"From Patent to Profit . . . the most
well-respected invention development book . . . "

Michael Neustel
U.S. Patent Attorney

"From Patent to Profit solves one of the greatest mysteries
that inventors have faced for decades.
It's about time someone wrote a book showing inventors
how to make money on their patents."

Don Kelly
Intellectual Asset Management Associates, LLC

"Inventing a highly marketable product is great . . .
but making money is another thing."

Andy Gibbs
CEO, PatentCafe

"Exciting, liberating . . .
a real eye-opener."

Phil Bourgeois
CEO, Studio RED

from
Patent
to
Profit

INVENTION DEVELOPMENT GUIDE

Bob DeMatteis

SQUAREONE
PUBLISHERS

Every reasonable effort has been made to provide reliable information and data, but the author, editors and publisher do not assume responsibility for the validity of any or all materials contained herein or the consequence of their use. The author is an inventor and not an attorney, and the content of this book is not to be considered as legal advice. If you require legal assistance, seek the advice of an attorney.

The author and publisher gratefully acknowledge PatentWizard, LLC and WaterPik for granting permission to reproduce their respective trademarks.

Cover Designers: Phaedra Mastrocola and Jacqueline Michelus
Cover Photo: Getty Images, Inc.
Typesetter: Theresa Wiscovitch
Editor: Barbara Tchabovsky

Square One Publishers
Garden City Park, NY 11040
(516) 535-2010 • www.squareonepublishers.com

Library of Congress Cataloging in Publication Data

DeMatteis, Bob.
  From Patent to Profit / Bob DeMatteis
    p. cm.
  Includes bibliographical references and index.
  ISBN 0-7570-0140-8 (pbk.)
    1. Inventions—Marketing—Handbooks, manuals, etc.
    2. Patents—Handbooks, manuals, etc.
  I. Title.
  T339.D45   2005
  608.773—dc22

                                                    2004005156

Printed in the United States of America

10 9 8 7 6 5 4 3 2

CONTENTS

As always . . . it's dedicated to
my favorite invention, my daughter, Lindsey,
an impressive inventor herself.

ACKNOWLEDGMENTS

I would like to thank those who have helped form the content of this 3rd edition and who have contributed to the abundance of inside information making it possible for inventors and companies to profit from their ideas and inventions.

Let's start with all the wisdom shared by Don Kelly, Andy Gibbs, Michael Neustel, and Phil Bourgeois. And let's include those invaluable contributions made by Matt Gilliland, Paul Lapidus, Mary Wollesen, the SBDC and SBA management team, the U.S. Patent Office, and last but not least, all the famous inventors and creators whose insight is shared throughout the book. This includes spectacular inventors like Thomas Edison, George Pullman, Eli Whitney, Yujiro Yamamoto, and Jerome Lemelson. It also includes sensational creators and marketers like Walt Disney, Steven Jobs, Steven Wozniak, Mo Siegel, and the formidable Bill Gates.

My thanks also go to the founding fathers of our country that created such an incredible patent system and made it all possible in the first place. Thanks to George Washington, James Madison, Thomas Jefferson, and perhaps one of the most incredible inventors/entrepreneurs of them all, Ben Franklin.

My appreciation would not be complete without thanking my publisher, Rudy Shur, his typesetter, Terry Wiscovitch, and an outstanding editor, Barbara Tchabovsky, who helped make the content remarkably clear and precise.

However, above all, I want to thank my family for being there for me through all the hard work and long hours it has taken to compile and write this book over the years. My wife, Dilcia, gave her support during long hours of research and editing. My daughter, Lindsey, with her remarkable inspiration, kept me going while perfecting the contents. In a way, you can say I did it for them. I love them dearly. Last, I have to give thanks to God for bestowing me with my creative mind and the ability to follow through, since it would never have been possible otherwise.

PREFACE

It's a Whole New World!

Invention in the twenty-first century is reshaping the world . . . and at a dramatic pace. Never before has invention had such an astounding effect on our lives as it does now. It's no wonder patents and intellectual property are being referred to as the "new money" in business throughout the world.

In just the past few years countless inventions and patents have influenced our lifestyles and workstyles and have created a whole new world. What we once thought of as science fiction just a few years ago is now reality. Just a few years ago, who would have thought that personal computers would be in every home in America, let alone outnumber televisions? Who would have thought that cell phones would be commonplace and become essential to our lifestyle? Who would have even imagined that the Internet would have evolved so quickly and become the primary communication system, business network, and information source on the planet?

The interesting fact is that many of these phenomenal inventions are being created by independent inventors and small businesses . . . just like Thomas Edison, George Pullman, Orville and Wilbur Wright, and Walt Disney did years ago. Just like Steven Jobs, Steven Wozniak, Mark Andressen, and of course, the remarkable Bill Gates did recently. And just like thousands of other successful inventors have done with their thousands of other not-as-well known inventions in recent years. I'm one of those "others" and never did I believe I would become an inventor . . . until one day it happened.

That's what this book is all about. It's all about the methodologies that I've learned over the last two decades. It's all about how I turned over 20 U.S. patents and several foreign patents into profit. It's all about how my inventions and subsequent patents have had an impact on well over $1 billion in sales.

It can happen to you too.

Let me tell you it was not easy to start out. I made many mistakes and fortunately I made some excellent decisions that brought about an early success. From those initial successes I developed a stream of successful inventions protected by patents. In fact, all my patents have resulted in profit even though I don't manufacture or market any of the products. Today, I license or partner them all. However, the methodologies to do this are not something you'd learn in school or at a university. It's strange but true. The research and time it took me to learn the invention development process was enormous. But I was determined and dedicated to achieving the outcomes I desired.

Over the years my inventions and patents created a whole new world for me. It gave me the best job I ever had—five years developing and launching the self-opening plastic grocery sack in America. It gave me the opportunity to teach Total Quality Management (TQM), which gave me tremendous insight into how manufacturing and marketing departments interrelate. More important is how these two key components interact in order to launch a new innovation. It also gave me the opportunity to create my own future as an inventor and to be self-sufficient.

It was not until March 23, 1994, when my daughter was born that I decided to quit my job and focus my efforts on developing my own inventions. Frankly, it was a somewhat scary decision, but seeing her born into this world, so brave and yet so fragile, I made the debatable choice. You can say it may have been based more on emotion than common sense, but I was lucky, because it was probably the best decision I've ever made. After all, my wife faithfully supported that decision too.

My new life started with a foundation consisting of a few existing licensees, which has since grown into an enviable portfolio. But there was another compelling passion I wanted to pursue, teaching. I had enjoyed teaching Quality Management and decided I would like to continue teaching in some form. After some thought, I selected the field of inventing and patenting over TQM. After all, it was what I enjoyed most.

Before I began developing a curriculum, I contacted three colleges and universities and inquired if such a course would be desirable. The response was a resounding, "Yes!" So, over the next eighteen months I worked to develop materials for the courses.

As always when creating something new, I came across a stumbling block that was not going to be easy to resolve. You see, I didn't want to teach a course on patents and patent law, or teach others about licensing or even about inventing. I wanted to teach would-be inventors and businesses how to do what I knew best and that is *how to make money on an invention*. All I could find were books written by attorneys on patent law and how to write a patent application, but none of them contained real life experience, or for that matter, common sense. I also found books hyped with "how to make millions on your ideas," but for the most part all they contained was hype, which lacked qualified methods of accomplishing the ultimate goal of making money. The last thing I wanted to do was to send someone down the road of filing for costly patents with a questionable—and most likely unsuccessful—outcome. Or to build false hopes that would result in a letdown later on. There simply was no book I could use in my courses to teach invention development from the standpoint of making money. No text existed showing inventors how they can be assured of success by following proven methods used by successful inventors. Not until now.

This is what ultimately compelled me to write this book. I didn't intend on being an author, I just wanted to teach others how to be successful inventors and how to profit from their innovations. However there was no choice, so I began writing my own class materials. The result was three small books, *The Art of Inventing*, *The Art of Patenting* and *The Art of Licensing and Marketing*. These books served their purpose for two years of invention and patent workshops.

It was in 1997 that I combined my teaching experience and much of the content of those first three books into the first edition of *From Patent to Profit*. From that first edition, it has continued its transformation over the years and today has grown in size and content and now contains over 400 pages of valuable information for innovators and small businesses. From that first printing it has also expanded with tremendous insight used by successful inventors over the years to parlay their inventions into valuable patents and at times, immeasurable profits. Everything you want to know about how to make money on your inventions is included in this, the third edition. I'm sure you'll find it will be your single most valuable asset in your invention development activities.

Today, I don't have to teach, I could be living on a Caribbean island enjoying the warm waters, drinking coconut milk, and fishing. But just like an inventor, I have to create, because that's what I do best. Most of my time is still spent on my invention activities, with a substantial amount of time dedicated to my books and workshops. Oh yes, a lot of time is also spent playing basketball, skiing, and hiking with my daughter and explaining how the world works. It's all been made possible because I took that first step to pursue a dream and then worked hard as it turned into reality. Never would I have imagined when I received my first patent that today I would be a successful inventor with over twenty patents licensed internationally.

It is with my sincerest wishes that you can use and apply the principles in *From Patent to Profit*, and that you too will become one of the many inventors bringing about change to our planet and making it a whole new world!

INTRODUCTION

From Patent to Profit is an invention development guide for entrepreneurial small businesses, independent inventors, engineers, and product developers.

Whether you are just starting out, want to expand your company's product line and protect the expansion with patents, or have been struggling with the development of an innovation, *From Patent to Profit* will show you how to make it happen. Whether you plan to go into business or license your innovations, you'll have all the tools you'll need to succeed and to launch your innovations. Our emphasis on marketing and licensing innovations from a practical standpoint, from a moneymaking perspective, is unmatched. It may sound trite, but inventing and launching new patented products is really all about money. Of course, your inventions need to have some unique, inventive, marketable attributes, but if your development effort does not reward you with sales and profits, you'll be discouraged and abandon the project.

Frankly, no inventor should ever give up. If your creations are not earning money for you, you have to find out why. Then you have to make all the right decisions and change, modify, and improve your inventions so they do make you money. *From Patent to Profit* is based upon this simple principle:

> *There are no failures in life, only outcomes.*
> *If you do not get the outcome you seek, continue*
> *to make changes and improvements until you*
> *get the desired results.*

Ideally, you should apply the tenets in *From Patent to Profit* early on in the invention development and patenting process. You or your company do not want to invest thousands of dollars and thousands of hours in products and technologies that are not highly marketable. You want to develop only those that have verified marketability. *From Patent to Profit* shows you how.

It is no wonder that the U.S. Patent Office says that 97% of patents never make any money for the inventors. Why? Few, very few, inventors apply the basic tenets and common sense business principles required to make money from patenting and developing their inventions. Plus, unfortunately thousands of individuals spend thousands of dollars on invention promotion companies that have pathetic track records—less than $\frac{1}{10}$ of 1% ever make any money for the inventor. The invention promotion companies can bring much harm to an inventor by causing pre-market exposure and complete loss of patent rights. Exposing a methodology that doesn't work also sheds light on the methodology that does work.

At the heart of the successful invention development methodology is the simple fact that you, the inventor, must pilot the effort yourself. You cannot rely on a middle man to do this for you. The creators of inventions often think of them as "offspring"—like they do their own children. Would you rely on some middleman to raise your children to become prosperous, outstanding citizens? Of course not. Someone from the outside cannot have the same foresight, vision, and passion for their children—or inventions—as you do.

Just as you raise your children to become worthy adults in society, you will be the one that guides the development of your inventions. And just like your children will need help as they grow, you will want partners to help you in the development of your invention.

In a way, you can say there are no secrets about how to develop an invention successfully and that all you have to do is apply the practical laws of raising your children in today's world. It's a paradox of a sort. But there are tenets and secrets to raising successful children, and there are tenets and secrets to developing successful inventions. *From Patent to Profit* will unmask these for you, so you can be successful and make money on your inventive endeavors and not fall into the 97% failure category. In fact, you really can be successful 100% of the time once you learn the process.

The *From Patent to Profit* process is easy to track, too. Once you read Chapters 1 and 2, you'll be in position to track your invention's progress by using the Strategic Guide chart. This chart divides invention into two phases, the developmental phase and the launch phase. Even the most inexperienced product developers can follow the process through to completion.

From Patent to Profit is not theory. It is based on years of experience in product development, patenting, and licensing. Contributions from many successful inventors, manufacturers, marketers, and patent attorneys are unsurpassed and cumulative, representing hundreds of years' experience. All the contributors have one common goal: making money on inventions now and in the future.

More importantly, *From Patent to Profit* includes all the most recent methods and technologies. It includes the newest laws and gives you insight into the future by showing you how to use the U.S. Patent Office's new 21st Century Strategic Plan to your advantage during invention development and financing. *From Patent to Profit* is the only book that does this. It will save you a small fortune as you proceed in your endeavors, and it will show you how to leverage your finances and development budget like never before.

Use *From Patent to Profit* as a manual. Use it to guide you in your efforts and you will be able to achieve the outcomes you desire. It is the textbook for the Patents in Commerce™ (PIC) training series, and we encourage you to apply the PIC model as you read about invention design, patenting, manufacturing, and marketing. We also encourage college faculty to use the Strategic Guide as a format for teaching students about the invention development process.

With all this said, we'll make one promise to you. Follow our time-proven tenets and you'll be successful, make money, and have the time of your life creating the future. Now go out there and get your motors started. The race is on to invent the future of our country and our society, and you're going to be in the lead!

PART

1

INVENTING

INVENTION DEVELOPMENT AND DESIGN

Inventing is all about money. It's also about problem solving, opportunity, and a different way of life. It is fun, exciting, at times challenging, and best of all, it can be very rewarding.

Prosperity in America is founded upon invention. Invention is right at the heart of having a better way of life. The pursuit of modern, efficient, convenient innovations, protected by U.S. patents, is one of the most sought out ideals of entrepreneurs and small businesses. That's what America is all about.

For the past two centuries inventors have literally created the future of our country. Invention in America has stimulated prosperity and commerce that have created an unimaginable world and hold the promise of an unimaginable future. Think about it. Mankind has been on this planet for one million years. But in just the last two hundred years, mankind has transformed a relatively primitive world into one bustling with industry and manufacturing and blossoming into the technological society we are today.

Eli Whitney's cotton gin made fabrics economical to produce. George Pullman and George Westinghouse made trains a viable means of transportation and opened up the West. Thomas Edison introduced a myriad of electrical products ushering us into a new era. Henry Ford ramped up mass transportation, and Orville and Wilbur Wright did the unbelievable—they actual flew!

Today, everywhere you look, you see inventions that have molded your way of life. The cars you drive, the computers and cell phones you use, the televisions you watch, the tools and machinery employed in industry all started out as inventions. The personal computers you use, the software created by the visionary Bill Gates, and of course, the Internet, were all unheard of just a few years ago. These are just a few of the remarkable inventions that have forged our history.

A WAY OF LIFE

We live in a country essentially created by inventors. America's economic strength and global competitiveness depend more than ever upon the successful harvesting of the great ideas of this gifted group of people.
–Don Kelly, Intellectual Asset Management Associates, LLC, former Director, U.S. Patent Office

Inventors have made fortunes from their inventions too. Edison and Westinghouse created entire industries. Ford launched what may be the most important industry in the history of our planet, automobile manufacturing, and Bill Gates created software from nothing more than his imagination and a lot of tenacity and determination. Young Mark Andressen created Netscape, the first Internet browser, and Yujiro Yamamoto created the telephone answering machine and cordless telephone. I created the self-opening plastic grocery sack, and Al Gore created the Internet (well, maybe former V.P. Gore didn't really invent the Internet). On a simpler level, a 19-year-old high school dropout, Betty Nesmith Graham, created White-Out, and Art Fry created Post-it® note pads. What did these individuals have in common? What were their objectives? How did they do it?

Keys to Success

What is invention and product design all about? What makes some products sell and others never make it. Whether you are a small business, an entrepreneur, or an independent inventor, you need to follow a certain methodology to ensure that your inventions make money. Frankly, it's a lot of work, but with the right plan you can achieve your objectives and save a lot of time and money. Entrepreneurs and fast, flexible, small businesses can almost always beat larger, cumbersome corporations to market. The methodology we reveal in *From Patent to Profit* is that methodology.

If you're one of those small businesses or inventors and you've got a good idea and want to act on it, this manual is for you. While Americans unquestionably lead the world in innovating and inventing, there are still a surprisingly large number of good ideas that never get to market. The chief reasons are really quite basic: many entrepreneurs and small businesses simply do not understand how to develop an idea and get it patented, mass-produced, and sold, producing a profit.

Successful entrepreneurs-inventors-innovators know that the world of invention and innovation is a lot of hard work. But they also know it can be a lot of fun and very rewarding. Those who take action and follow all the right steps will be successful. They will also have valuable patents protecting their inventions.

Having a good idea is the first step and the easiest part of the process. Next comes the more difficult part, which is learning how to turn your idea into a reality. Then, you can enjoy the benefits and rewards of your creative ability.

Patent Prowess

More than 200 years ago, in 1793, George Washington created the U.S. Patent Office. It became the impetus that made invention and innovation in America blossom. After all, what better way could there be to promote commerce than to grant the inventors the exclusive rights to manufacture, use, and sell their inventions?

It's also no wonder that Benjamin Franklin was the first Commissioner of the Patent Office. One of the most inventive humans who ever lived and having a keen business sense, he was the perfect individual to set the stage for patents in commerce. A powerful concept and a powerful system, the U.S. patent system would not only change our lives forever but would also become the envy of the world.

There is no doubt that the U.S. patent system has contributed to the success of America's prolific inventors and small businesses. There is also no doubt that most major corporations today were the small businesses of yesteryear. Independent inventor Thomas Edison founded several small companies and industries that become General Electric, Edison, and RCA. Perhaps your next endeavor or your improved product or system will create the focus of a new or expanded company and contribute to your personal wealth.

What is most important for inventors and small businesses to know is that the U.S. patent system gives them incredibly powerful rights that can become extremely valuable. In a world based largely on receiving monetary rewards for creative efforts, what more can you ask for?

Take Action

Before you or your company pursues an idea and embarks on a new product development project, there is one important point to remember. Your success is directly related to your commitment. You and your development partners, such as your company's manufacturing and marketing departments or external licensees, must make a substantial commitment to invest the time, money, and effort to see the project through. In other words, you must take action, create a team effort, and be determined to follow through. You must solve problems and overcome obstacles until you have attained the outcome you want. Invariably, this requires a significant amount of patience.

The best news of all, however, is a little-known secret: *it doesn't have to cost a lot of money to do it!* That is what this manual explains. You are about to learn how you can do it in a fraction of the time and at a fraction of the cost of what it takes large corporations. You are also going to learn how to protect your ideas before filing for patents and to secure patents in a timely manner. You will learn how to evaluate your invention's potential, how to find manufacturing and marketing partners, and how to master the art of inventing and patenting.

If you're ready to embark on an exciting journey, fasten your seat belt. Your commitment, determination, and patience to learn all you can and make all the right decisions will take you on a fast-paced, exciting journey. Make all the right moves and all the right decisions and your inventions will sell. You'll create profit in the process, and you'll have a whole lot of fun doing it!

Today's Perfect Business Climate

The time to invent and create new products may never have been better. We are living in one of the most creative, inventive, opportunistic times in history. Creative individuals all over the world have been developing ideas and inventions into new patents and products at a record rate. As we enter the twenty-first century, more than 200,000 patent applications are filed annually at the U.S. Patent Office, and more than 125,000 patents are granted. True to America's entrepreneurial spirit, more than 25% of these patents are from independent inventors and small companies.

It's a well-known fact that many successful innovations have been invented by independent inventors and small businesses. They are more focused and flexible and know how to get their concepts to market faster. Many major inventions were created by inde-

DON'T RUSH

Don't rush to patent your idea too early. You must keep in mind that patenting can be expensive if you don't know what you're doing. There are many steps you can take that will save you time and money. Patents are not a validation that your product will be a success in the marketplace. First, you must go through all the background and research necessary to adequately define and validate your new product idea. When it's time to patent, learn all you can, take all the right steps, and you'll be successful.
–Paul A. Lapidus, IDSA, New FunTiers

AN INNOVATIVE NATION

The words still ring in my ears. The President, holding a press conference in the White House Rose Garden just days after September 11th, 2001, offered the reassuring message that America remains strong and will overcome those horrific attacks. *"We're still,"* he said, *"a nation of entrepreneurs and small business vitality . . . still a nation of innovation."* What an interesting choice of words—a message that spoke volumes about America's high regard for its most valued resource, invention.
–Don Kelly, Intellectual Asset Management Associates, LLC

pendent inventors throughout history. The list includes Thomas Edison, Walt Disney, Elias Howe (Singer sewing machine), Murray Spanger (the Hoover vacuum), Jerome Lemelson (components of audio cassettes and the VCR) and many, many more. These inventors, just like Bill Gates today, started with a single concept and then formed some sort of small business to exploit their creations or license them.

Many independent inventors have prospered from the licensing of their creations. You may be unaware that Walt Disney got his initial capital to ramp-up the Disney Corporation and Disneyland through licensing his intellectual property. Lemelson, holder of more than 500 patents, licensed his inventions, which include hot wheels and several manufacturing-related systems used by major corporations such as Ford and General Motors. Yujiro Yamamoto licensed his inventions of the telephone answering machine and cordless telephone. Would you like to do the same?

According to the U.S. Census Bureau, there are more than 22 million businesses in the United States. Of these, about 70% are home-based businesses and another 20% are small businesses with fewer than 10 employees. Entrepreneurs with new innovations and inventions represent a large and growing group of home-based and small businesses. With corporate downsizing, job elimination due to robotics and computers, and the elimination of entire industries that have moved overseas, it's no wonder that much of America's creativity is a result of the inventive, entrepreneurial efforts of this segment of commerce.

Many talented workers, engineers, innovators, and marketers forced to leave their employment with larger corporations want to set out and create their own future just like Lemelson, Disney, Edison, Gates, and many others did. This shift in just the past ten to fifteen years has caused an unprecedented number of individuals to become "re-employed" in entrepreneurial home-based businesses and new start-up companies.

What is the result of this enormous shift in employment of talented individuals? A comment in the Wall Street Journal sums it up:

"The U.S. economy is the envy of the world . . . "

Since the inception of the U.S. Patent Office in 1793, entrepreneurs and inventors have changed the landscape of American business and the world's industries forever. American invention literally spurred on the industrial revolution *en masse*. After the trials and tribulations of the 1900s, with two world wars and major advancements in transportation and communication, we have entered the twenty-first century. We are now on the threshold of technology breakthroughs that not only affect our lifestyles and the workplace but also offer the ability for millions of inventors to work from their homes. Just like two hundred years ago, we are a country filled with independent inventors and small businesses with new ideas and inventions that once again will shape the future of the world.

Today's business climate is excellent for another reason. Small- and medium-sized businesses need proprietary products to compete against large generic competitors. The best way to create such products is through invention, innovation, and patented technologies. Since many of these smaller companies do not have research and development (R&D) departments, they are inclined to form partnerships to develop new ideas and new technologies, either high tech or low tech, conceived by creative individuals and businesses like yours. You just need to know how to protect yourself, your inventions, and your interests, so you can do it.

If your intention is to start a business or expand a present product line, you will find customers may pay more for innovative, high performance, patented products. If you are an independent inventor who wants to license your ideas and inventions, you can find quality companies willing to pay for your money-making ideas and inventions. There is more opportunity than ever before to find partners who are willing to listen to your good ideas and license them.

The principles that you should apply to develop your concepts are really just good business sense, common sense when you think about it. Differentiate yourself and focus your effort as this manual explains and you'll learn how to:

- Protect your ideas
- Make the necessary changes to ensure success
- Secure bulletproof patent protection
- Make money without spending a lot of your own

After all, good ideas don't need a lot of your investment dollars. Of course, you have to make some investment, but if you plan to partner your innovations with other established companies that may manufacture, subcontract, and perhaps sell those innovations, remember that they have to invest their time and money as well. Frankly, they'll be willing to do so. This teamwork approach is the method that almost all successful inventors have applied over the years.

From Patent to Profit shows you the various options available so you may apply them and turn your ideas and inventions into a reality and profit, while protecting them with strong patents.

Laws, Court Cases, and New Legislation

You can put yourself in a position of power from the onset by knowing four important legal factors that are having a positive impact on invention and innovation in America. They are:

1. **Provisional patent applications.** As of June 8, 1995, inventors and small companies may file a "provisional patent application" that sets a priority date and allows the inventor to legally post "patent pending" on an invention. The filing fee for a small entity, a company with fewer than 500 employees, is $110 as of 2009 (check with the U.S. Patent Office at www.pto.gov for fee updates). With a provisional patent application filed, you have patent pending status for an entire year. This gives you valuable time to further develop your concept, run up sales, or secure partners and licensees. If the invention is subsequently modified, which is frequently the case, a new provisional patent application, covering the newly discovered inventive matter, may be filed for the same inexpensive price. Or, if sales are poor, you may abandon the development without incurring major expenses. Last, if existing suppliers in the industry show no interest in licensing the technology, the provisional patent application may provide you sufficient protection to make the necessary changes and improvements so that the invention will be of interest. Any way you look at it, you can save a lot of money by filing a provisional patent application before filing the more expensive, regular (or permanent) patent application.

FACT

Numerous record judgments have been awarded to patent holders in the past few years. From his famous intermittent windshield wiper patent, Bob Kearns has been awarded tens of millions for patent infringement by Ford and several other automakers. More recently, Ron Chasteen, an Arizona inventor, was granted a $57-million award from Polaris Heavy Industries of Tokyo for stealing his electronic fuel injection system for two-stroke engines.

A WORD OF CAUTION

One problem we've encountered with inventors who have filed provisional patent applications is that once applied for, and after public disclosure, the clock starts ticking. A year goes by very fast. If an inventor has not yet found the right opportunity in which to successfully market or license the new product idea, whether as an entrepreneur or as a licensor, he/she may be forced to spend money on what may turn out to be a premature patent, or worse yet, abandon the patent application and the opportunity.
–Paul A. Lapidus, IDSA, New FunTiers

2. **The Doctrine of Equivalence.** This law, struck in 1952, states that if a product functions "essentially the same as that which is claimed in a patent," it infringes. This Doctrine was tested in the U.S. Appeals Court with the Hilton Davis case. In March 1996, the case went to the U.S. Supreme Court and on March 3, 1997, Supreme Court Justice Clarence Thomas handed down a decision. In essence, his ruling upheld the impact of the Doctrine and further clarified its meaning, in favor of the patent holder. More recently, the FESTO case also challenged the durability of the Doctrine of Equivalents, which has thus far survived.

3. **Recent legislation and the 21st Century Strategic Plan.** In the late 1990s legislation originally slated to help large corporations that may have had a negative effect on independent inventors was amended in favor of independent inventors and small companies. In 2003, the U.S. Patent Office initiated its 21st Century Strategic Plan that includes upholding those laws and also includes a proposal to significantly increase patenting cost, with some exceptions. In summary, the new Strategic Plan allows all entities to continue to file inexpensive provisional patent applications. It may increase fees for small and large entities but has a new provision to allow independent inventors to file at greatly reduced rates. It also contains provisions that allow a business to extend and defray filing and patent review costs over a longer time period. This can be an important strategy for inventors and small companies to conserve cash flow. After all, it is far more important to spend money on marketing and development to produce sales in the earlier stages. That's what produces profit, not just "having a patent." This is discussed in more detail in Chapters 9–13.

4. **Court Cases.** Hundreds of court decisions in recent years have been ruled in favor of patent holders. It is commonly understood in the world of intellectual property, that 60% to 70% of all court cases on patent infringement are being ruled in favor of the patent holders. If you separate out computer- and Internet-related patents, the percentage jumps to about 80%. But the best news of all is that this has forced most infringers to settle out of court to avoid the high cost of litigation. Most legal experts agree that 90% of these out-of-court settlements are in favor of the patent holder. Patent infringement is not bad news. In fact, it is usually just the opposite—if you are the patent holder. You may employ strategies that will cost you very little to pursue infringement. These strategies are discussed in Chapter 8.

The real benefit of these developments is that patents are more respected than ever. As a result, patent values have soared. Most experts say that values have increased 20-fold to 100-fold or more.

There is, however, a downside to all this—you have to make the commitment to stay in the game and weather the storm. You have to make sure your patents are well thought out with bulletproof claims or you'll have no protection. And you certainly want to make sure you have the best, most efficient, and cost-effective technology so the product will endure long-term market trends.

In Chapters 5 through 8, you will learn all about patents and patent strategy and more about these topics and how to use them to your advantage. Build strong patent protection around your innovation and you'll have a formidable opportunity!

The Best and Worst First Steps

The best first step in inventing is exactly what you are doing right now—namely, learning how to turn your inventions and patents into profit. The more you know and the more you understand your invention and how it fits into the field, the closer you are to getting it launched and making money.

There are two worst first steps you can take. The first is to run out and hire a patent attorney. Even worse is to respond to the slick TV and radio ads of the invention promotion companies that charge inventors thousands of dollars for worthless information and support.

You should use patent attorneys in a timely manner, and they are usually a pleasure to deal with, but they are not usually the best "first step" an inventor should take. You can't rush to file patent applications on your ideas and inventions based upon emotions. You have to objectively verify that your inventions are marketable. You'll see how to do this.

As for invention promotion companies, we know of none that provides qualified assistance. In fact, they can harm the inventor by prematurely disclosing an invention. In most cases, the supposed "invention research" work they do is boilerplate, using the same words and sentences lauding an invention, whether it is a simple hair styling implement or a high-tech electronic device. Invention promotion companies have extremely low success rates. Fewer than $1/10$th of 1% of inventors earn more money than they've paid the companies. Most of the companies have been repeatedly indicted for fraud. Read their literature and you will see that their success rates are infinitesimal.

Development Objectives

Inexperienced inventors may think that getting a patent is a main objective. Patenting is only one aspect of the big picture that brings about success. Other inventors may think that making a big hit quickly is a main objective. This, too, is misleading and is usually the philosophy for a hit-and-run fad. Rarely does it apply to long-term trends and long-lasting inventions.

There are a myriad of inventions and a diverse population of inventors, engineers, and product developers, but for all, the four primary objectives of the invention development process are essentially the same. They are:

- **Make money.** A monetary reward is what 99.7% of inventors want in return for their creative efforts.

- **Protect your future.** Patents are a wonderful, powerful way to protect you and your company's future. What could be better than having the guarantee of the U.S. government that you'll have a legal monopoly to manufacture, sell, and use your invention?

- **Become the leader in the niche or field.** With marketable inventions and patent protection, you'll have a jump on competitors and the ability to carve out your own niche and/or be the leader in your field. With that comes the ability to create new opportunities in the field.

- **Have fun.** You also have to have fun doing what you do; if you don't, you won't maintain the necessary interest and drive to be successful.

THE INVENTION PROMOTION COMPANY

The most disheartening thing about inventors who have been taken advantage of by scam companies is not that they've lost money, but that they may have lost something much more valuable . . . the spirit of inventing.
–Stephen Paul Gnass

TAKE YOUR TIME

What is the single most important rule for all inventors? Do not rush. Take your time!
–Michael S. Neustel, U.S. Patent Attorney

SELECTING A FIELD

When you kick off your invention career, make sure it is in a field you enjoy. Once you have developed some special relationships with your manufacturing and marketing partners, you will find it will be a lot easier to launch additional inventions in the future. Thus, from the perspective of having fun and developing relationships, try to stay focused in a single industry. To jump from one industry to another is like starting all over again.

One of the worst approaches to making money is to expect patent attorneys, marketing professionals, and design engineers to deeply discount their services to you because "you don't have any money." If you want the best hired services to help you succeed, budget for it. If you're an independent inventor, start a dedicated invention savings account and in a year, you'll have more than enough to begin your invention project the right way."
–Andy Gibbs, CEO,
PatentCafe.com

DESIRE VS. NEED

Sometimes desire can become need. A compelling design may specifically relate to lifestyle- or workstyle-enhancing products. When a product can relate to someone's specific style, then desire becomes need. That is why someone will buy an SUV who never takes it off road, puts it in 4-wheel drive, or even sees snow.
–Phil Bourgeois, CEO
Studio RED

If you understand your invention development objectives, you'll be in a position to make all the right decisions. Let's elaborate, because the need to succeed is driven by these objectives.

Make Money

There is nothing more compelling than making money. Since money has been established as the means of conducting interstate and international commerce, people should obviously be willing to pay you. To be rewarded for your efforts, you will need to evaluate your inventions early on and repeatedly throughout the invention development process to ensure that you are on the right track. How to assess your invention is discussed throughout this book.

When assessing your invention, the first and most important criterion is to determine its moneymaking potential. Sales generate money, which equates to profits or royalties for you or your company. Bear in mind, however, that patents do not necessarily generate money. Neither does the manufacture of a product generate money. The product must be sold before profits and royalties are generated.

Use sales potential and marketability as your guide to making money. Don't make the common mistake of assuming that your ability to get a patent is a guide to how much money you are going to make, and, don't make the mistake of thinking that finding a manufacturer willing to make the product is any measure of the product's sales potential either.

The amount of sales your innovation will generate is based on three economic principles:

1. **Is there a strong need or desire?** Need usually results in higher sales than desire. We take care of our needs first. We have to eat. We have to have shelter. We don't have to have the newest, most modern sports utility vehicle, however. In our advanced society, most Americans' needs are met. Thus, the driving principle behind most inventions today is satisfying the psychological desires of consumers. For instance, people really do want SUVs, and they do sell well. In fact, consumers not only want them but they also want them to be luxurious with automatic cruise control, climate control, heated leather seats, four wheel drive, and so on. It also goes without saying that if there are opportunities to fill a void, such as meeting new mandated environmental laws, there would be a compelling need for your invention. These could be exceptionally good opportunities.

2. **What is the price point?** Can you sell your product for more than the current generic products on the market? Or does it compete with commodities and should it be priced close to existing competitive products and why? Many inventors and new product developers unknowingly proceed developing their inventions with the assumption that consumers will be willing to pay a lot more. This is usually erroneous. Brand loyalty, confidence with existing products, and plain old habit can keep customers buying the established brand for a long time, even though it may be inferior. Many new products are sold at higher prices than existing products, but as volume expands and competition increases, the prices tend to decline. This is commonly seen in the computer products trade. However, if your invention is in an industry that has little price elasticity, you may not be able to get a premium price, which will result in lower profits or a non-existent market. For instance, attractively

printed designer trash bags that cost just 5% more than existing brands were a flop because users would just not pay the extra for pretty flowers printed on the bags. The price elasticity threshold on trash bags was only 2%, and 99% of consumers were unwilling to pay 5% extra to have flowers printed on bags that were discarded. Would you have paid more? Determining the price point is one of the first criteria you'll need to examine.

3. **What is the market size?** You can't assume that you're going to capture "an entire market." It just doesn't happen very often. Your market size should be determined by how much of it you can capture. A 5% niche is substantial, but if that represents only $100,000 a year in sales, it probably is not enough. However, if the invention is quick and easy to develop and launch, it may be. Market size is also determined by the concept of market durability. In other words, will the new product endure year after year, or will it be vulnerable to knock-offs and design-arounds next year? This is where the importance of patent protection and patent scope come into play.

Can you answer these important questions now? If you are an expert in the field of invention, you probably already have reasonably good information through networking with potential customers and sales associates. Or, like many large corporations do, you may conduct an inconclusive and relatively expensive beta test. Remember, however, that beta tests may result in inaccurate, misleading reports, and frequently, they do not clarify what you need to fix in order get the outcome you want.

One of the best and most inexpensive ways to determine marketability is to discuss sales potential with qualified, experienced sales/marketing experts in the field of your invention who have a desire to sell your innovation to the trade. Through their experience, you can usually get a good idea of sales potential. Sales potential determines earnings potential, which is the objective you want to validate. Working with sales/marketing experts is discussed throughout this book.

Secure Your Future

Market durability is important. No company or independent inventor wants to spend a lot of time and money on developing a new product line that lasts only a year or two. Whether you intend to market your inventions through your company, or license them to an existing supplier in the trade, you'll want to protect your ideas and inventions in a timely manner with patents. You will also want to establish other intellectual property rights, such as trademarks, trade dress, and copyrights in the process. There are ways to do this without incurring large expenses. With patents, your focus becomes one of securing long-term income, not short-term profit.

In addition to verifying your invention's marketability, you must consider whether you can hold onto the business. You have to consider seriously the following factors and competitive responses:

* **Is the invention patentable?** Will you be able to get a patent (better yet, more than one patent) to help protect your long-term security? Will the patent give you broad enough protection to keep others out of the marketplace? If you don't have patent protection, it may be easy for competitors to enter the market. You will learn how to develop strong patents in Part Two of this book.

PERCEPTIONS

When evaluating the potential sell price of a new product, keep in mind that you are dealing with the consumer's perceptions—not the product's cost plus mark-up. If you cannot zero in on the price a consumer is willing to pay, it will be difficult to sell or partner your idea to anyone. Once you have a general understanding of what the market will bear, then your objective becomes figuring out how to get it made for the right price.

LICENSING VS. ENTREPRENEURSHIP

Give some thought to whether you're going to license your product on a royalty basis to an established company or bring the product to the marketplace yourself. In the beginning, these approaches have parallel tracks requiring much of the same work from concept to first "proof of concept" model. Once a concept is well defined however, the approaches are quite different, and the up-front costs of licensing are considerably less than taking a new product all the way to the marketplace yourself.
–Paul Lapidus, CEO, NewFuntiers

- **Is the invention worth patenting?** The beginning of the patenting process is relatively inexpensive. However, securing a patent with adequate coverage and protection will eventually require an expert patent attorney and usually costs several thousand dollars. Will future sales be sufficient to justify the costs?

- **How will competitors respond?** If you capture a noticeable niche in the market, are competitors going to stand still? Are they going to lower prices, trying to drive your price levels and profits down? Will they try to develop a new competitive version and design around your patents? Will the competitors take a chance on infringing on your patent? Or, will you be willing to license your patent rights to them? You will read more about competitive responses in Chapter 2 and licensing in Chapters 14 through 16.

- **Is the invention a fad or a long-term trend?** If your invention is a short-term fad, is it worth pursuing in the first place? Is it worth patenting? If your new invention is based on having instant success, remember, that success can fade away just as quickly. If your invention is a short-term fad, like the Pet Rock™, do you really want to spend a lot of time and money on product development and patenting? No, it is a waste of your or your company's resources. You want long-term trends that represent long-term security and profit.

There are other barriers to market entry, such as the inability to purchase raw materials and components, to attain or license other technologies, or to overcome environmental or safety hazards. Later, we will assess these secondary influences so you can be sure you're on the track to making money and securing long-term protection.

Own Your Niche, Lead the Field

Owning your niche or being the leader in the field of your invention puts you in the enviable position of being able to create new improvements and to literally invent and patent future generations. One success should lead to the next. Each time you improve an existing product, or add a new one to the product line, you are that much closer to making a breakthrough discovery that could really result in some huge profits.

The importance of learning all about your industry's niche and all about patents in the field (and about patenting in general) cannot be stressed enough. This knowledge can extend the life of your inventions and earnings for years. It will also allow you to make good decisions and the right contacts much faster. Don't feel intimidated if you don't have a lot of knowledge in the field right away; just be willing to learn and willing to go out there and start meeting people.

The theory that "knowledge is power" is true when it comes to inventing. Once you have knowledge in your field, you will be able to pilot your inventions and innovations to success far faster and with greater certainty of success.

You should always be in a learning mode as you carry out your invention activities. After all, inventing and innovating represent a process of learning and discovery. The world of patenting is also based on discovery and learning. Ask yourself the following questions:

- Do I have adequate knowledge in the field to make my invention successful?

- If not, am I willing to invest the time and expense to get the knowledge and information I'll need? What resources are available to me? They should include people, schools, books, trade magazines, the Internet, etc.

- Are any barriers blocking me from learning all that I need to know?

You must be willing to learn all you can and do all the right things while you develop your innovations. You must be inquisitive and constantly aware of what is happening in the invention development process. Be flexible in your approach and always consider and even create new options along the way.

Have Fun!

Frankly, if you focus on the means, solve the problems you encounter along the way, get your patents in place, and your invention launched, believe me, it'll be a lot of fun! Then, you can continue to do what you do best: invent!

You really must enjoy what you are doing. If you don't, you will lose interest or end up doing something you hate. Either will result in a poor performance and not much financial success.

Here are two important questions:

- Is the field of your invention one that you really enjoy? Would you like to spend most of today, and the next ten years, creating new ideas in this field?

- Are you really going to have fun talking to and working with people in this industry? Will you enjoy working with them as your invention is developed and launched?

Making good money can make just about anything fun, but you have to enjoy the day-to-day process as well. If not, your profits will probably be short lived.

Ongoing Evaluation and Research

As you proceed with your invention's development, you will be continually evaluating its potential to fulfill your four primary objectives. One of the first things you should do is to assess your invention. This assessment should be done at the earliest possible date so you have a reference point and you'll know what improvements or modifications may be required.

You must be able to evaluate your invention honestly and accurately. You do not want to chase a dream that is going to cost you a lot of time and money. Sometimes, the emotion surrounding a new idea can cloud reality. You must be realistic. You cannot expect to capture 20%–30% of an existing market, unless you already own a substantial percentage. Established competitors are not going to let you run over them and then close their doors and go out of business. It just doesn't happen very often; in fact, it rarely does.

If you are active in the field of your invention, you probably already have a good idea of its potential. For instance, if yours is an established business, you may already know that your idea will help you protect current sales and allow you to capture additional markets. Obviously, pursuing and patenting a new improvement will help secure your future.

However, if you plan to start a new business or a new product line or to license your invention, you need to evaluate it in more depth before committing your resources. This

NO BUSINESS PLAN REQUIRED?

Business plans are a waste of time unless you are going into business. If you plan to license your inventions, the licensees may draft their own business plans. You can't do it for them. Your goal should be to do unbiased evaluation and research, which you can then present to potential licensees. The information you gather can be put up in a simple one-page New Product Summary and a more in-depth prospectus, like those in Chapter 9.

Regardless of how thorough your evaluation and prospectus might be, a potential licensee or partner will still show due diligence and get their own facts for themselves.

is particularly important if you are not familiar with the field. You must do some solid research and assessment before and during development.

Your company may already have distributors and customers lined up to purchase your innovation. Or, it may have a good idea of those companies that will be interested in using or marketing your invention. As an entrepreneur or independent inventor, you may possibly have some companies targeted for license, manufacture, and sales. You'll want to be conservative and realistic in your evaluations. An inadequate, invalid assessment can cost you a lot of money and a lot of time.

Usually an invention begins because your intuition or instincts tell you that it's a good one. If you know the field well, you can count on your intuition or instincts to be accurate. In fact, in many cases, you'll probably be more accurate than a massive 82-page business plan accompanied by expensive marketing research studies.

Business plans on new inventions are for companies and individuals who are either going into a new market, plan to secure financing, or plan to go into business. It serves to help identify those many unknown factors that impact entry into a marketplace. If your invention is truly a new concept, such as the discovery of radio or the discovery of the personal computer, how can you project sales on something that has never before existed? You can't, but this is a rare occurrence. However, it's easy to see that certain concepts such as these intuitively would sell and would sell well. Nevertheless, in the unimaginative world of bank financing, a track record is still required. As an inventor, entrepreneur, or established company, the only way you might be able to provide a bank with a track record is to use some of your past invention successes.

With all that said and done, there are ways you can assess and evaluate your inventions. In the chapters ahead, you'll discover ways of tracking and assessing your inventions during development.

Traits of a Successful Inventor

Before setting out on your invention development course, take inventory of yourself. You may call yourself an inventor, innovator, engineer, or product developer, but all of these terms are essentially the same as far as invention development is concerned. The dictionary defines innovator as "someone who begins or introduces something new." An inventor is described as "someone who conceives a previously unknown device." We know that either definition describes an entrepreneur and that both describe product developers.

BUT CAREFUL EVALUATION NECESSARY

Successful inventors follow the proper steps in evaluating, protecting, and marketing their inventions. Unsuccessful inventors typically skip the first step of evaluating their invention and charge forward with blind emotion.
–Michael S. Neustel, U.S. Patent Attorney

Inventors and innovators tend to be naturally creative, intuitive, flexible, problem-solving people. Whether you call yourself an innovator or inventor, you will use your natural abilities to develop your inventions. Develop these skills and you will achieve the outcomes you want.

Here are the most important traits of successful innovators and inventors:

- **Commitment, persistence, tenacity.** Frankly, business people like individuals who are committed, persistent, and tenacious. You just have to do it in a nice way. In other words, smile, be polite, and let others know you're committed to obtaining the right outcome.

- **A sense of urgency.** Almost invariably it takes longer to launch new products than we would like. But there's no doubt, it would take an awful lot longer without a sense of urgency. When problems arise, you want to fix them *now!* When opportunities arise, you want to move on them *now!* In all you do, you want others to "do it *now*."

- **Intuition and then analysis.** Inventors tend to know how things work without explaining it. You want to be sensitive and use your intuition to identify and solve problems. Then, analyze the possibilities and make decisions. You simply cannot guess or let your emotional bias affect your decision making.

- **Always learning, thinking ahead, anticipating.** Inventing is learning. It's discovering. When you focus on using your intuition, you'll be able to stay ahead of the game, and better yet, develop a keen sense of anticipation. Just like in professional football, the smart defensive back doesn't react to what the other team's quarterback and wide receiver do. He *anticipates* what they'll do. A sense of anticipation makes you well-prepared for even the toughest obstacles you might encounter.

- **Thinking in terms of customer benefits.** Almost every successful new product is developed based on satisfying the customer. Later, you'll learn how to key on various customer-driven attributes. In essence, they are customer benefits. They drive every aspect of invention development and the invention's performance.

- **Fast, flexible, focused (and analytical).** Successful inventors want to engineer their products, solve problems, and make prototypes fast. They can't wait around. They are also flexible and able to make changes and adapt to new discoveries that are made along the way in the invention development process. They stay focused in their efforts, solve the various problems they encounter, and keep their projects on track.

- **Enthusiastic and positive attitude.** You really must show a positive attitude and enthusiasm in your effort. It'll rub off on everyone else and is invaluable to your team effort. Can you think of a better way to be determined and tenacious than by doing it in an enthusiastic and positive manner?

Last, but not least . . . the only trait shared by all successful people is:

- **Continual improvement.** Of course you want to continually improve your inventions since breakthrough concepts usually come later on. But you should also improve yourself and your company's methods in every way you can. For instance, inventors who do not use the Internet are at a marked disadvantage compared with those who do. The inventor who improves his computer skills and uses the Internet is soon prospecting international business partners.

The more you think and act like an innovator-inventor, the sooner you will become a successful one. In many respects, a positive attitude will ensure you'll have the outcome you want because frankly, there's no other alternative. No matter what you do, you must continually improve and change until you get the outcome you want.

Take a personal inventory and ask yourself what are your strengths and weaknesses. Use your strengths to your best advantage. But don't forget your weaknesses; be aware of them and sensitive to them, so you can improve or at the very least compensate for them.

Positioning Yourself

If you establish yourself as an innovative leader in your field, your customers and licensees will always be interested in seeing what you're developing. They'll always be

FACT

The success gained by the inventor will be in direct proportion to his/her commitment, ability, preparation, and most importantly, his action and tenacity.
–Stephen Paul Gnass

FACT

One reason why our company has never deployed any of the new ideas we have developed here in Studio Red is because it needs substantial dedication. It needs a champion that will live, eat, and breathe the idea into fruition.
–Phil Bourgeois, CEO Studio RED

FACT

What comes first, an "understanding" of your invention or a "commitment" to making it a success? While it seems to make sense that a good understanding would come before commitment, it generally doesn't. If this were so no one would ever get married, because we could never truly understand our spouses. It is similar with inventing. Smart inventors know that if they make a commitment to learn all they can during the invention development process and make their decisions based upon what is best for the customer, the end result is going to be a good one.

looking for new improvements and ideas that will increase sales. Best of all, when you are ready to introduce new innovations, they will provide a built-in customer base, ready to help test and market the invention.

You can change and improve your attitude and your position just like you change and improve your products and systems. After all, you are an inventor who is bringing change to the world. So why not bring change to yourself as well? Just remember . . .

The weak indulge in resolutions, but the strong *act*!

Making Money From Inventions and Patents

When an invention is sufficiently developed and patents are granted, there are essentially five alternatives to consider. If your patents are being developed specifically for your business, you probably already know your intentions. If you are an independent inventor, you may also have some ideas. If you're just starting out, here are your choices:

1. **Sell the patent.** A reasonably strong patent should be worth at least $100,000 to the patent holder. But values are very difficult to determine. To do so, you may take sales projections for 20 years and then evaluate them as you would evaluate real estate rental property. But purchasers of patents generally assign a much higher risk to patents than they do to real estate.

   As an example, you may project sales from your patent for the next 20 years as $40,000,000 (averaging $2 million per year). With a typical royalty rate of 3%, your total income would be $1,200,000. However, it may take years to establish annual sales of $2 million. Besides, would a purchaser who has sufficient cash to purchase your patent have the same sales and profit projections as you have?

   For this and many other reasons, the outright sale of a patent is uncommon unless it occurs several years after the patented product has been on the market and sold. The exception is a company with vision that sees the potential of the patented product. If this is the case, it will approach you about purchasing the patents.

2. **Create a job.** One excellent outcome for an inventor is to create a job with an established company. Together, you and the company commercially develop the inventions and secure patents. There are many alternatives regarding your compensation for the value of your invention and your know-how. Royalties can be exchanged for stock (or stock options) and of course, steady income. The income from the job should be enough for you to live comfortably while you pilot the development and ultimate success of your invention.

   The chief advantage to this approach is that you have the invention know-how to make it happen and the company has the money to develop it correctly. With almost any new idea or invention, a company expecting to market it with a high degree of success will have to appoint or hire a product development manager anyway. Why not you? With the promise of the marketing success of your product, you can have a secure future with new invention opportunities as well.

3. **Go into business.** This approach is for those who have financial resources to tap. You build a business based on your inventions and patents. While this approach can

be extremely profitable, it is also the most time consuming. You will need lots of energy and stamina, since building a business and developing new inventions and patents is like having two or more full-time jobs! Suddenly you'll be an inventor, sales/marketing expert, manufacturing expert, and administrator. You'll need substantial financial resources to hire people and do it right.

4. **License the patent.** Since trying to arrive at a value for an outright sale is difficult, licensing is generally more acceptable to manufacturers and marketers. Licensing is much safer for the licensee than buying the patent, since the licensee does not have to worry if it has purchased a pig in a poke. For the licensor, the profits will most likely be greater than if he had sold the patent outright. Licensing is similar to a leasing plan. Through leasing, a business can conserve cash while investing money on equipment, raw materials, inventory, marketing, etc. Through licensing, the patent holder can typically earn from 2%–7% on sales, without the hassle or the liability. As you'll read later on in Part Four, there are several variables to licensing that can determine the royalty percentage and terms.

5. **Form a Joint Development Project.** This is an approach used by major corporations throughout the world. Frankly, it makes even more sense for smaller companies. Having financially sound partners finance your developments just makes a lot of sense. A Joint Development Agreement (JDA) is really nothing more than two or more companies agreeing to invest their money to research and develop a concept. Those involved obviously must have something to gain. For instance, a JDA with a raw material supplier makes sense if your innovation represents a substantial increase in the supplier's raw material sales. So, if you're already in business and want to launch a new frontier, this may be the approach for you. You'll learn more about JDAs as you create your development teams and in Parts Three and Four.

Deciding if You Are Ready

If you are ready to take action now, keep in mind that success can happen quickly or may take years. There are many determining factors, such as the amount of time you invest, market readiness, and the time it takes to find a licensee or development partners. At times, it's just a numbers game, looking for the one company that says "yes."

If you are a first time inventor, don't quit your job. It is probably going to take you a while to develop your contacts and build a team. You will start out part time and hopefully catapult your successes into a full-time occupation. As you learn the *From Patent to Profit* development system, the time frame for new introductions will decrease with each new concept. If your invention is in a field you're already active in, or you are creating a new product for your existing company, it will be easier to take action and build your development team.

In the next chapter, you will learn all about invention development strategy and the *From Patent to Profit* system and how to develop your ideas and inventions into a moneymaking enterprise.

AN EXCEPTION

The exception to the rule on the sale of a patent occurs when a company sees the future market potential based upon its future direction. If they are familiar with what your patent could mean to them and their industry, they'll pay. One example is a friend who sold a chemical tag that referenced a specific germ to a medical company for millions before ever coming to market. The reason was the company's familiarity with the difficulty in both that germ and the technology offered in the patent.
–Phil Bourgeois, CEO Studio RED

DO WHAT YOU DO BEST

Many highly successful inventors found out early on that they should keep doing what they do best . . . that is, invent and create. One of the most prolific inventors in U.S. history, Jerome H. Lemelson, licensed his 500-plus patents after a bad attempt at going into the manufacturing business. Yujiro Yamamoto, inventor of the cordless telephone, the telephone answering machine, the air purification system used by General Motors, and many other products, licensed his as well. Both Lemelson and Yamamoto knew very well to keep doing what they do best—invent—and leave the marketing up to the experts!

DESIGN AND DEVELOPMENT STRATEGY

Don't start out by spending a lot of money on patents . . . Research your concepts, develop a team effort, and let others invest in your success.

Invention development leading to a successful launch is rarely accomplished by a single individual. There's a certain methodology and system to follow. You'll learn that system in this chapter.

It is also rare that an inventor or company will embark on the development of a new concept and it will go unchanged throughout the entire process. Almost invariably, inventions go through a metamorphosis and are continually upgraded, improved, and literally re-invented. There are simply too many problems that will require solving before the invention will be ready for market.

New product development also doesn't have to be costly. In fact, properly done, it shouldn't cost much at all, as the costs should be shared by those who stand to gain.

So often an inventor or small company burdens itself by investing its money and its resources on product development instead of using various partners to make an appropriate contribution. Or, a small company may only be thinking in terms of purchasing all the various elements and materials needed and bearing the entire load.

Be prepared to look at invention development in a new perspective, one that is refreshing, will enlighten you, and will be kind on your pocketbook. We're also going to take it to the next level and show you how you can leverage your intellectual property and know-how so others will eagerly want to be your development partners.

PLAN AHEAD

Getting the patent is only part of the plan. Inventors have to look down the road and see where they're going and know they will actually make some return on the investment in a patent.
–Don Kelly, Intellectual Asset Management Associates, LLC

Three Keys to Developing Inventions

The system and methodology used by experienced inventors and product developers includes three elementary aspects that you must employ in order to be successful.

First, you must pilot your own inventions. Second, you want to form a team that incorporates the four essential processes of product development. Last, you want to continually update and improve your inventions.

First, Pilot Your Inventions

You have to be in charge of your own development project, your dream. You can't turn it over to someone else and expect the same results. You will team up with other experts along the way, but you must always maintain focus and control of your invention's destiny. Once you have other partners on your team, little by little they will take on more responsibility. Once your invention is launched, you'll have much less to do (other than collect royalty payments or share in the profits).

As you pilot your inventions, you will learn all about how various nuances can change or improve the marketability of your creation. You will learn all about how to get it patented, mass produced, and sold, thus producing income. Like a navigator, you, the inventor, will pilot your invention on a journey to success. When you have achieved the outcome you want, you will feel a tremendous sense of accomplishment. You'll also own a piece of history in the process.

Second, Monitor Four Processes

Once your idea, your dream, begins to unfold, and you have made a commitment to take control of invention development, you'll want to develop a plan. This plan includes identifying manufacturing and marketing partners and legal counsel. Combine the expertise of these partners with you at the helm of the inventing process, and you have four primary interrelated processes.

There may be other secondary processes you'll have to monitor, such as external engineers and package designers or qualifiers such as the Federal Drug Administration or local government agencies. You might also have companies involved that may be pioneering certain manufacturing equipment, just to make your invention. In other words, manufacturing processes and technologies necessary for the production of your invention may not yet be available, and thus have to be invented.

The four key processes that drive your development are:

1. Inventing
2. Patenting
3. Manufacturing
4. Marketing

Each process is headed up by a single individual, with you, as the project leader, in charge of the inventing process. The processes you will manage are illustrated in the Strategic Guide, which you may use to follow the project to completion.

Third, Continually Analyze and Improve

When following the Strategic Guide chart, you'll be continually analyzing and evaluating your invention. You simply cannot rely on your own opinion or the opinion of fellow workers or friends and neighbors.

A wide array of end-user customers must test the invention to determine what they

PARENTAL GUIDANCE

If you are a parent, you know best how to raise your child. You can't pay someone else to do it for you. Remember, inventions are like children: you want to raise them properly so they will become mature, desirable products. The difference between inventions and children is that inventions can make money, while children cost money!

BEAT A PATH?

No one knows what you want to accomplish better than you. Too many inventors I've prototyped for have spent the time and money necessary to develop an idea only to sit back and wait for the world to "beat a path" to their door. This isn't going to happen. Drive the car. You know where you want to go. The car can't get you there by itself—it needs direction and control—that's up to you.

–Matt Gilliland,
author Microcontroller
Application Cookbooks

like and dislike and what needs to be changed. You'll also have several primary buying influences that will have to be met, with the most important one being the end-user customer. And there'll be several secondary influences, many of which may have nothing to do with the inventive matter itself. These influences are addressed by applying customer-driven innovation (CDI) as discussed in detail in Chapter 3.

The Strategic Guide System

The *From Patent to Profit* system is illustrated in the Strategic Guide, which shows you how to qualify and accomplish your development objectives. Figure 2.1 summarizes the Strategic Guide. A complete version is in the appendix. Use the Strategic Guide chart to track your progress on an ongoing basis as you develop your products. Write in it or copy it as you work. You can literally check off each small step you take on your invention development journey.

FACT

An invention is never "done." There is always room for improvement, sometimes GREAT improvement. Keep this in mind when you think you've created the best invention in the world, and every day, look for ways to make it better.

–Andy Gibbs, CEO, PatentCafe.com

Figure 2.1 *From Patent to Profit* **Strategic Guide – A Simplified Model**

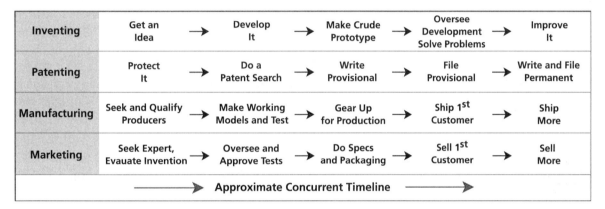

Inventing	Get an Idea	→	Develop It	→	Make Crude Prototype	→	Oversee Development Solve Problems	→	Improve It
Patenting	Protect It	→	Do a Patent Search	→	Write Provisional	→	File Provisional	→	Write and File Permanent
Manufacturing	Seek and Qualify Producers	→	Make Working Models and Test	→	Gear Up for Production	→	Ship 1st Customer	→	Ship More
Marketing	Seek Expert, Evauate Invention	→	Oversee and Approve Tests	→	Do Specs and Packaging	→	Sell 1st Customer	→	Sell More

→ Approximate Concurrent Timeline →

You'll soon find that invention development is not a single, linear, step-by-step procedure whereby you start at step one, end at step ten, and presto you are rich. Each invention has variables. Some projects may already have some team members in place, others may not. Sooner or later, you'll have to assemble this team in order to complete your objectives and achieve the desired outcome.

The Strategic Guide is broken down into four processes. Success in inventing and patenting is based on your piloting these four processes. Each process is led by one of your four primary team members. These team members must buy into the project if there is to be any chance of success.

The four processes are illustrated in the Strategic Guide by four overlapping, continuous, and somewhat flexible timelines. You are the pilot, the team captain. The four processes you pilot are:

1. **Inventing.** This process refers specifically to activities relevant to your invention. It includes your initial evaluation, prototyping, and the development of the idea from an invention into a saleable product. Along the way, your monitoring of the process will include several assessments and improvements that will overcome potential obstacles and will also give your inventions emotional appeal. You manage this process.

THE PLAN

Unless you don't mind wandering around aimlessly, it's always best to develop a plan so you and your company know where you're heading. It's just as important to follow that plan. This is especially true when it comes to developing new products and bringing them to market. So, make a plan. Keep it dynamic so it can be adjusted as needed and revisit it often.

–Don Kelly, Intellectual Asset Management Associates, LLC

2. **Patenting.** This process starts when you learn the basics about patenting, protecting your idea with an invention disclosure, conducting a patent search, creating the best possible patent strategy, and then writing a patent application (or more than one). In most cases, the first application will be a provisional patent application, followed by a regular application a year later. Once filed, the regular application is prosecuted and hopefully granted. Last, you may have to prepare for others' infringing your patent, which usually results in good news! The first stages of the patenting process you will do, but you'll eventually hand it over to a patent attorney, who'll then manage the prosecution of your patent applications.

3. **Manufacturing.** New inventions must be manufactured in a cost-effective manner, with the right quality, great service, and timely deliveries. Reaching those goals is not quite as easy as it sounds. To generate sufficient sales and garner sufficient profit or royalty income, you must find the very best fit between your invention and a manufacturer. If you work for a company, this partner may be the manufacturing department, or it may be an outside subcontractor or importer. You'll read how manufacturing fits into invention development in the following chapters and will see how to find and evaluate manufacturers in Chapters 9–11. One individual, such as a plant manager, will lead or oversee this process.

4. **Marketing.** Far too many inventors fail miserably with marketing. In fact, some inventors tend to go through the entire invention development process without any influence from the marketing partner. This is a huge mistake. Inventors who think that having a good idea and getting it protected with a patent is enough are usually in for a rude awakening. If your invention doesn't get sold, it will not earn any money. If you work for a company, your marketing partner may be your internal marketing department. If you are an independent inventor, either you have a specific company or sales team in mind to market the new product line or plan to find one to license your product. The primary focus of every principle disclosed in *From Patent to Profit* focuses on this perspective—your invention's marketability and thus its earning potential. You simply must get these experts on your team as early as possible. Again, one individual, such as a national sales manager, will lead this process and oversee the marketing effort. This individual plays a central role in your profit-making enterprise.

If you find that your invention activities are well-developed in the inventing and patenting processes, but you have not considered manufacturing or marketing, you have a lot of catching up to do. You have to have manufacturing and marketing specialists to validate your invention's cost effectiveness and sales potential.

Use the Strategic Guide

In your invention development activities, you will be piloting and monitoring all four processes *concurrently*. The overlapping timelines for the four processes are flexible and approximate, but try to stay on track as your ultimate success depends on the interrelationship between the processes and the successful completion of each one. For instance, trying to make changes and modifications without the input of your

---

NEVER BE LOST

Making sound business decisions in a timely manner is what the Strategic Guide is all about. This is right at the heart of making money on your inventions. If you are uncertain about what your next step might be, the Strategic Guide is there to direct you. Follow it and you'll never be lost.

FIRST STEP

At a *From Patent to Profit* seminar, one participant asked what would be the smartest first step to take in the invention process. One thought came immediately to mind and after some thought I knew this was the most desirable first step an inventor can take. The answer is, "Get a strong marketing partner on your team as soon as you can!"

marketing partner may result in an unsellable product. Or, modifying a prototype without a manufacturing expert's input could result in an invention that would be costly to produce or worse yet, would have to be assembled by hand, driving the cost skyward.

At the top of the Strategic Guide chart located in the appendix are the numbers, 1 to 10. Your objective is to complete each task under each number in the four processes before proceeding to the next number in the timeline.

Your journey has most likely already started at the upper left-hand corner of the Strategic Guide chart under the invention process, with "You have an idea." From there, you want to complete the other three items in column 1, "Learn about IP," "Learn about manufacturing processes," and "Learn invention marketing essentials" before proceeding to column 2.

As you can see, you'll take several small steps to protect your idea, evaluate it, and then go out and make things happen. The four parts of this book correspond with the four processes just discussed and the alternatives you may encounter along the way.

Monitoring and applying the Strategic Guide may appear a bit overwhelming at first, but relax; it gets a lot easier as you move forward. After a short while, you will suddenly find you are well advanced and closer than ever to completing your objective.

It's a Team Effort

An invention invariably starts as an idea born from a single inventor, but sooner or later, it will require a team effort to make it happen. Bringing a new product to market requires a tremendous amount of effort and resources. If nothing else, you simply will not have enough time or expertise to do it all yourself.

Ask yourself if you are an expert in all facets of your invention. Of course you're not! The *From Patent to Profit* system is about building your team, so you can pursue and enjoy the fruits of your success, without guessing what the market or your customers want or what the market will bear.

Your team consists of your four primary team members: you, a patent attorney, your manufacturing partner, and a sales/marketing partner. The four of you work together as a committed team to do whatever it takes to get the project launched. Later on, in Chapters 9–12, you will learn how to find and evaluate these partners if you don't have your team assembled.

Your team members are crucial to your success. You want a smart patent attorney who can help you secure bulletproof patent protection. You want a manufacturer who can give you the quality, service, and price you expect. You want a sales/marketing partner who can get your new invention sold and producing income. The better marketing experts may even get commitments for orders before the product is manufactured.

With an expert and committed team, you will be able to shorten the development time frame significantly and save an enormous amount of money. Think about it. It could take you six months to get an appointment with the decision-maker of a major retailer, just to make a sales presentation. You'll also have the expense of flying to a faraway corporate office. A marketing expert who has the right contacts can do this in a matter of days or weeks and without any cost to you. He'll also make a far more professional presentation than an inexperienced inventor.

TRADITIONAL APPROACH = FAILURE

An older traditional approach to inventing talks about five steps to success. They are: 1) protect it; 2) do a patent search; 3) file a patent application; 4) make a prototype, and 5) try to market it. This 5-step approach will usually guarantee three outcomes. They are: 1) You will spend a lot of money on a patent you have absolutely no idea how much money will make; 2) You will probably design around your patent by the time you have finished prototyping, and 3) You will have wasted a lot of time and emotion in pouring money into a dry hole and may never invent again. The reality is that this approach puts you into the 97% failure category.

DO IT RIGHT THE FIRST TIME

A sales representative I hired for one of my companies taught me a great lesson, "You never have the time to do it right the first time, but you always have time to fix your mistakes." How true. Inventions typically take three to six years before they produce revenue for the inventor—so don't rush. You have plenty of time to do it right the first time.
–Andy Gibbs, CEO, PatentCafe.com

Analyze and Follow Through

Experienced inventors and product developers know that an initial idea or invention is rarely marketable from the onset. Invariably inventions go through a metamorphosis that identifies the best manufacturing processes and the best marketing attributes.

The best way—perhaps the only way—to discover the best, low cost, high-quality manufacturing processes and marketing attributes is through continual testing and evaluation and analysis. It's easy to say that you have tested an invention thoroughly, but your strategy must include unbiased end users testing the invention thoroughly as well. After all, "you" are not the market. Your customers are. That's where profit is made.

Throughout the development of your invention and throughout this book, you will find areas that will challenge you to evaluate and validate your invention's potential. When you find problems, you have to fix them. You might have an invention that is unique and exceptional in ten ways, but one simple negative can destroy a product's market potential.

For instance, you may have discovered a trash bag design that is very easy to use and requires no dispensing; this is accomplished by nesting the bags, one inside the other. But the inherent problem is that there are no available technologies to nest the bags. Thus, the cost of such an invention would be ten-fold the present cost of trash liners since nesting would have to be done manually. Bear in mind that with this example, the price sensitivity of such a trash bag is only 2%. In other words, no one or no company will pay a price higher than 2% of the existing market prices for like products. So, your market at ten-fold the present cost is about nil.

Throughout this book, you'll learn the methods successful businesses and inventors apply to assess and analyze their inventions and make the appropriate changes. As in the preceding example, you may discover facts that will make you want to reevaluate your position.

Frankly, the price objection with the previous trash liner example is an extremely tough one to overcome, because any mechanized methodologies used to nest large trash bags, like the tapered flower sleeves commonly seen in supermarkets, are all too costly to overcome the 2% price objection. The best choice may be to abort the concept and to pursue another one that is more promising. Or, it may be to avidly pursue some single niche market that may have an interest in expensive nested trash bags, albeit we don't have a clue what that market may be. Do you?

The more involved and experienced you get with development and testing and assessing your inventions, the more you'll be able to anticipate problems and obstacles. Fortunately, when you become good at anticipating, you will already have thought up an assortment of solutions.

Seven Benchmarks to Profit

As you follow the Strategic Guide timeline, there are seven benchmarks you can use to track your invention development progress. You may have already reached one or two. When you reach these benchmarks, you can be assured that you have answered some important questions pertinent to a successful launch or offering for licensing, partnering, and so on. You must overcome any hurdles during this development phase because your development partners and/or customers won't usually do them for you.

R, R AND R

The three R's of inventing are Research, Research and Research.
–*Michael S. Neustel,*
U.S. Patent Attorney

You'll also notice that the Strategic Guide is separated into two phases, the first is the development phase and the second is the launch phases. All seven benchmarks are in the first development phase. So, you can say that satisfactorily reaching these benchmarks is tantamount to reaching the second phase, the launching of the new product. In many respects, once an invention enters the launch phase, it is no longer an invention, but a new product.

The benchmarks do not necessarily have to be met in the exact order listed below as there may be some variables, such as when patent applications are filed. Nevertheless, the benchmarks are presented in the Strategic Guide in the most efficient order so you can transform an idea into an invention, then into a legitimate new product, and then patent it and make profit.

The seven benchmarks are:

1. **Protect your idea.** One of the very first things you should do is write an invention disclosure. The United States is a "first-to-invent" country, not first-to-file country, so you want to establish your "date of original conception" right away. The second criterion, which is of more importance, is "reduction to practice." This means following through with diligence and showing that the invention works the way you say it does. Frankly, that's what you're doing by developing your invention and maintaining logs and other records. By doing this, you are validating your rights. This first benchmark objective is by far the least time consuming of all. Invention disclosures are explained in detail in Chapter 6.

2. **Conduct and evaluate a thorough patent search.** You should do this next, unless substantial patent protection is not of interest to you or your company. Frankly, even if securing patent protection is not of interest to you, you should do it from the onset just to make sure you are not infringing some other person's or company's patents. It is illegal to make products covered by a patent, unless you are the patent owner or have licensed it. This benchmark may also be met after you have done some initial assessment of the marketability of the product. However, make sure you do not talk seriously to others about licensing or partnering until a thorough search has been conducted. Patent searching is discussed in detail in Chapter 6.

3. **Do an initial marketability assessment.** You should do an initial market evaluation early on in order to determine the potential of the concept. You may start by discussing your concept with business associates, or even talking to family members, friends, and relatives. Next, you should talk to marketing experts who actively work and sell in the field of your invention, your own marketing department (for instance, your national sales manager), and customers and retailers if they apply. You'll see how to do this without revealing the inventive matter in Chapters 3 and 12. Later on you, your company, or a licensee can conduct beta testing, if desired. With strong results in your marketability research, you're on your way.

4. **Qualify manufacturing potential.** After a marketability assessment, you need to make sure the product can be cost-effectively produced. Your marketability research should give you some price points you'll need to meet; these will have an impact on how and where the product will be produced. In many cases, you'll know approximate manufacturing costs before you've finished the preceding marketability assessment. In Chapter 10 you'll learn about manufacturing cost scenarios you can apply to your inventions.

PATENT PROTECTION

It is unimaginable that anyone would want to proceed on a project with poor or marginal patent protection. Why would you want to spend a lot of time and money on an invention that anyone can make afterward or easily design around?

EARLY QUALIFICATION

One of the biggest advantages of having qualified experts evaluate your invention early on is not getting stuck developing an idea that is going to be a loser. Pursuing an idea and prototyping it without the input of qualified marketing experts usually results in an unsellable product. Pouring a lot of money into patenting and prototyping before properly evaluating the product's money-earning potential usually results in an unsellable product. The outcome in such cases is an emotional attachment after years of expending a huge effort and substantial expense.

5. **Prototype the invention to a reasonably functional state.** Every invention goes through various prototyping stages that will start out by illustrating basic functionality and appearance and proceed through to prototypes that can be used for testing and consumer trials. To reach this benchmark, you will need reasonably functional prototypes that can be presented to potential development partners, licensees, or customers. Early-stage prototypes won't cut it. Prototypes should look close to the real thing, albeit they don't always have to be. In certain cases, computer-generated designs may be preferable to an actual functioning model. For instance, inventing a new million dollar machine would best be illustrated through virtual prototyping—computer generated, a series of engineering drawings or, at times, conceptual sketches.

6. **Secure marketing.** This is the one area that many inventors never reach, in part because one or more of the previous benchmarks has been overlooked. For instance, the inventor may have neglected a consumer-driven market evaluation or not have determined that the product is too costly to manufacture or that it has poor patent protection. However, if you have been continually evaluating your invention during the course of its development and making all the right changes and improvements, this should not be a problem. Good ideas will always have willing marketers. Sometimes the only issue is to find the right ones. If marketing experts in the field of your invention continually show disinterest, then there's something essentially wrong with your approach or invention, and you need to find out what it is. Finding the right marketing expert for your team is probably your most important benchmark, and if you're not finding one, you have to know why and then fix it.

7. **Write and file patent applications.** You will want to start writing at least one provisional patent application during the development process, after you have made sufficient progress to portray your discoveries and their unique attributes. Rushing to file rarely accomplishes anything other than spending a lot of time and perhaps money on filing on unqualified inventive matter. Unless your sole interest is to accumulate a portfolio of worthless patents, you should reasonably validate market potential beforehand. You can do this effectively by using confidentiality agreements, which are an important part of an inventor's toolkit. Once you understand the process, you should file a provisional patent application in a timely manner. Later on, you may file additional applications if new discoveries and improvements are made. Ultimately, within one year of filing a provisional application, you'll begin filing the permanent one. This strategy is discussed throughout the book.

When you've reached your seventh benchmark, you're on the road to profit. Not much can stop you at that point. The only question will be, "How large will sales be and how much profit will be made?"

Based on these seven steps, it may be more appropriate to describe the Strategic Guide process as "how to turn your ideas into profits and then into patents." Frankly, that is one of the keys to the *From Patent to Profit* methodology. You will find that by mastering the four processes in the Strategic Guide, you can begin earning money before you incur any major expenses. In fact, if you follow the guide properly, you can avoid almost all costs you might otherwise have to bear. This includes costs to patent, prototype, license, test market, etc.

Last, keep in mind that most breakthrough ideas occur in the later stages of development, sometimes even long after a product is launched and on the market. It is also rare that an inventor's first invention is a smash hit. It is usually sometime later that an improvement will substantially propel sales upward. The good news is that future improvements take far less time to develop.

Thus, once the seven benchmarks have been reached, a smart inventor will focus on continual improvement. He/she will continually watch for future trends and customer-driven advantages that the present product doesn't have. These future improvements will not only improve sales potential but may also result in additional patent protection and a longer product life cycle.

Benchmark #1: Protect Your Idea

Begin the invention protection process by first preparing an invention disclosure document describing your invention. The content of your disclosure is directly related to the potential scope of your eventual patents and should be as accurate and complete as possible. Since writing an invention disclosure is directly related to patent scope, you should learn the basics about patents and patent scope so you can include all the inventive aspects of your invention, not just "the product" itself. This is a common mistake.

You'll soon find that you can probably secure additional patent protection on the method of use, various individual elements, and perhaps even the process of manufacturing. For this reason, instructions on writing an invention disclosure are illustrated in detail in Chapter 6. When the disclosure document is completed, check the appropriate box on the Strategic Guide chart.

Your invention disclosure usually starts the paper trail that is created as you change your idea into a product, thus establishing your first-to-invent status. This paper trail becomes important in your product development activities for a variety of reasons you will read about later.

Maintain Logs and Other Records

All of the subsequent steps you take are part of your paper trail. In a log book, such as the Scientific Journal (see appendix), record your invention and patent-related activities. Logs are the

Figure 2.2 Scientific Journal Sample Log Page

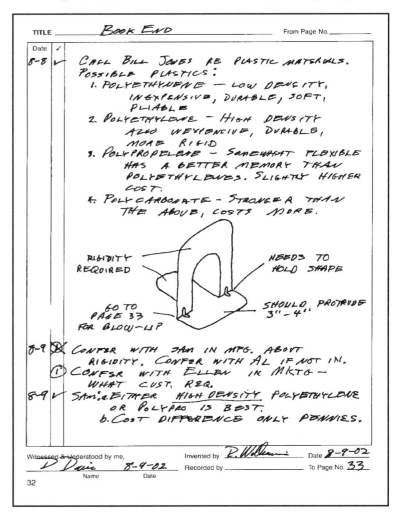

WHO INVENTED THE
TELEVISION?

Actually there were three important inventions leading up to the mass marketing of the television. First, in 1916 William D. Coolidge invented the vacuum tube and was granted U.S. Patent 1,203,495. Then, in 1930 Philo T. Farnsworth invented the first television system and was granted U.S. Patent 1,773,980. However, television was not a feasible mass-market product until Vladimir K. Zworykin invented an economical means of producing images as we commonly see on televisions today. In 1935, he was granted U.S. Patent No. 2,021,907.

primary means to establish the fact that you are reducing your invention to practice, hence securing your first-to-invent status. If substantiation is ever needed to prove that you are the original inventor, or to prove the date on which you conceived the idea, logs can represent an indisputable paper trail if they are properly maintained. Without logs and records, you could lose your first-to-invent status and your claim to an invention.

Your invention records will consist of logs, sketches and drawings, letters, computer files and CDs, prototypes, samples, and any other materials related to your invention. You can keep these records in envelopes, file folders, or cardboard cartons. Give your invention a nickname (if you haven't already) and write it on the notebooks, sketches, files, and containers you use.

Another positive aspect of maintaining logs is that as you record your day-to-day activities and discoveries, you are also developing a reference guide that you can use to quickly review what you have learned. Inventing takes time; over the weeks and months, you may forget certain important test results or discoveries. With a well-maintained log, you will have a valuable resource when you need it. It can save you hours and will assuredly help speed up invention development.

The best approach is to use a dedicated log for each invention. It should be sequentially numbered, with a glued binding (stitched is not important), with ruled or graphed paper and a defined left column. A glued binding prevents someone from fraudulently trying to insert pages at a later date. The log book page in Figure 2.2 is from the Scientific Journal (see appendix) and illustrates how a typical page would be recorded and signed.

Your log book should include a left column where you can date your entries. As you begin the day, you may start out with a checklist of things to do, and then you can check them off when they are accomplished (illustrated in the second vertical left side column). The topics you can include in a log typically are:

- Those to whom you have talked and what was discussed

- Discoveries about your invention

- Solutions to manufacturing problems

- Simple concept sketches

- Test results from experiments

- Ideas for new inventions . . . related or unrelated

- Ideas on new uses of the invention

- Marketing ideas on the invention

- Problems you may anticipate

At the end of the day, one of the best methods to verify content is to have the page signed and dated by someone who understands the subject matter.

Try to keep the log with you at all times. You never know when the creative juices might flow. Sometimes the "video in the brain" plays something important that you will want to remember when you are driving. You may want to pull over and make a quick sketch. If you do not have your log, use whatever is available, even a placemat in a restaurant, to sketch your ideas.

Maintain Sketches, Drawings, and Notes

Beginning on the first day you come up with the idea for your invention, date and sign all sketches, drawings, and handwritten notes. Seeing your signature on sketches and drawings (engineering or otherwise) may be important if they are used later on with a subcontractor, supplier, or potential licensee.

If a sketch is later corrected, maintain both the correction and the original. If you are ever challenged, a paper trail with serially dated sketches showing the metamorphosis of your invention will help substantiate that you are the original inventor.

Maintain File Folders for Invention-Related Matters

In file folders write your invention's nickname on the tabs and keep the following data:

- Incidental sketches and notes, even if they contain errors
- Correspondence to suppliers or others involved in the project
- Technical data from suppliers

Maintain Separate File Folders for Patent-Related Activities

In these folders, maintain the following data:

- Your invention disclosure
- Your patent application drafts (sequential application drafts improve the paper trail) and drawings
- Correspondence with your patent attorney and the Patent Office
- Correspondence with potential licensees

It is not uncommon to create three or four file folders on a single invention in a short time period.

Store Records

When your file folders become full, you can store them in cardboard cartons. Make sure to keep the cartons in a place where they will not get wet or damaged. In addition to the overflow from file folders, keep the first prototypes of your original invention, regardless of how crude they may be. These original prototypes can be important in proving reduction to practice. Write the invention's nickname on the outside of the boxes for easy identification.

Benchmark #2: Conduct a Patent Search

While your patent search will almost always have a significant impact on the patentability of your invention, it also plays an important part in its development as well. A patent search usually occurs sometime after you have done some initial invention development, at least enough in order to understand the inventive matter. Patent searching is also key to knowing if your innovation may be infringing an existing patent, to understanding patent scope, and to writing applications. All this is covered in depth in Chapter 6.

When reviewing related patents, you very possibly will find your invention does not infringe any others. While this is good news, you might also want to check the related

---

WARNING!

It used to be that only stitched journals were recommended and that a line or page should never be skipped. But sometimes this is difficult to do. If you're actively developing your concept, a log book is only one of many items that become part of a paper trail. Thus, what may be written in a log will almost always be corroborated in other materials. For instance, a simple sketch gets converted into a CAD drawing. Keep in mind, that ink date testing is easy to do so it's not difficult to legally determine when an entry was written. Someone trying to commit a fraud by back-dating a journal won't get away with it. What is more important is that you take action, move forward, protect your concepts, and get them sold!

SEARCH TIP

Good: Search U.S. patents at http://www.uspto.gov for free. Better: Search global patents at http://patentsearch.patentcafe.com Best: hire an attorney to conduct a patent search and provide a legal opinion of patentability.
–Andy Gibbs, CEO, PatentCafe.com

THE IMPORTANCE OF
SEARCHING

Do a patent search yourself or through a patent attorney. It makes no sense to reinvent the wheel if it has already been patented. It's a waste of time and money. Although a patent search is not guaranteed to bring up all the prior art (patented products) that predate yours, it's a necessary investment of your time and money. It is very embarrassing to an inventor to introduce a concept to us that has already been patented by someone else!

–Paul Lapidus, CEO, New FunTiers

art, or patented products, that predate yours, and ask yourself if any of them have been successful. If not, why? Try to use this to your advantage. Is the industry dominated by just a few companies and the barriers to market entry too great? Are the manufacturing processes related to the patents too costly? Most new inventions fall flat because of poor marketing plans, which, of course, are unrelated to patentability.

You may also find that the inventive matter in your invention, or something very similar, is already patented. You don't want to proceed if your invention is going to infringe another patent, at least not without the patent owner's permission. If something similar is patented, this narrows your patent protection. Would you want to proceed knowing your invention may be easy to design around? These are serious decisions. Here are some alternatives to consider:

1. If it's already patented, do you want to try to license it from the owner?

2. If it is already patented, do you want to try to design around it? Create the next generation?

3. If something similar is patented, will the scope of your narrow patent protection be strong enough to accomplish your objectives? To license it? To protect your company's investment?

4. Do you want to drop the project altogether?

From the perspective of creative brainstorming, designing around an existing patent can be challenging and fun. If this happens to you and your decision is a design-around, read the section on problem solving in Chapter 4. Then get your creative juices flowing and try to figure out a design-around that can make money. In doing so, you just might make an important discovery even though the scope is considered narrow.

The key to designing around an existing patent is to try to imagine the next generation of the product. Ask yourself what the next generation is likely to be. Answering that question qualifies you as a skilled inventor in your field.

Benchmark #3: Evaluate Marketability

This is an important initial step that should not be postponed. Usually marketability opinions start with a small core of insiders, such as employees at work, or even family and friends. But to acceptably provide evidence of an invention's potential marketability, you'll need the input of some expert marketers in the field. You also may need to research the sales volume potential. Otherwise, you'll just be guessing whether or not your invention will ever sell, let alone earn sufficient profit.

Making this initial marketability evaluation is not a substitute for thorough, in-depth studies with consumer testing and so on, but it is a means to substantiate whether there is market appeal for your invention and whether the market is large enough to justify the investment. It can also provide you with some feedback that may give you a clearer picture of what may be appealing to consumers.

Methods for Exploring Marketability

You have several means you can use to explore potential marketability while maintaining your rights to file patents. Three methods do not require having a patent application filed, and three do. Usually the best methodology to assess marketability initially is to use confidentiality agreements. It's also the method most commonly used by experienced entrepreneurs and established companies.

These following three methods don't require a patent application to be filed. They are discussed below in the general order of desirability, with the most desirable being the first.

1. The best way to evaluate the merits and sales potential is with a marketing expert you know and trust. If your invention development is for your company, this is easy for you to do, since it is most likely you'll be consulting your company's national sales manager. If he gives the green light, you're on your way. If the individual works for a potential marketing partner or licensee, use a confidentiality agreement and discuss it. The beauty of discussing and assessing the inventive attributes early on with trusted experts—potential licensees—is that frequently it will lead to a license agreement or some form of JDA.

2. If you plan to license or partner your invention and know some potential candidates, you may start out with a New Product Summary (described in detail in Chapter 10), which does not reveal the confidential inventive matter. If you elect, you may follow-up with confidentiality agreements (also discussed in Chapter 10) as needed. If you receive a favorable response from sending out a simple New Product Summary, you're probably on the right track.

3. If you do not know potential licensees or marketing partners who can help substantiate marketability, you are at a disadvantage. Obviously, you'll need to have this partner on your team soon so sales can be made. Ultimately, this is going to be your sixth benchmark, the "Commitment from a Marketing Partner." If you fall into this category, and assuming you want to maintain confidentiality, you may try to qualify marketability in some other ways. One way you may try to evaluate an invention is through retailer or user surveys, using "walk-in" testers or third-party surveyors. Retailer surveys may be based on the use of a New Product Summary without a confidentiality agreement. This may be a more costly approach, as it usually requires some testing agency or university to conduct the surveys. Consumer surveys may also may be conducted with actual user tests and, by using confidentiality agreements. The results can be used effectively to attract marketing partners.

The next three approaches you can employ assume that you want to maintain your patent filing rights and require that you have a provisional patent application (or the permanent one) already filed. These methods are not usually as preferable as the three approaches discussed above, but they have their place.

4. File a provisional patent application and start manufacturing and selling your invention right away. In doing it this way, you are actually putting benchmarks 4, 5, 6, and 7 ahead of benchmark 3, and this may not be the most desirable approach. It's rare for an initial invention to be the "final model." So, be prepared to incur some additional costs for product modifications should it be required.

5. File a provisional application and overtly seek out marketing partners without using confidentiality agreements. You'll want to aggressively pursue candidates in the field because within one year of filing the provisional application, you'll have to file the costly permanent one—and that one year can pass by fast. When using

NEW OR OLD TECHNOLOGY?

Did you know? Herman Hollerith revealed the first computer-related invention in his U.S. Patent No. 395,782 in 1889. It was entitled "Art of Compiling Statistics." In 1892, William Painter invented the bottle cap and was issued U.S. Patent 468,226. Thomas Edison invented the motion picture projector in 1893 and was issued U.S. Patent 293,426; however, it was not until 1897 that Edison was granted U.S. Patent 589,168 for a motion picture camera. Willis H. Carrier invented air conditioning 100 years ago and was issued U.S. Patent 808,897 in 1906.

this approach, make sure you don't discuss any confidential trade secrets pertaining to the patent application, sensitive marketing studies, engineering drawings, and so on.

6. File provisional patent applications and continue development and sales and file new provisional applications until someone says, "I like it."

Oh well, the last one isn't all that appealing, but who knows, it is used. With a provisional application filed, you can more openly discuss marketability with your potential marketing partners. The downside of filing a provisional application early on is that you will incur the higher cost of patenting a year later. Even with a committed marketing partner on your team, one year may not be sufficient time to develop an invention, complete the packaging, and so on. If you are licensing your innovation, you may ask your licensee to pay these costs, as discussed in Chapters 14–16.

Do Your Own Preliminary Survey

Here are four methods you may use to conduct a preliminary survey yourself. Before you do this, however, read Chapter 3 on customer-driven innovation and review the section in Chapter 10 on how to write a New Product Summary. Use it to prepare your surveys and to be able to talk about your invention without revealing the inventive matter. The four methods are:

1. **Survey retailers.** If your product is to be sold to retailers, call a typical store that would carry the product and talk to the store manager or department manager. Tell him/her that you are looking for a certain type of product and ask if they carry it. For instance, act as though you are a consumer and ask, "Do you carry a biodegradable paint solvent?" If he/she responds affirmatively, you know it has already been developed. If the answer is "no," ask if he/she knows where you can purchase such an item or if it can be ordered. This will further indicate if such a product exists. You can also ask the manager if he's had other inquiries for such an item. You can even ask him if he thinks something like that would sell. Listen carefully, because an experienced manager will have heard a lot over the years. Then, duplicate this inquiry with other retailers to see if the results are the same.

What you may find is that there really is not much demand for your invention. You may also find that there are biodegradable paint solvents on the market, but when you analyze the contents, you find that they are toxic or very high priced. In contrast, you know that your product is not since it's made from citrus oil (for instance).

2. **Survey commercial/industrial users.** If your invention is for industrial or commercial use, you can contact the appropriate manager in a plant and determine the interest level. Generally speaking, you should talk to someone in the actual operations department and not to someone who does the buying. For instance, ask the vice president of operations in a manufacturing facility, "Would you be interested in a new electronic device that automatically monitors hydraulic press safety hazards?" If the answer is "yes," you know your invention might have potential. You can also find out if there have been many safety issues concerning their hydraulic plant operations and what the current safety steps are. Likewise, if the manager responds with "no," find out why. There may be little demand because there are few hydraulic-related prob-

---

RETAILER SURVEYS

The method of asking a department supervisor for information about a new product is a method taught by Phil Henry, President of Gray Electronics. Phil has several successful electronics-related inventions. Phil also uses this method of screening as what he calls a "Poor Man's Patent Search." In other words, if the expert working for the retail outlet says that a product is not available and Phil wants to pursue the invention, he feels that is reason enough to have a thorough patent search done.

lems in the industry. Make sure you talk to others in the industry in order to confirm your findings. Don't rely solely on one person's opinion.

3. **Survey distributors.** The sales manager of a distributor of your new product should know what customers are asking for and what might sell. For instance, you call a local distributor and ask the sales manager, "Do you carry disposable heat-resistant gloves for baking?" If the answer is "no," ask if there is a demand or "the possibility of a demand" for such a item. Then listen and you'll hear all about why it may or may not be desirable. You may also discover the potential sales volume and where the price point should be. You may find that there are disposal gloves on the market, but they do not sell well because the price is too high. This is invaluable information because now you can ask the begging question, "What would people pay?" Keep in mind, however, that distributors are only order takers and do not necessarily know much about their customers' true needs.

4. **Survey manufacturers' representatives.** Talking to manufacturers' reps is similiar to talking to a distributor's sales manager. You can ask essentially the same questions, but you'll tend to get more qualified answers because the manufacturers' reps handle only a few products, understand them better, and tend to have a much better grasp of end-users buying habits. They may arrange to have your innovation sold *en masse* to the distributor, but the market demand for any new product is ultimately generated by pull-through from the end users. For instance, manufacturers' reps may go directly to the district manager or operations manager to create the demand. In contrast, the distributor sales reps will tend to call on the buyers instead and take orders as placed. Buyers rarely make the decision to change or to add new products. It's usually done by their bosses, the managers.

Frankly, by talking to several candidates who will either use or sell your products at the retail or industrial/commercial level, you are building a network of potential customers who can help in the start-up stage. They may be perfect test markets for your invention. As you will see later on, they may also be able to help you find manufacturing and marketing partners.

In your phone surveys, if you receive a lukewarm response, don't give up right away. Try to field as many objections as possible to better understand them. In Chapter 3, you will learn some of the secrets to making your inventions more desirable and more saleable to end users. In fact, what you will learn in Chapter 3 will become the central focus in all your invention-related design activities.

Guiding Principles for Evaluating Marketability

There are several other assessment principles you can use to help determine your invention's sales potential. The following are not substitutes for evaluating marketability, but are additional qualifiers to any marketability surveys you may do.

Look for Trends

The best innovations to develop are those that create long-term trends. If your invention is only a fad, you may waste a lot of time and effort without receiving a proper return

---

DESIGN TRENDS

Design trends in vacuums in the twentieth century tended to follow scientific developments in transportation. The first ones with the little headlight resembled a locomotive. Then, when the jet age arrived, the new canister versions resembled jets and rockets.

GO FOR IT!

If your idea is not a long-term trend, but a short-term fad and you want to try to market it, then forget patenting and go out and try to sell as much of it as you can, as fast as you can! Just go for it before others start copying you.

A MILLION $ QUESTION

Finding the next trend is the million dollar question. In the technology industry we call it the next "Killer App," or killer application, meaning what will be the next market maker like the cell phone or PDA.
–*Phil Bourgeois, CEO, Studio RED*

on your investment. Use your intuition, instincts, and just plain common sense to determine if your idea is a fad or trend. Don't make an emotional response to this issue. Qualify it if you are unsure.

Corporate America often looks for short-term successes, short-term profits. But this is relevant to existing product lines. With new products and product lines, this approach would be similar to a company president asking, "Why don't you sales/marketing people go out and look for new gimmicks and fads for us to sell?" This is in direct contradiction to the philosophy of the same entrepreneurial corporate CEO who says, "We want long-term proprietary products." Looking for short-term profit is usually not compatible with inventive, innovative philosophies like those that have built General Electric, Ford, IBM, and more recently, Cisco and Microsoft.

Remember that if your innovation-invention is based on having instant success, that success can fade away just as quickly. Look for the long-term trends that may slowly develop. They'll give you the long-term security and income you deserve for your efforts.

Study Probable Competitive Responses

Another consideration with new inventions is to make certain your competitors cannot respond by trying to drive prices down or drive you out of business. This can happen regardless of how wonderful your new innovation may be. If your invention is a threat to their existing technology and your sales are soaring, they will most likely be determined to fight.

TRENDS MAY TAKE TIME TO TAKE HOLD

If you have confidence in your invention's success, but it is not being readily received, be patient. If you have carefully thought out all the product's attributes and you know that sooner or later it will have broad acceptance, a *trend* may ultimately develop. You just may be a little ahead of your time. It is ironic but true, that it sometimes takes years for a trend to develop, even with the best ideas. But when a trend does take hold, it sticks around for the long term. Some excellent examples are Ziploc® bags, PC computers, and even small cars! Always try to innovate with future trends in mind.

Competitors that are vertically integrated can create serious problems. Vertically integrated companies—those that are the supplier of raw materials and the intermediate components used in the product's assembly—tend to have a cost advantage over those companies that are not. If you are a threat to a company such as this, you may not be able to secure consistent raw material or parts supply. This can be particularly distressing in a market that is being ramped up, as supply tends to fall short of demand anyway. On the other hand, when old-fashioned industries are being replaced by new technologies, there may be little challenge to your new product launches. There is much to learn in new emerging industries with great opportunities for new products and innovations. A few examples of new industries replacing old are:

- PCs and software replacing typewriters and adding machines
- Plastic bags replacing paper bags
- Plastic bottles replacing glass bottles
- Lasers replacing surgical scalpels
- Robotics replacing people
- Small cars replacing large cars
- Email replacing snail mail and scanning replacing faxes

At times creative new products may not elicit much of a competitive response, yet at other times they will. It's best to qualify this upfront. Foreign automakers entered the United States with the small car market easily because they had been building small cars long before the U.S. producers. They had experience and quality. The U.S. car companies simply could not respond quickly. Similarly, plastic grocery sacks replaced paper grocery sacks and the competitive response of the paper industry was virtually non-existent because the paper industry had no expertise in manufacturing plastic bags and worse yet, paper prices were soaring, so they couldn't respond by dropping prices. Even if the paper industry had responded, it would have been for the short term because paper costs had been steadily increasing over the last fifty years, whereas plastic resin prices have been more or less stable during the same time period.

You'll also find that trying to license or partner your innovative ideas to the dominant existing suppliers and their defensive-minded management is very difficult. For instance, IBM declined entering the PC market long before Apple launched it. You will learn in Chapter 9 of Part Three that old-fashioned industries that think in terms of high-volume generic products and protecting their existing market share find it very difficult to embrace invention and innovation.

Compare Your Invention to the Industry's State of the Art

You can also rate your invention relative to the present state-of-the-art and the existing suppliers in the industry. First, evaluate the present state-of-the-art in your industry. To give you an example, when plastic grocery sacks began replacing paper sacks in supermarkets, it was fairly easy to see that the current state-of-the-art mainly consisted of a number of large, inflexible, paper bag manufacturing conglomerates. These mega-corporations had little desire to do anything but protect their existing market share.

The paper industry's state of the art was a hundred years old. With recent introductions of abundant high-strength, high-density polyethylene plastic, the plastic grocery sack was destined to ultimately cost far less than paper. It was clearly an opportunity for an innovative company. If you or your company wanted to invent new product opportunities in the grocery sack market in the late 1980s, would it have been a paper bag product or a plastic bag product? Obviously, the future here would be with plastic bags. It was unique and bound to be highly cost effective in the long run.

STRENGTHS AND WEAKNESSES

Just as you have considered competitive responses in your ability to enter the marketplace, you will want to consider carefully competitive responses to each unique attribute of your invention. One of the keys to inventing and innovating is to declare your niche and be prepared to move forward and take it. You want to make sure it is a niche you can protect. The more difficult the competitors' ability to respond, the better your opportunity. Even if you erred and did not anticipate their responses, being first to market can give you the competitive edge to maintain your newly found niche. Just keep inventing and stay ahead!

Figure 2.3 State-of-the-Art Positioning

According to the chart in Figure 2.3, the large paper bag producers and the state of the art of their product line would be rated between 1–4. The emerging, high-volume, low-cost plastic grocery sack producers would be rated between 5–8. Generally speaking, highly unique products such as the thick stand-up, handled, plastic or paper shopping bags and the canvas bags would be rated at 9–10. These markets are highly specialized and may or may not be patentable. In this particular case, the paper and canvas bags would probably not have been, but the plastic ones would have been. Use the chart in Figure 2.3 like this to get an idea of where your new invention is positioned.

1. Mark the approximate location of the state of the art in your industry as a whole.

2. Next, mark the position of key competitors' products.

3. Last, mark your invention's position.

By positioning your invention on this chart, you may also be able to get an idea of your product's potential sales volume. Is it in the emerging unique, high-volume location? Or is it in a highly unique, more specialized niche?

Identify Unique Attributes that Sell!

Is your invention a major innovative improvement with unique attributes or is it just a minor variation on the same old-fashioned theme? Generally speaking, you need a decided advantage to entice customers to buy your innovation. If you're uncertain whether or not your invention has significant advantages, don't worry just yet. The next chapter lists several attributes and methods to improve your invention's saleability. You can use these customer-driven "potentials" to direct development.

Benchmark #4: Learn About Manufacturing

For those who are inexperienced in the field of invention, this step is one of the most important steps of all. You simply cannot expect "whatever you've dreamed up" to be easy to manufacture. Too often, what may seem like a good invention, an easy thing to make, may be nearly impossible.

For instance, placing a coupon inside a plastic grocery sack during the manufacturing process may seem like a simple inserting operation. However, what is overlooked is that plastic grocery sacks are manufactured out of a continuous, closed tube (called a layflat tube), which begs the question, "How are you going to insert thousands of coupons, one-at-a-time, every 21 inches or so, inside a continuous, closed tube? Second, when the bag is formed from the continuous tube, it is sealed at both ends first and stacked, and then 50 to 100 bags are cut out simultaneously forming the open bag mouth. So how are you going to insert 50 coupons in 50 thin plastic bags stacked one atop the other, to the height of about $^3/_8$ of an inch? It can be done only by hand, which is too costly. What is also overlooked is that adding a single coupon inside a plastic bag during the manufacturing process would require new shipping cartons. Depending on the thickness of a paper coupon, it may double the cost of shipping.

All this is surprising, but true. The cost of putting a coupon in a plastic grocery sack during manufacturing would probably drive the cost from about 1.5 cents up to about $.15. There are a lot cheaper ways to put coupons inside a plastic grocery sack, such as having the checker or bagger insert it.

Even though you or your company may not be manufacturing the product, you have to have a good understanding of the cost-effectiveness of the manufacturing process. It would be a travesty for an inventor to pursue such an item as the coupon-in-a-grocery-sack concept, wasting hundreds of hours and thousands of dollars in patenting costs. Whether the end product was planned for your company or you intended to license it, no one would be interested in taking on the project.

You must determine two important aspects early on. They are:

1. **What is the price elasticity of products in the field of your invention?** This essentially is finding out how much more consumers will be willing to pay, if anything at all, for your invention, whether it is an improvement over the existing products or processes or is entirely new. Generally speaking, it is only about 2% for high-volume generic product lines. Brand loyalty and force of habit can even make 2% unreachable. For instance, using our previous plastic grocery sack example, the cost of plastic sacks is an expense item, the number one expense item in a supermarket. If you are asking a major chain to pay only 2% more for its plastic sacks, you may be asking it to spend an additional $250,000 to $500,000 per year! Purchasing agents get fired over such decisions. With a small, privately owned supermarket, the 2% increase may cost the owner an additional $200 per year. In these days of tight budgets, consumers would most likely be reluctant to spend even 10% more on unique new products in just about any field.

2. **How will your invention be made?** You probably have a good idea how it'll be made if you are developing a new product for your company in an existing product line. Or, if you have experienced friends or business associates, you'll have a viable resource to tap. Generally speaking, you can consult with these existing contacts and better understand the various manufacturing processes that may be required to build your invention. If you are unsure, you need to find out. One of the best ways to find out how your invention would be manufactured would be to inquire with your local Small Business Development Center (SBDC). The SBDC is a division of the Small Business Administration (SBA) and almost all have a wealth of knowledge in the manufacturing arena. Appointments are free.

As a product developer, you simply cannot go off half-cocked. If your innovation costs 2% more to manufacture and requires a 2% larger shipping carton, then more warehouse space will be required to stock the item. If there are costs for dies and molds to make the new product, this too must be amortized into the profit picture. Just these simple changes can result in a 5% to 10% increase in price. If you are offering the invention for license, your royalty may, in effect, drive up the cost to 10% to 15% over existing products. Or, shall we say, it'll drive up the cost well outside of the 99 percentile, where 99% of consumers will consider the purchase.

If you cannot answer at least these two important questions early on, you're not ready to proceed with patenting or marketing of your innovation.

Benchmark #5: Prototype Your Invention

Your invention will go through a metamorphosis as you substantiate its marketability and learn about related manufacturing processes. This initial input almost always has an effect on your prototyping process. It will direct the development of salient attributes and generally establish the foundation for prototype design. Your prototyping may be done concurrent with your marketability and manufacturing activities. But generally speaking, you want a good understanding of what would make your invention highly marketable, and how it would be manufactured, long before you get into the final prototyping stages.

PATENT MODELS

With the Patent Act of 1790, the U.S. Patent Office established the requirement that a working model of each invention be produced in miniature. In 1880, the model requirement was deemed impractical and models were then only required when the Patent Office Commissioner thought it was necessary. By 1880, 246,094 patents had been issued and about 200,000 models had been submitted and stored at the patent office. For interesting facts on patent models, visit the Rothchild Petersen Patent Model Museum web site at http://www.patentmodel.org.

In the later phases, prototyping may include in-house testing, customer trials, and eventually larger, expanded test markets. Your manufacturing and marketing partners can help you tremendously during these later phases. If you can get them involved early on, do it.

If your invention is a sophisticated machine or a complex electronic device, you will pursue a different form of development. For instance, machinery is usually prototyped as it is built and its function may be illustrated through computer-generated design. With electronics, you'll most likely need a competent engineering or prototyping company if you don't have in-house capabilities.

The Five Phases of Prototyping

We know that a prototype is usually necessary for you to properly turn an invention into one with all the right customer-driven innovation (CDI) attributes you'll learn about in Chapter 3. Prototyping is not a simple one- or two-step process. But as you will see, it can be done quickly and inexpensively if you take all the right steps.

A prototype invariably begins as a crude model and then transforms itself, little by little, into a cost-effective working model with the proper marketing attributes. It is from this perspective that your prototyping will probably not be entirely completed until after you have assembled your manufacturing and marketing partners. Your final prototype must be what your marketing expert says is "saleable" and your manufacturing expert has determined that, "it can be cost-effectively manufactured."

What follows are the prototype phases that a typical simple invention follows. You may make only a few before you get one that can be presented in sales presentations. Subsequently, the final working models will be manufactured and put through final testing before being released to the public.

Phase 1. The first prototype is usually simple and may be very crude. It is only to get an understanding for size, shape, and basic design. It may be nonfunctional or function crudely. It does not have to be made of the desired material. For instance, it may be carved in balsa wood or molded from clay instead of being injection-molded in plastic. If your invention is easy to prototype, you may go through a few crude, first-phase prototypes just to determine size.

Phase 2. The second prototype phase can be used for testing functionality and should more closely resemble what the true invention should look like and the desired material. You should personally try to do as much testing as possible, regardless of redundancy. Try to validate the appearance and functionality with marketing experts with your company or others in the field of your invention, but use a confidentiality agreement.

Phase 3. Testing your second-phase prototype will expose the need for some improvements. In the third phase, you'll incorporate these improvements and hopefully you will also be able to put some "dazzle" or "sex appeal" into the prototype. In this stage, you will continue to be a primary tester, but it is important to have the product tested by others in a test environment. Ideally, you want end users to be those testing your innovation. When you test and observe, be superobservational and hypersensitive to functionality. Try to "crawl inside" your invention and see what is really going on.

When others try it, always listen to their comments, positive and negative. All negative comments are good news and represent great opportunities in this phase of development.

This phase 3 is important. The phase 3 prototype should be good enough to show others in the industry and to use for making presentations to marketing experts or even to certain potential customers. You can use it to secure marketing partners or to get order commitments. You'll see how to do this a little later.

Phase 4. This may very well be a first article, a true working prototype manufactured from a test mold or from an actual manufacturing process. It is tested in actual working conditions by real end users who do not necessarily know that they are "guinea pigs." Don't call them that, by the way. Remind the testers that they are the lucky ones who are the first to use the new improved product or innovation. You (or at the very least, your marketing expert) should be present during tests. You should listen to all the "ooooo's" and "ahhhhh's." Learn all you can about the testers' opinions regarding size, shape, style, etc. You can casually ask what they like and what they don't like. Never assume. Watch the actions and movements of the testers carefully. After extensive observation and listening to feedback, you will determine what final improvements and alterations, if any, are needed.

Phase 5. The final prototype phase should be production models that are used in a substantial test market. Instead of a single test application, use several; instead of one store, use an entire district. Unlike the previous prototypes, the final prototypes are usually sold to the customer (the end user), thus generating some profit.

Creative Prototyping Methods

Inventors are inherently creative. Use your creative ability to create and make your first-phase prototypes. Here are several suggestions:

- Purchase or use a similar product and modify or adapt it.
- Carve it out of wood and paint it with chrome paint to resemble steel.
- Try to use computer-generated graphics to illustrate the first phase.
- Use a craftsman's file and sandpaper and form the prototype out of a block of solid plastic.
- If it is an electronic device—for instance, a remote control—make the general shape out of balsa wood or a block of plastic, but connect it to wires that go to the functioning electronics in a separate box. At an early stage, it is not important to have the electronics neatly presented with chips, microcircuitry, and so on.

Be creative and you'll be able to figure out ways to get some of your first prototypes made so you may make presentations to manufacturing and marketing partners.

Establish Prototyping Partnerships

Almost always you will be required to complete the first two or three phases of prototyping. However, being a successful inventor/product developer requires fast, economical prototyping and testing in the later, functional phases—phases three and thereafter.

Your final prototypes must have all the right attributes and you cannot be guessing. The best way to do this is through a team effort with your manufacturing and marketing

---

THE TRUE PROTOTYPE

Phase 3 more accurately may be called a true prototype. In our industry, a prototype is exactly representative of the manufactured product but is made without the tooling or final production process. Other than that all things are the same as the production model and can be used in FCC, CSA, CE compliance and some environmental testing.
–Phil Bourgeois, CEO Studio RED

T1, T2 AND T3

To industrial designers Phase 4 is called "first article" (off the mold) or T1 (which stand for tooling 1 or the first shot off the mold. Then T2 is the second shot which includes modifications to fix any errors in the mold. There even might be a T3 for complex plastic parts with huge interrelated functionality.
–Phil Bourgeois, CEO Studio RED

RAMP UP

Again for us, Phase 5 is called production ramp up or beta round.
–Phil Bourgeois, CEO Studio RED

FOAM CARVING

There is a new material called Ren shape which is a dense foam, carves much better than wood because it has no grain, can be painted to look like plastic, and machines well. There are many densities of foam like 10-lb and 20-lb, which is easier to carve and sand to make shapes. We use the 10-lb foam during our early design studies.
–*Phil Bourgeois, CEO*
Studio RED

PROTOTYPING

If you plan to license your invention and it's one that requires highly technical, expensive components and research and development, look for a special relationship with a licensee. For example, instead of spending $100,000 of your money on an electronic telephone alarm paging system, seek out a partner that will confidentially work with you to develop the desired results. They can pay you a contract salary until the product is launched and royalties thereafter. Even with this relationship, you will still need to have some expertise in the workings of the invention and have pre-qualified patentability. You can't just "take an idea to them" and claim inventorship. You have to show them how it works to be the real inventor.

partners. With the right partners, you can create prototypes with the attributes you want and with little out-of-pocket expense. You'll also be able to get the prototype completed far faster with a team approach.

Following are two examples of how you may work with your team members in the final stages of prototyping an innovation and actually secure commitments for orders. These kinds of sales approaches are used all the time by expert marketers.

Industrial and Commercial Prototyping Applications

Inventions dealing with industrial applications are usually directly related to problem solving and result in saving customers money, time, or space. So it is usually not difficult to get an audience with a potential customer if your idea or invention falls in these categories. You are also in an excellent position to get a commitment to buy before large sums are spent on the later phases of prototype development. When dealing with industrial and commercial applications, develop a working partnership with your internal departments or with your external manufacturing and marketing entities. This team effort should focus on problem solving in the customer's facility where the invention will be used.

The team effort includes four members who will contribute to and benefit from this prototyping partnership:

1. **The inventor.** You will contribute your time and know-how (you invented the product or system in the first place) and will work directly with the end-user customer and marketing people during testing. Remember that you are the pilot navigating the ship.

2. **The sales/marketing partner.** This is your company's sales manager, your outside marketing partner, or perhaps the national sales manager of a potential licensee. This individual contributes his/her time to set up initial customer presentations. The outcome is to secure a commitment from an interested customer to test the invention. This customer is one that will either strike an order upon completion of testing or will give a commitment to purchase in the near future. Depending on the nature of your product, the salesperson's relationship with each company, and the willingness of the customer, you may be able to use crude (even handmade) prototypes in the initial presentations and tests.

3. **The manufacturer.** Once you and your sales expert have a committed customer, you present the opportunity to the manufacturing department or to an outside manufacturer or subcontractor. The manufacturer does not have to play a part in the initial customer presentations. If you are working with a manufacturer who may become a subcontractor or licensee, it must prove its worth. It can do that by providing the prototypes at its expense. This is done all the time, by the way, so don't think you're asking for something unreasonable. If the producer wants to get in on the ground floor, it will gladly invest its money to make these late-stage prototypes. Besides the manufacturer needs to prove that it can make the product with the expected quality and performance. Thus, in many instances the manufacturer must commit to *make or pay for the prototypes* that will be used in testing. Otherwise, you should seek another supplier. In many cases, these prototypes will be sold and billed to the customer anyway. That makes it even easier for a manufacturer to justify the cost of pro-

totyping. In the long run, the manufacturer obviously benefits from having the new business. In fact, they will most likely try to strike some form of exclusive manufacturing agreement with you.

4. **A customer.** A willing customer must contribute the time necessary for you and your partners to complete the in-house testing. With new industrial products and innovations, you will typically begin with one application to "get the bugs out" and then expand the applications within the same customer's business as required. The benefit to the customer is obvious.

If your innovation may be sold in the retail or the industrial/commercial markets, usually your best choice is to pursue the industrial/commercial marketplace first. It simply requires a lot less investment in time and money.

Retail Prototyping Applications

Retail applications are not as simple as industrial or commercial applications, since a relatively fine-tuned prototype (including representative packaging) may be required. However, it is not impossible to secure prototyping assistance for retail applications.

The format here is essentially the same as with an industrial-commercial application, but a little more work is required. With retail applications, you may also need an additional expert to get your prototype prepared for market. Here is the scenario:

1. **The inventor.** It starts with your piloting the invention and working with the other team members.

2. **A sales/marketing partner.** This person should have excellent ties with a retailer of sufficient clout to make a purchasing commitment. For instance, if your invention is electronic, the retailer might be Radio Shack; if it is paint related, Sherman Williams; or if kitchen related, Target Stores. The sales expert schedules an appointment with his/her good connection, such as a purchasing director or district manager, to make a sales presentation on the soon-to-be-released product. This presentation may be made with a previously made prototype, computer-generated pictures, or at times, none at all. The sales expert gets a conditional commitment from the retailer to test market the product in a given district—usually 20–50 stores. The conditional agreement could be to guarantee sales of all units. At this phase, the product's pull-through is not of consideration to the retailer; it just wants to be the first one to exploit a new opportunity and profit from it.

3. **A manufacturer.** In receipt of an order from a major retailer the manufacturer should be eager to gain a new customer of this magnitude. The manufacturer will go out of its way to get the final prototyping done and get the packaging finalized as well.

4. **A retailer.** A willing customer who trusts your sales expert or the company will be necessary to get a purchase order struck conditioned upon guaranteed sales. It will be even more appealing to be the first customer if it can have an exclusive for a certain start-up period, which it may ask for. This is how Direct-TV started. RCA was first and others, like Sharp and Sony, were added later. These producers probably bought their position or made certain other start-up guarantees. Similarly, because retail is so competitive, a good salesperson can secure an order on the come if the new product represents new profit potential. For instance, could you imagine if only

ELIMINATE PROTOTYPING COSTS BY PLACING AN ORDER

This statement is particularly true in Asian countries like Hong Kong and China where molds and dies are inexpensive. You may save thousands of dollars on these costs by placing a substantial order for goods instead. In such a case, you'll be asking them to amortize the mold or die costs into the first order. The downside to this approach is that shipping from overseas takes longer and at times communication may be difficult. When using this approach, don't expect an overseas company to spend $20,000 on a mold with an order commitment of only $40,000. More realistically, you'll have to order $60,000 to $80,000 worth of merchandise or more. On the other hand, U.S. manufacturers may be able to use a temporary mold to accomplish the initial start-up objective. The more that's at stake, the more a U.S. supplier will be willing to invest in your project.

Toys 'R' Us had the first release on all new Barbie® dolls during the first year? Your innovation may not be a Barbie doll, but you can make those kinds of deals.

5. **A packaging expert.** This expert may or may not be needed. The benefit is that prototype packaging with great graphics can be quickly created on a computer. The cost to do this is typically born by either marketing or manufacturing. At times, prototype packaging that looks awfully close to the final printed version may be computer generated.

As you have discovered, the key to free (or virtually free) prototyping is to get an order or a commitment first. With an order in hand, it is not difficult to convince a manufacturer or marketer to pay for the later phase prototypes.

Making Presentations With Prototypes

There are several points to consider when using prototypes in sales presentations. First, these meetings are usually arranged by a well-connected salesperson in the field. The presentation can't be made to just anyone; it must be made to the right decision makers. Here are some key points:

1. **If the invention is an industrial or commercial application, meet with a Vice President of Operations, *not a buyer!*** Buyers are rarely authorized to make changes and test new ideas and systems. Think about it. In an industrial application, will buyers understand labor savings, productivity or timesaving? It's usually not their field of expertise. Even if they understand the concepts, they're usually not in a position to authorize change.

2. **If the invention is a retail product, contact the Vice President of Merchandising, a District Manager, Purchasing Director, but *not a buyer!*** New products can be introduced quickly with the support of a VP of Merchandising, who usually directs the allocation of shelf space, oversees advertising, arranges district meetings, and determines how store managers merchandise new products. All of this is not the buyer's job. Speak to the person who can authorize change or who can approve new product purchases. Since most inventors are not salespeople, it would be wise to have your well-connected sales expert do this. It's what they do best. They'll know what to do and how to handle objections.

3. **If you accompany your salesperson, speak in terms of benefits.** Speak in terms of benefits to the customer and the retailer, not benefits to you. You can identify certain existing problems and then talk about the benefits of customer-driven innovation.

4. **Use a partnering approach.** Get everyone to be a partner in the launching of the new product. The industrial-commercial customer gives you the ability to test, monitor, and fine tune the invention. The retail customer allows you to go into stores and help merchandise and set up counters. Your marketing partner will follow up on the shipments, call on out-of-the-area store managers, etc. The manufacturer also partners by being ready to make any necessary last-minute changes and improvements required to successfully get the final-stage prototypes perfected and the working models shipped.

5. **Always ask for an order . . . even if it is conditional.** Remember that you are developing a partnership with the customer. Guarantee them success by saying, "We will

do whatever it takes to make this a success for you." Then *do* whatever it takes! In retail, can a customer be given an exclusive sales arrangement for the start-up period? Some free publicity? What will it take?

You don't need a huge first order or commitment when using these methods. Manufacturers take in a lot of money and part of their budget is based on purchasing new dies, molds, etc. It should be easy for them to justify prototyping expenses if they have a legitimate high volume customer on the hook. Under the circumstances, many manufacturers would rather gamble thousands of dollars on prototyping a new product than blow it at a Las Vegas casino at the next trade show. In return for the initial start-up expense the manufacturer should be guaranteed a supply contract for a certain period of time.

Your sales expert should be eager to help. A sales manager wants sales and a new profit center is always appealing. If your sales/marketing expert is an outside rep agency, it will want an exclusive sales agreement based on performance. If the sales expert is your licensee, it too will most likely ask for some start-up exclusivity.

When You Can't Get a Commitment

Sometimes prototypes must be extensively developed and tested before orders can be secured. Other times the cost for a prototype is more than you can afford. Yet other times, the cost of development may be impractical for you, but may be practical for a licensee or JDA partner. There are alternatives to consider.

1. **Submit sales projections to manufacturing.** Have your marketing expert (your internal marketing department, outside partner, or potential licensee) provide some sales projections based on the crude concept and/or first-stage crude prototypes. Ask your manufacturing department or outside subcontractor to pay for the prototyping based on being the dedicated supplier. The more costly the prototype, the longer term the commitment. But it has to be based on performance. If the marketing expert is convinced, is solidly behind the project, and has a track record of previous successes, he/she should be able to persuade the manufacturer to pay for prototyping.

2. **Offer incentives.** Offer a potential licensee an exclusive license based on performance, for a limited time or for the term of the patents, in exchange for prototyping. Or, you can consider passing up part or all of the initial license fee (up-front fee) in lieu of the prototyping expense.

3. **Rethink the invention.** If you are not getting the interest you expected, it could mean that your invention is not ready. It may not be a significant enough improvement over existing products; it may be priced too high; or perhaps it doesn't function quite right or has any number of other problems. Be sensitive to what you see and hear. If those you are talking to indicate that your invention just doesn't have sizzle, it's time to find out what you need to do to improve it. If you have to, invent the next generation instead.

4. **Abandon it.** You may want to consider abandoning the project altogether. If you cannot get interested parties to invest modest sums of money for product development, it may mean that the benefits you perceive in your invention are not being well received. You may want to use your creativity for other inventions and opportunities.

PROSPECTING, TOO

Even when making initial presentations to prospective marketing partners and licensees you may use crude prototypes as long as they illustrate and convey the right effect. These smart managers know the prototypes will need perfecting and will be willing to work with you.

KEEP SEARCHING

If you are have a difficult time getting commitments from marketers and manufacturers, don't give up. The fit may not be the right one or may not be able to justify the return on investment (ROI). If you've got a great invention, someone will step forward and want to get in on it.

In Chapters 9 through 12, you'll read in depth about finding, interviewing, and evaluating manufacturing and marketing experts and how your product development relies on their input.

Prototyping Machinery and Complex Devices

When prototyping machinery and complex devices, it would be very difficult for a small company or independent inventor to spend hundreds of thousands, even millions of dollars, on prototyping. In many industries, every machine sold is considered a prototype by its very nature because the technology is changing so rapidly that the next machine that's built invariably contains many improvements over prior machinery.

The key to developing patentable machinery is to do it with an established company in the field of the invention. You protect the new patentable process in the normal way, and then have the company's engineers engineer the concept. They may need to have a committed customer to purchase the equipment, but if the new process you are creating is a substantial improvement over existing ones and customers see the merits, it should not be difficult to garner interest. Then, let the machinery manufacturer provide its technological expertise to build the machinery and pay a royalty to you on sales.

Chances are the new innovative machinery can be used in many other future applications leading to new technologies and breakthroughs. The key is to let a dedicated manufacturer of machinery in the present industry invest its money, know-how, and contacts to get it made and sold.

Electronic Prototyping

If your invention requires the use of electrical or electronic circuitry, you will probably need to hire a company or individual with knowledge specific to this field. There are many ways in which circuitry can be developed and tested, making it very important to clearly communicate your requirements to the developer early on in the process.

You don't need to know how the circuit operates internally; however, you do need to know what functions are to be performed. If at all possible, enlist the help of your designer to determine the appropriate types of electronic devices that you wish to use.

There are a series of questions you and the developer should address up front in order to expedite the design process and avoid costly revisions. Is your invention going to be battery operated? Will it require a motor? Will it need torque, a reverse direction, or should it accelerate and decelerate? All these requirements must be communicated as clearly as possible to avoid unnecessary work and expenditure.

For example, say your invention is supposed to turn on a motor every time a switch is activated. The designer needs to know what type of switch you're using (toggle, momentary, low-voltage, etc.) as well as the specifications of the motor (AC, DC, high voltage, capacitor-start, etc.). If your developer is unaware of your desire to create a battery-operated device, he would probably use some inexpensive "power hungry" transistors to save cost for your production units. However, since the circuit runs on a battery, it may only operate for an hour, instead of weeks like it should.

Does the circuitry need to be placed on a printed circuit board (PCB)? Will it be single, double-sided, or multi-layered? Should the PCB be flexible or rigid? Do you need to get Underwriters Laboratories (UL) or Canadian Standards Association (CSA) certification for your device? If so, be sure that the design firm has a track record of seeing their designs through this process.

ALEXANDER GRAHAM BELL

In April 1875, U.S. Patent No. 161,739 was granted to Bell for his discovery of a method and apparatus for transmitting two or more telegraphic signals simultaneously along a single wire. This was accomplished by the use of transmitting instruments, each of which sends electrical impulses at different rates from the others. Referred to as multiple telegraphy, it became the foundation of the telephone communications systems we use today.

Digital electronics has enabled the creation of many supersophisticated, high-tech devices that enrich our lives. For instance, cell phones and "key-chain" pets—complex circuitry in a small package for a cheap price.

The tendency to go digital sometimes creates tunnel vision in electronic circuit development. Many developers try to solve *everything* with a digital circuit. However, if your device is relatively simple in operation, it's possible you'll be able to use an analog circuit to accomplish the same task. If you get proposals from more than one designer, each will probably come up with different methods of accomplishing the same task. By evaluating different design solutions, you might discover new features, enhancing your product.

Of course, your prototype will cost much more than the final retail price, but remember there is a lot of engineering that goes into making a reliable, customer-friendly device. Soliciting multiple proposals from various companies not only helps you determine what your development costs may be but it may also show you a different way to look at your invention.

With the advent of low-cost easy-to-use microcontrollers, it's now possible for those who are inexperienced in electronics to create real-world, functional circuits that can yield proof-of-concept prototypes. If your circuit is relatively simple, and you have the time and are willing to learn, you can save significant money by developing an early version of the circuitry yourself using microcontrollers.

Choosing appropriate and reputable electronics developers is crucial to your product's success. You should select an electronic prototyper who is experienced and available for troubleshooting when needed. Some prototypers who are moonlighting may not be available for future circuit revisions when (not if) the prototype circuit fails.

MASTER CREATOR

This section on electronic prototyping was created by Matt Gilliland, known as one of the top microcontroller inventors on our planet. He illustrates in great detail how microcontrollers can be used in prototypes and products in his Microcontroller Application Cookbooks series. Matt is a consultant to Hewlett Packard and the inventor of its new method of blueprinting in color.

Virtual Prototyping

This is gaining more and more popularity. Believe it or not, you can have almost anything prototyped in a virtual environment. It can be demonstrated with computer-generated graphics that illustrate functionality and style. This type of prototyping can be invaluable in the early phases of prototyping to show your marketing department or potential marketing partners how it works.

Using this procedure, you may quickly find out if there is sufficient interest in your invention. It also opens the door to discuss with the potential partner what additional product development is needed and what the associated costs will be. It is a perfect time to dovetail these discussions into a licensing agreement, with the licensee paying the development costs.

It's also possible to prototype many electronic circuits virtually, on a computer. The schematic is created on screen; then, through simulation, the circuit's operation can be observed. Although not a real-world prototype, it's a useful means to develop a circuit's concept and functionality without going to the expense of a printed circuit board and chip-set design in the early stages of the development process.

Benchmark #6: Secure Marketing

Most entrepreneurial companies will have marketing commitments in place before patents are filed. In other words, they have confirmed and secured a commitment from their marketing departments to go sell the heck out of their new products early on before the legal department or their patent attorneys are called in.

If you are an independent inventor looking for licensees or a small business looking for outside marketing assistance, then securing marketing is probably going to be your most time-consuming task. This will become a central focus in your strategy and the most important one of all. Never forget, nothing happens until something gets sold. And no one can do that better than a smart sales/marketing team.

Whether you have an internal marketing department that has signed off on the project or you have to secure outside assistance, there really is very little you can do until this crucial step is fulfilled.

The testing of your invention starts in-house with you and a few associates or maybe even friends and eventually moves outside your sphere to end users. The best way to set up end-user trials is through the marketing team. That team should be arranging and overseeing those tests with you. After all, the results will illustrate marketability, which the marketers will use to compile a list of features and benefits for future sales calls.

Part Three of this book is dedicated to finding, evaluating, and hiring marketing partners.

Benchmark #7:
Write and File Patent Applications

One of the most important aspects of having a successful invention and attracting development partners is to have substantial patent protection. Strong patent protection increases your chances of securing financially sound, industry-respected partners. In fact, without it, you may completely lose your ability to partner or license your creations altogether.

Rushing to file patent applications will usually NOT accomplish this objective. However, if you dovetail your patent-writing activities with your product development activities, you'll be able to file in a timely manner and with broad coverage.

You will want to start writing your first application—most likely a provisional application—sometime during the development phase. Based on pure strategy, initially writing a regular patent application instead of the provisional application makes little sense, unless at least one of the following factors prevails:

1. **Infringement is imminent.** In this case, you will want to secure your patent protection as soon as possible, so you can plan your patent infringement strategy.

2. **The sales window is short.** If your product has a short window of opportunity and you want patent protection to ward off competition, it makes no sense to file a provisional application first.

3. **You have money to burn.** If so, don't waste time on filing provisional applications.

There are a lot of good reasons to start out with a provisional patent application, which you'll learn about in detail in Chapters 4–6. But the most prevalent reasons affecting your development strategy are:

• They are a great way to preserve cash flow and use that money to pay for prototyping and marketing instead.

SWITCH BENCHMARKS #6 AND #7

An alternate and somewhat common approach you may use in your invention development activities would be to switch benchmarks #6 and #7 and file a provisional patent application first. Next, openly start searching for your marketing and manufacturing partners or licensees with your prototypes. When taking this approach, you are committing yourself to the content of the inventive matter in the provisional patent application and to filing a permanent patent application within one year.

DREAMS COME TRUE

The U.S. patent system has worked well for more than two centuries and is often referred to in the context of the American Dream.
–Don Kelly, Intellectual Asset Management Associates, LLC

- They save a lot of time, thus allowing an inventor, engineer, or entrepreneur to focus on product development and improvement.

- If the new product is designed around, or improved upon, during development, a new application can be filed quickly and the old one may be abandoned. This happens all the time.

Today, provisional applications are widely used and can be an excellent means to divert the problem of interference. Interference occurs when two persons file patent applications on the same subject matter at the same time (within one year of granting). While it can happen, it is rare, and is reportedly becoming more uncommon each year. You shouldn't be obsessed with this possibility and should learn how to make provisional patent applications an important tool in your development arsenal. This is a central theme in the *From Patent to Profit* methodology.

Writing Patent Applications and Invention Development Strategy

The United States is a "first-to-invent" country that establishes the true inventor by recording the date of original conception followed by reduction to practice. Thus, this tells us that rushing to file patent applications doesn't set priority anyway. Frankly, few experienced inventors rush to file, unless there are specific reasons to do so, such as imminent public exposure of the new concept. At times, we may solve a critical problem with a simple breakthrough concept that's patentable and is begging to be tested or presented in a public forum.

Over the past ten to fifteen years there has been talk about changing the United States to a first-to-file system, so we would "be in concert with the rest of the world." This will most likely result in generally inferior patent applications and probably won't affect the speed to which companies, engineers, and inventors file their patent applications anyway. This change would probably not have much effect because provisional applications can be filed for $110 for small entities, and patent-filing rights are preserved up to a year. If you use this strategy, you'll most likely be filing your applications long before any other competitive filings occur.

Smart inventors and product developers write their patent applications while they are developing their inventions. There are very good reasons for this. First, besides being protected by first-to-invent laws, the true nature of the invention and clarification of the inventive matter will unfold. It is during development that inventors will be able to identify, understand, and subsequently write patent applications blanketing the subject matter.

By developing your ideas and inventions thoroughly and then writing your application after some initial testing and proving, you will have a sufficient understanding to portray your discoveries and their unique attributes. Filing too soon frequently requires you to file an additional patent application (or more) on new inventive matter as it is better understood, in order to broaden the all-important claims.

Based on the provisional patent application laws, you should be able to file applications quickly, based solely on the fact that you will have been writing your patent applications as you have been developing your concepts. Chances are you'll be first to file even if someone else had been developing the same invention simultaneously, solely because your inventive matter will be more clearly developed and thoroughly stated and qualified. You patent application would tend to be much broader, based on your more thorough development and understanding.

The tremendous savings of a provisional patent over the cost of filing a regular application (at $4,000 or more to write and file) can be applied to your invention development. Later, within one year of the provisional filing, you should have some sales generated on your invention, which will make the cost of the permanent application affordable.

However, an inventor should not think that he/she can wait forever while the invention is being developed. Patent laws say that you cannot claim your first-to-invent rights if you wait more than one year after the granting of an identical invention to another party. Nor can you claim first-to-invent rights if the development of your invention had been considered abandoned. You'll read more about this subject in Chapter 6.

Last but not Least . . .
Improve Your Invention

In life, the one common element that leads to success is continuous improvement. This single principle almost guarantees a person will become successful. This principle holds true with innovations too.

Breakthrough inventions are rarely your first ones but are usually some improvement on the initial concept. Thus, it is important to continue to develop your invention and look for opportunities to improve it. This facet of invention development must be avidly pursued if you plan to have big success.

In terms of innovating and inventing, improvements may be small in comparison to the original idea. However, they are easier to incorporate into the product's design and to launch as well. With a manufacturing and marketing team in place, you just follow the Strategic Guide one more time.

Figure 2.4 Continuous Improvement of Your Invention

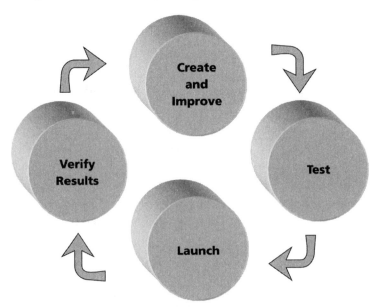

From the perspective of continuous improvement, the successful invention development process is a circular, continual process. Figure 2.4 shows how the ongoing invention development process becomes your *modus operandi*.

By continually improving your invention, you are reinforcing its ability to be a continued success. New patents will evolve, giving you added protection against competitors, and sales will continually climb, providing more profit or royalties.

What may be more important is that you are becoming the expert in your particular niche. As you continue to develop your niche, taking it further into the future, you are distancing yourself from the other me-too products.

Apple Computer became a hit quickly, but sales did not really take off until certain improvements were made—namely, the use of a mouse. This and other new software additions suddenly made the Apple usable in everyday business applications. The addition of the mouse sent Apple Computer to the next level altogether. Once it became the industry giant and the creators sold their interests, Apple slowed its new product development and Microsoft took the lead.

HISTORICAL FACT

To every inventor, I always say, "Don't give up, ever." But I also say to inventors, "Hug your families every day. The invention process is a long road and you need those people behind you. On every balance sheet you should list family as an asset."
–Don Kelly, Intellectual Asset Management Associates, LLC

CUSTOMER-DRIVEN INNOVATION

What makes some inventions sell while others don't? Why are
some inventions quickly licensed, yet others just can't get attention?

It goes without saying that inventions people want to buy will make money. But what makes consumers want to buy certain products and not others? What are the attributes of moneymaking inventions and how can you incorporate them into your invention?

Before you consider your invention's attributes, first focus on the state of mind that can help you determine which ones fit best. This is referred to as "customer-driven innovation." It is the central theme in your invention-related design and engineering activities. It will even impact and validate your patent's claims and scope.

The Customer-Driven Philosophy

The way in which products are made has changed dramatically since the beginning of the twentieth century. The beginning of that century marked the end of the era in which ***individual efforts*** of a single person or a small group resulted in individually made products. In the early 1900s, Henry Ford ushered in the production line. The United States, followed by other first-world countries, became a ***production-driven*** country. By the 1950s, the emphasis in manufacturing had changed to a ***sales-driven*** philosophy—one in which sales promotions became the driving force. Compare Henry Ford saying, "You can have any color you want as long as it's black" (a production-driven mentality) to General Motors' sales-driven approach that offered five models of cars in a variety of colors. The result is that GM overtook Ford as the sales leader in the automotive industry. In spite of these approaches to business, the United States did not begin thinking in terms of customer's satisfaction until the 1980s. Figure 3.1 shows the driving forces behind manufacturing in the United States and the world.

FACT

The customer-driven innovation philosophy is what drives Scott Cook, co-founder and chairman of Intuit, makers of Quicken®. Competing in a market against huge competitors, Intuit has outperformed the likes of Dell, Microsoft, Intel, and Wal-Mart over the last five years.

51

Customer satisfaction is a key component in the Quality Management systems (QM) or Total Quality Management (TQM) used by most of the world's leading corporations. Unfortunately, in the United States, the Quality Management transformation began decades after the systems had been adopted, and proven highly successful, in Japan. In the twenty-first century, customer satisfaction is more important than ever to the survival and future success of any company.

Take it to a Higher Level . . . *Customer-Driven Innovation*

Today, highly successful entrepreneurial companies of all sizes know that if there is one underlying key to success and their future security it is their customer-driven philosophy. Yet most companies confine their customer-driven philosophy to better service and faster customer response. Few companies really consider a customer-driven philosophy when designing, manufacturing, or developing future products and

Figure 3.1 History of Manufacturing

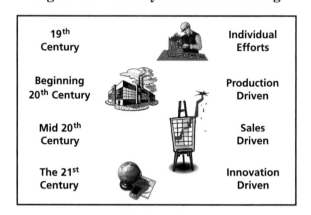

innovations. A focus on ***products that customers want to buy*** should be the central theme of every company's philosophy.

Emphasizing product innovations and inventions that appeal to your customers' ***real needs*** and ***desires*** is a winning theme for finding new customers and keeping old ones. This theme defines customer-driven innovation (CDI).

As an inventor, engineer, or product developer, CDI is your invention design focus. It will make or break your invention's success.

Focus on Niche Products

All inventors and businesses can use CDI principles and create niche products that can be protected long term. Meanwhile, many high-volume competitors will try to continue promoting and expanding market share of their generic product lines. From this perspective, their strength shows their weakness. High volume, market-dominating companies generally do poorly developing lower-volume niche products.

However, virtually every new entrepreneurial effort and product introduction begins as a niche. In time, some of these niches become high-volume generic products. Even major worldwide industries started as niches. For instance:

* Electricity replacing candles
* Automobiles replacing horse-drawn carriages
* Airplanes replacing trains
* Jet engines replacing propeller-driven engines
* Calculators replacing manual accounting practices
* Computers and software replacing calculators and other business machines

- Cell phones replacing hard-wired phones

Once a niche has grown into a dominant market, it is typically split out into many new niches. For instance:

- Neon and florescent lights replacing incandescent lights
- Dozens of herbal teas competing with Lipton® tea
- Small cars and multiple styles replacing the basic family sedan
- Ziploc® bags replacing fold-top plastic bags and plastic wrap
- PCs and software tailor-made for a myriad of applications—graphics, publishing, industrial controls, web design, and so on.

Apple Computer, Honda, and Intuit are just a few examples of companies that were able to create innovative, customer-driven products and niches in existing high-volume markets. Even though all these companies started out small, they beat out their giant competitors to huge new markets! Apple beat out IBM, Honda beat out Ford, and Intuit beat out Microsoft . . . yes, that's right.

Today, there are many small, inventive, and entrepreneurial companies that are destined to be tomorrow's giants because they are introducing customer-driven innovations and carving out new, even smaller, niches. As an inventor, focus on a new niche. If it later grows into a marketing monstrosity, you will have surpassed your fondest dreams.

Consider Generic vs. Value-Added

Successful innovators know the advantage of differentiation by selling value-added, customer-driven products. Being value-added is usually synonymous with better profit margins, too. If you focus on value-added benefits for your customers, you will discover the secrets to products they will want to buy!

The "Total Product Concept" flower (Figure 3.2), developed by Ted Levitt at Harvard, helps us to better understand product focus. Most manufacturing-driven entities concentrate their efforts on the two inner rings. It is here that products can become overly engineered and more and more generic. Many manufacturers (especially larger ones) concentrate their efforts on making products cheaper, instead of better. They focus solely on cutting sizes, gauges, raw material requirements, and scrap rates and reducing downtime to compete in a generic marketplace. Instead, they should be inventing and creating value-added products that can garner higher prices, profits, and customer satisfaction.

THE GREATEST NICHE EVER?

In the mid-70s Jobs and Wozniak invented/created the first mass-marketed PC that used software . . . the Apple computer. What they really created was an entirely new industry . . . an entirely new way of life. In 1997, computers outsold TVs. And when you think about it, how many jobs today are directly related to PC computers, software, and their offshoot, the Internet? What they started just a few years ago may become one of the greatest innovations affecting the history of the world. And, it all started out as a small niche.

INNOVATIVE?

During WWII, innovative toy companies went to work for the war effort. Marx toys started making bazookas, and Fisher Price began making bomb crates. Before WWII, metal was the preferred toy material. But since it was in scarce supply, the outbreak of war spurred the use of plastic in toys, which has dominated since.

Figure 3.2 The "Total Product Concept" Flower

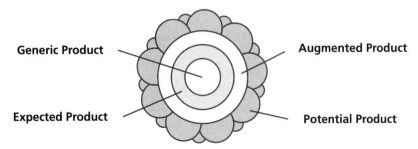

Generic Product

Augmented Product

Expected Product

Potential Product

At a recent conference, I heard a futurist, James Canton, PhD, talk about the products that would resonate with the consumers of the future. He was describing products, that might be the next "killer app" (Killer Application), which is a term describing a new product that takes center stage from all other technologies around it. He indicated his research suggested that products must be "lifestyle- or workstyle-" enhancing. That means the product should make a person have a better life or work experience. When designing a product, be very careful to address the users' experience with it—how they see, carry, use, store, or dispose of the product. This could also apply to how they view themselves when using or carrying it, which starts to be encompassed under the banner of fashion.

–Phil Bourgeois summarizing the words of futurist James Canton

Innovators tend to focus their efforts on the two outer rings. Innovative companies such as Apple, Intel, Intuit, and Microsoft have become industry giants in their field by doing so. They focus on the augmented product and the potential benefits to their customers.

To expand on this concept, the long-term strategy of traditional, large manufacturing-oriented companies focuses on generic products and looks like the flower in Figure 3.3. In contrast, the long-term strategy of innovative companies looks like the flower in Figure 3.4, with the focus on the augmented and potential product.

**Figure 3.3
Focus on Generic
Products**

**Figure 3.4
Focus on Augmented
and Potential Products**

Keep in mind that it is difficult for large and cumbersome established companies to change their production philosophy. It usually takes years if it'll ever happen! From the examples in Figure 3.3 and Figure 3.4, it is easy to see that there are excellent opportunities for small, flexible manufacturing and marketing companies to cash in through innovation.

Where is Your Mindset and Where is Your Invention Positioned?

Are your inventive efforts being directed simply to make cheaper products or are you focusing on creating new products without any regard for cost? Safety? Patentability? Market position? Competitive response? Or, do you view your creations as a small CDI niche that can be captured and protected? Are your concepts founded in "not-very-different, me-too" inventions or do your ideas carry some quantifiable value to the customer? Using Figure 3.5, you can determine your current philosophical position. To be a successful product developer of niche products, you need a philosophical rating of 8 or higher.

Figure 3.5 Determining Your Current Philosophical Position

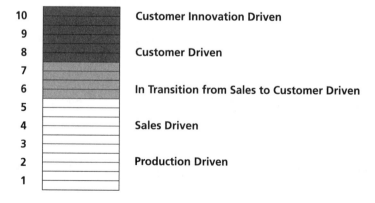

Your challenge is to seek and create niches you can protect. The most profitable ones with long-term security are those that are unique and specialized customer-driven innovations with excellent patent protection.

Applying CDI

There are several ways you can apply CDI to your invention development processes. In fact, it should pervade all your decisions, how you think, and how you approach problem solving. Besides, applying CDI principles makes it a lot easier to be able to focus on the most important issues that will make customers want to buy your products.

Think in Terms of Customer Benefits

Start thinking of your invention in terms of its CDI potential and how it directly benefits your customers. To get an idea of where you are now, answer these questions:

- What features and benefits do customers expect from the existing products? Are they getting them now?

- Do you know the true underlying needs of customers who would buy your innovation? Do they want improved productivity? Lower labor costs? Greater convenience? Ease of use? Durability? Can you list them?

- Have you evaluated end-user needs? Do you have or can you get connections with the all-important operations and merchandising decision makers?

- How do your products enhance the lifestyle and workstyle of your customers? What are the compelling reasons they will want to buy your products?

Answering these questions is a good starting point. Now carry it over into the design and development of your invention.

Make Your Philosophy "Flexible"

In the process of developing and fine tuning your inventions, be flexible in your approach. Don't be so fixed on your ultimate objective that you ignore possible changes and alternatives that might improve sales or lower manufacturing costs. Likewise, don't let manufacturing costs dictate price either.

Adapting your innovation to existing manufacturing processes may be a challenge. You want your invention to be cost effectively manufactured, so it can be sold at a reasonable price. But it must also have as much dazzle as possible. That's what gets attention and makes new products sell. You'll see how to apply this concept shortly.

By being flexible, you may also discover new opportunities and related inventions, such as a new production process that may be patentable. This will give your new product further security. If you are contracting all or part of the production, persuade your subcontractors to adopt your CDI approach and be flexible, open-minded, problem-solving partners. They have as much to gain as you do!

Get Everyone Involved and Communicating

In a team effort everyone must be involved in the CDI commitment. If you have an innovation that is going to benefit certain customers, everyone must know what those benefits are and why they are important. You will not be successful if everyone in the

DESIGN FOR THE LICENSEE

If you plan on licensing your invention, you must satisfy TWO customers—the person who will ultimately buy your invention and the company that will license your patent. If you don't spend as much time addressing your licensee's requirements (like market share, manufacturing costs, product line expansion, to name a few), then your invention will probably not succeed.
–Andy Gibbs, CEO, PatentCafe.com

HOW TO MAKE ALL THE RIGHT DECISIONS

If you make your invention, marketing, and manufacturing decisions based on what is best for the customer, you will almost always be making the right decisions. It is a foolproof approach to decision making! Use CDI when making decisions and you won't make a wrong one.

OUCH!

Be excruciatingly specific about what you want accomplished. Write it down, communicate clearly, and maintain a spirit of open-mindedness to new enhancements. Realize the driving force behind the product is you.
–Matt Gilliland, author Microcontroller Application Cookbooks

communication stream—from top management to manufacturing to sales to distribution and a retailer, right on down to the end-user customer—is not informed.

Good communication starts with you and then goes to managers, supervisors, and the workers who will make the product. It is your responsibility to make certain that communication is flowing down to the very last employee who will be making your invention. Through open communication, the importance of the customer-driven innovation (your invention) will be conveyed. Every employee can (and must) take part in ensuring that the innovation is produced on schedule and with the desired outcome.

Think about it. Without adequate communication down to the very last person, making changes in products is difficult. There will always be employees who will revert to the old way of doing things, because "that's the way they've always done it and that's the way they were told to do it." It happens all the time and can be one of the most frustrating experiences in product improvement you'll have.

Figure 3.6 Responsibilities in an Innovative Company

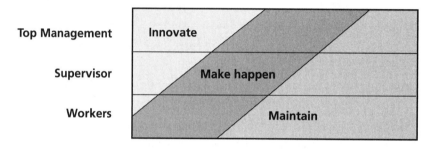

Figure 3.6 shows the areas of responsibility for an innovative company. Most of the responsibility for product innovation rests in the hands of top management. Today, innovation even trickles down to the workers . . . a shift from just five years ago. The shaded area containing the responsibility to "make [things] happen" is perhaps more important when it comes to launching new products. The "make happen" arena is where supervisors and workers become more involved in the effort, as it becomes everyone's responsibility. Frequently, it is in this "make happen" arena where the lack of communication delays and thwarts innovative efforts.

For innovative efforts to be successful:

*** All barriers to communication must be broken down to allow immediate feedback from one person to another . . . from management to supervisors, from one department to another, and from one shift to another.***

Get Super Close to Your Customers

Start by taking the time to get to know your customers (or potential customers) extremely well. This should include all buying influences from the top level management of retailers (or suppliers in an industrial/commercial application) right down to the all-important end users. One of the best ways to locate innovation opportunities is to be supersensitive to problems your customers—end users, distributors, and retailers—have with existing products or the lack of an existing product fulfilling their needs. Your customers' problems are your greatest opportunities.

With industrial/commercial applications, buyers, operations people, and managers do not always have the answers. This is why you must gain as much access as possible to a customer's operation. Remember that *you are more likely to be the expert in your field and are in a better position to discover new innovations they will want . . .* and perhaps cannot live without. If you understand your customers' needs and problems, you will have the opportunity to create some brilliant inventions and improvements that will be well-received.

In industry, decisions to change are made at upper management levels. Decisions to add new products and product lines with retailers are also made at upper management levels. Presentations of new CDI inventions and products to top management are usually well-received, providing they have perceivable benefits. Upper management can provide valuable qualifying input and pave the way to test markets and in-house trials. It is usually fairly easy to get appointments with top executives to introduce new problem-solving innovations or new profit centers . . . if you just ask.

Once your innovation is in place, continually improving it by focusing on additional customer benefits is a surefire way to secure long-term, profitable relationships with your customer base.

Think Small

Too often in America we think in terms of large volumes, big dollars, and huge markets. Even in terms of innovation, we have a tendency to think in terms of major changes.

To be a successful, innovative entity, it is far better to think small. The Japanese have learned this lesson well at the expense of many American companies. Thinking small means not only thinking in terms of smaller (niche) markets but also of smaller innovations. You don't need to recreate the wheel overnight. It is best to recreate it one or a few spokes at a time.

Besides it is unlikely that your first invention will be a smashing success. However, it can represent a significant door-opening opportunity. With a little experience, you will be in position to introduce future improvements that will become breakthrough opportunities. With a series of small innovations, you will either reinforce your current market niche or be able to capture new ones. Either way, you are warding off competitors, expanding your sales volume and profits, and securing your future via more and more high-quality patents.

Without question, once you become experienced in developing inventions, the time it takes to bring them to market will dramatically decrease. You will be able to bring new products to market with lightning speed that dazzles your competitors. Of course, your competitors will be continually struggling to keep pace with your last dazzling innovation!

Use Intuition (or Instincts) to Guide Your Innovations

Experienced innovators know how to use their intuitive abilities. The more you use your intuition, or instincts as the case may be, the smarter you become in the field of your invention. There is a difference between the two. "Intuition" is defined by *The American Heritage Dictionary* as "the act or faculty of knowing something without the use of

---

ASK THE CUSTOMER

When Waterpik decided to redesign their Aqualink temperature control system for pools and spas, they turned to their existing customer base for answers. They listened to issues posed by existing users, retailers, installers, and service people. The needs they identified were the specific directives during the new design called "One Touch"™.
–Phil Bourgeois, CEO, Studio RED

DEVELOPING CLOSE RELATIONSHIPS— THINKING SMALL

When I think about launching the sales of a new packaging product, I start out with a small test run. I can usually get the manufacturer to make these at no cost to the customer. This is an easy way to get a new product tested. It behooves the manufacturer too, as it shows the customer they're willing to go out of their way to satisfy them. Then, working closely with the customer, when tests show positive results, I ask for blanket, truckload orders to cover an entire year and keep the competitors out.
–Marketing wizard, Joe "Truckload" Bowers

rational processes." *American Heritage* defines "instinct" as "the innate aspect of behavior that is unlearned, complex, and normally adaptive." Which one fits you best? Whichever fits, you can use it to your benefit, if you learn how.

Myers-Briggs, the co-inventors of the Myers Briggs Type Indicator personality test, differentiate between two personality types. "N" types generally prefer to use their intuition. "S" types prefer to use their instincts. Inventors are more likely to be "N" types, but "S" types can be dynamic innovators as well. Leonardo Da Vinci, Benjamin Franklin, and Thomas Edison were certainly "N" types. Lee Iacocca and General George Patton are perhaps two well-known "S" type innovators. Either trait—intuition or instinct—can serve you well in the invention development process.

Use it to:

1. Determine the merit of existing products or new ones you want to develop.

2. Determine the changes or improvements your inventions need to be successful.

3. Help identify the underlying causes of problems hindering existing products or thwarting the development of your inventions.

4. Discover, even by borrowing at times, new solutions.

5. Pilot your ongoing invention developmental processes.

6. Direct your future innovations.

One of the key aspects of using intuition effectively is you must follow up and be analytical as well. Use the thinking side of your brain—not the feeling, emotional side. The combination of an intuitive and thinking mind can come up with a myriad of brilliant inventions. Just like Franklin, Da Vinci, and Edison, this planet's history is littered with the Einsteins and the Bill Gates of the world. "Anything is possible" . . . and you can rest assured, just about anything is!

Once you've become the expert in your field, your intuition will drive your inventive efforts and you'll know without qualification what needs to be done and why. You'll owe no explanation—you'll just know! Take the time to develop this important facet of your personality. Ask yourself, "Does Bill Gates wonder about what his next product development decision will be?" Or course not, and nor will you wonder about yours.

An undeveloped intuition that is out of touch with reality or one that does not follow up with analytical measures can result in pie-in-the-sky products too far advanced for the current market or impossible to manufacture at competitive prices. Likewise, instincts without seasoned experience and a clear understanding of the future can result in ineffective, misdirected efforts.

Find the Champions to Your Cause

In every innovative effort there will be leaders, movers and shakers, who rise to the top to help move the development project forward. Look for those who take charge in your project and give them the leeway, encouragement, and power to make things happen. You send powerful messages to the work force when the champions are given the power to push forward (and occasionally ramrod) an innovative effort. These champions may ruffle some feathers and cause some problems and disagreements, but they must be supported and rewarded. The message must be crystal clear: nothing stands in the way of innovation. After all, the company's future may depend upon it!

FOUND THE CHAMPION!

I was working with General Motors on an air filtration project and was having difficulty trying to keep it on track. I was having to fly to Michigan far too often and it was wearing me down. On my next visit to Detroit I decided to apply this one bit of advice—"Find a champion to your cause and let him lead." It worked! It not only helped my effort and saved me immeasurable weeks of travel and a lot of money—but it greatly enhanced General Motors' ability to quickly conclude the project for an immediate launch.

—Yujiro Yamamoto, world renown inventor

Zero in on Your Target Market

Your invention's attributes should target a specific target market that you can identify and use as a central focus to your initial marketing thrust. Once you have determined the target market, you'll want to continually ask yourself what it specifically wants and needs. What specifically does the target market want and what is required right now in order for the market to approve, and ask for, your invention? Think about this:

*If you make all your decisions based on the needs of
and benefits to your target market, you will make
only excellent decisions.*

Your target market is your niche. It is where you will become the premier expert in your field. You will become well known in this field over the years. Don't lose sight of your target market and it will never let you down. The field of opportunity is yours to create.

Focus Your Efforts on CDI

Can you clearly state the true needs of your customers? Which problem(s) are you trying to solve? Frankly, your customers do not always know their problems. In fact, they frequently don't! It is often difficult for customers and end users to articulate their true needs. Oh, they may know that they want to save money or they want a product to be more convenient to use, but they usually can't articulate exactly how to do that. Maybe you can.

Your chief task as an inventor is to uncover these problems. It is not necessarily an easy task. But if you think of your inventions in terms of potential CDI benefits to your customers and how your inventions may accomplish that, you will discover several good opportunities. These opportunities include the creation of new products as well as improvements on existing ones.

While you are developing your innovations, keep in mind this simple, but important, concept that will influence which attributes your innovations will have:

Cleverness can create innovations, but *character* sells them!

Putting character in your innovations should be your underlying theme as you strive to find CDI opportunities. You will learn more about this concept at the end of this chapter.

CDI Benefits Defined

The CDI benefits defined herein represent the potential, not the actual. Your objective is to apply as many of the concepts as possible. One of them may even represent a central development theme for your invention or your company. Inventions must have solid, verifiable, and observable customer benefits or they will have little success.

The remainder of this chapter is dedicated to the discussion of CDI aspects you can use to develop and evaluate the potential marketability of your inventions. You'll confirm certain positive attributes of your invention and you'll find others you've overlooked.

Incorporating CDI in your inventions is important, but it's not the only consideration. Don't forget there are other considerations equally important to having a well-received product, such as character and patent protection. CDI is the foundation of the design of your invention.

COMPETITOR'S STRENGTHS REVEAL WEAKNESSES

Just as you have considered competitive responses in your ability to enter the marketplace, you will also want to carefully consider competitive responses to each new unique attribute your invention will have. You must declare your niche and be prepared to move forward and take it! You want to make sure it is a niche you can protect. When evaluating each potential customer-driven innovation, you will always want to weigh them against how your competitors will respond. The more difficult their response, the better opportunity you have.

59

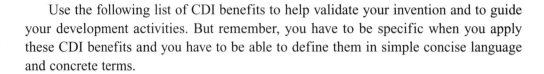

Use the following list of CDI benefits to help validate your invention and to guide your development activities. But remember, you have to be specific when you apply these CDI benefits and you have to be able to define them in simple concise language and concrete terms.

CDI Benefit #1: Inventions Good For Mankind

It might sound altruistic, but it is true. Your inventions should have a positive effect on the health and welfare of mankind. They cannot be founded in greed, but really should be designed to make some long-lasting impact on commerce in America and the world. After all, that's why the patent system was developed . . . to promote business and commerce.

For example, in the late 1800s and early 1900s, a huge number of inventions were targeted toward improving the ease of life for housewives in the home—washing machines, toasters, ovens, irons, and so on. In the mid-1990s, inventions in the electronics field made massive contributions to communications and to freeing time in the home. Inventions also contributed heavily to transportation and applications of raw materials such as plastics.

In the late 1900s and into the twenty-first century, a noticeable shift took place. Inventions had to have a positive impact on the environment and safety, else they would not be accepted by business or society. Oh, you might get a patent without a product's making a positive environmental or social statement, but rarely would it be of value in industry.

Today, new patented technologies in business and the Internet are also reshaping the way we do business and how we live. Business- and computer-related patents have inundated patent offices around the world. About 20% or more of all patents issued are in this field. Are they going to make a positive impact on mankind? Well, that may depend on how you view them. If you believe the Internet is the future of business, then, of course, many will make valuable contributions.

The key here is that over the years the various themes of the most important commercialized inventions were based on what was of benefit to mankind at any given age. Whether that benefit is to the housewife or to communications, transportation, the environment, how business is conducted, or how we live our lives, successful inventors fulfilled their obligation to improve the existing way of life and create new life styles.

While some of these innovations may have been ahead of their time, they can make an impact if you are patient. For instance, you may have developed a process that eliminates or neutralizes certain types of contaminants, thus having a positive impact on global warming. However, you may have to wait until legislation is in place before the process will be sought out. By monitoring proposed legislation for laws mandating environmental changes, you can improve your timing for a market release. But why wait? Take a leadership position and do it now and this can be your declared niche!

Your invention's niche (and the central theme of your company) may be on environmentally sound products such as biodegradable cleansers, with no phosphates, lye, or ammonia, like the orange-based or oxygen-based cleaners. The large existing, chemical suppliers can't quickly adapt their huge processing systems to make these new products, let along make them in small runs. So, it gives you, the inventor, an opportunity to carve out a niche that is attractive to many environmentally conscious consumers. Besides, environmentally sound products are not normally associated with corporate giants that dominate with high-volume generic products.

Making products better for mankind applies both to inventions and the many ways to improve existing products. When you improve the environmental qualities of an existing product your company may be manufacturing, or have licensed, you also improve your ability to maintain the current sales position. Here are some questions to consider in your invention activities:

- Can you alter existing formulas to reduce or eliminate contaminants?
- Can you lower the amount of raw material used, reducing waste?
- Can you make your product more durable, hence increasing the life?
- Can you incorporate additional uses of the product?
- Can you design it to improve lifestyles? Workstyles?

Take the time to evaluate how your invention will have a positive impact on mankind. More specifically, evaluate how it will improve the lives of consumers? The world of business? The safety and well-being of our fellow man? Be specific and see how many benefits you can identify. If you can't answer this, you will probably want to re-think your objectives.

CDI Benefit #2: Saving Time, Money and Space

Everyone looks for savings, which can come in many forms. The three most powerful savings you can provide your customer are:

1. **Saving time.** Time *is* money. In the current madness of a fast-paced world, everyone is looking for ways to manage their time better. If you can save your customers valuable time, you will find plenty of followers. You'll soon see how you can specifically accomplish this objective in almost every innovation.

2. **Saving money.** There are several ways to save money for your customers. Be smart and look for innovations that have a *total* cost-cutting impact. Cost-cutting means lower expenditures on the bottom line, not necessarily a lower price per unit. Anyone can create a lower price per unit by cutting quality. Per-unit costs can be very deceiving. Your challenge is to cut costs without cutting quality.

3. **Saving space.** Here is where thinking small can really pay off. Americans tend to do this poorly, while the Japanese understand this concept well. With the high cost of real estate today, space is at a premium. Large "anythings" that you can replace with small, efficacious "anythings" are in demand. Computers and cars are prime examples, but there are opportunities in virtually every field.

A good example of applying the savings concept is through reducing the cost of raw materials by using one that is stronger

CONSUMERS ARE INFORMED

Consumers are better informed today about available products with the Internet. You also need to know the products available to consumers and how your product can be differentiated from them.
—Michael S. Neustel, U.S. Patent Attorney

NEW, IMPROVED!

Your inventions and innovations can be in the form of improvements on existing products or completely new concepts. Sometimes we see them in supermarkets with labels reading "new, improved formula." You can use the many CDI benefits for both original ideas and improvements to existing ones. You can also look for other areas in industrial, commercial, and retail applications, in which your innovations can be "improvements." For instance, look for ways to improve down-time, change-over time, handling, shelf life, strength attributes, durability, assembly, ease of operation, speed, defect reduction, etc., etc.

and longer lasting. As stated in CDI Benefit #1, we know this would make a strong environmental statement while also possibly saving space and time (less material requires less space and less handling). It may even affect energy costs. For instance, if you could pack 24 items per case instead of 12, you'd be shipping twice as much product per truckload, hence saving the energy required to ship the product. These savings may indirectly improve an industrial or commercial customer's bottom line, regardless of the unit cost. By lowering shipping costs, you can also reduce the cost of the item.

CDI Benefit #3: Safety

DISNEY = SAFETY

Product safety is relatively new. Some of Disney's first toys in the early 1930s—figures of Donald Duck and Mickey Mouse—were made of eggshell thin celluloid, which is highly flammable. When touched by a match, they would explode into a brilliant display of colored flames. Disney became concerned and changed the material, leading the way for all manufacturers to make safe toys for children.

Several companies have had tremendous success by being the safety experts. Safety as a central theme will always have a buying audience in just about any product line. Mercedes Benz, Volvo, and the originators of seat belts and airbags come to mind. Even certain natural food products have a degree of safety when you purchase them. Many industrial products require certification by Occupational Safety Hazard Act (OSHA). But why not exceed OSHA requirements? Why not apply safety facets in every product, regardless of whether they are mandated by law?

Every producer has a challenge to design products that are inherently safe to use and safe around children. If you wait for government to require you to improve safety, you could be losing a competitive advantage. Strive to make all your products the safest to use by designing them to exceed existing safety standards. Or, you may even make this theme your niche. Every field can support a niche supplier that focuses on being the safety leader.

CDI Benefit #4: Convenience

CONVENIENCE

There is no end to innovations that strive to improve a product's convenience. This has a specific appeal towards lifestyle. Improvements can be made continually until the product works automatically . . . which is your ultimate innovation objective!

How convenient is "convenient"? Developing your ideas into products that customers find convenient is a never-ending opportunity. You will not exhaust the search for innovations that make your products convenient to use until you have made them totally automatic in every way imaginable. In fact, that could be your goal! When your customers tell you that they like your products because they are convenient to use, you will know your innovations are right on track. A few examples of convenient-to-use products are:

- One-at-a-time Kleenex® dispensers
- Remote control devices for TV, garage door openers, and auto door locks
- Automatic dialing for phones and voice-activated cell phones
- Internet access for email and banking services, travel plans, and movie rentals
- Automatic ice makers and water dispensers in refrigerators
- Talking computer calendars (They have them for electronics for the blind, but why not all computer calendar programs like Outlook™?)

Think about the future generations of these items. There is really no end to the possibilities. It's a way of life in America and will be in most of the world in the years to come.

CDI Benefit #5: Comfort

In many respects, we take comfort for granted in the twenty-first century. Yet, if we think about it, our homes have been comfortable for only the last several decades. As recently as one hundred years ago, most people had few of the comforts we enjoy today. These comforts include chairs, sofas, beds, plush carpeting, hot water heaters, baths, and showers. The list goes on and on. Comfort also includes other aspects of life, such as clothes, cosmetics, and personal items.

There are still exceptional opportunities to improve the level of comfort in many products, such as:

- Seats in airplanes, airports, and folding chairs

- Handle grips of all sorts, such as on luggage, briefcases, and bicycle handlebars

- Air conditioning that does not dehydrate, but humidifies and purifies the air

- Telephone systems (Isn't it amazing to think that we still hold most phones up to our ears? That's starting to change though.)

- Shoes and clothing (just ask Johnston, Rockport, and Hilfiger)

Adding comfort to an existing innovation may make a difference leading to success. For instance:

- A comfortable backpack, especially one that perfectly adapts to an individual's shape, sells better than one that is not . . . no, it's not available yet.

- Could you imagine a bicycle that was comfortable in every respect? A shape that supported your posture with a comfortable seat and handgrips.

- A tennis racket with a form-fitting handgrip that adapts to—literally molds to—an individual player's hand. Automatically, the player's hand grasps the racket in the perfect location every time. Would that improve a player's game?

- Operating machinery in industrial applications can be more comfortable, too. Have you ever used a jackhammer? For a real challenge, can you invent one that does not vibrate the body into numbness?

Look for comfort opportunities in your innovations, because many companies will not! If comfort is your central theme, then you'll have an exceptionally great career ahead inventing blissful inventions for your customers.

CDI Benefit #6: Entertainment, Toys, and Games

Americans like to be entertained. There are countless numbers of patented games and toys that have made fortunes for their creators over the years. The good news is, with the invention of the PC computer, there are more opportunities than ever before.

No longer does a game inventor have to depend solely on board game sales. Just look at Yahoo and nick.com to see the countless number of games being invented every week. As the creator of a new game, you don't need to rely on having a computer background; there are plenty of programmers who can assist you in creating just the right look. If you're a game inventor, you should seriously consider pursuing this avenue and apply as many other CDI benefits as you can.

With toys, there are also more opportunities than ever before. Small innovative toy companies are located everywhere in the United States and most all toy companies openly seek new innovative concepts.

Licensing agents like Paul Lapidus' NewFuntiers abound as well. They don't accept all ideas, but if you've got a good one, they know how to make the difference.

There's another avenue of toy and game development that can also be pursued. Some

---

PALM PILOT

The Palm Pilot is an example of marketing a system over the overly capable notebook computer that did much more but was not convenient to use when all you needed was a phone number or to determine if you had time for an appointment. Palm succeeded in creating a market, which was not responsive to the Apple Newton a decade earlier.
—Phil Bourgeois

COMFORT AND LIFESTYLES

Comfort is right at the heart of lifestyle-related inventions. Look at all the new beds and sleeping systems. Exercise equipment is more comfortable and convenient than ever. Look at the effect of air conditioning on population dispersal. Today's cars make yesterday's cars look incredibly uncomfortable.

great selling points for toys and new products are the characters that promote them. For instance, a neon-glowing magic wand might sell OK, but it would sell much better if it were tied to a Walt Disney character such as Tinkerbell® or Minnie Mouse®. If you have an invention that has such a possible tie-in, you can propose the development and production to one of the several companies that market these kinds of products to the trade. They'll arrange the secondary licensing for you.

At home, we enjoy entertainment, but there may be another arena far more ripe for development: our workplaces. While computers address this in some respects, there are many other possibilities. With the many Internet and virtual reality concepts available, the future of workstyle-related inventions is limitless.

Can you imagine creating new and entertaining machinery and processes for business? For instance, why can't machinery process controls be both informative and entertaining? For instance, can thermometer readings on a monitor be illustrated with colored markings that clearly show green when in the safe zone, change to orange in a field of caution, and to red when too hot? It could have a person's face that smiles at you and an arm the reaches out and gives the OK sign when the proper temperature is maintained. When the temperature moves out of the designated safety range and needs to be adjusted, the face starts to frown and finally whistles at you if it is dangerous. At high temperatures, steam shoots out of its ears! Might be fun to watch while the machine blows up? All kidding aside, it would get attention fast.

In the years to come, machinery could take on transformations that make them entertaining. Doing this should increase productivity and decrease absenteeism. Just as teenagers are entertained as they skillfully manipulate Nintendo® and Gameboys®, workers can skillfully manipulate their machinery.

Only your imagination can limit the possibilities. Providing entertainment potential to an otherwise boring work environment can boost morale. Improved productivity, less waste, and fewer defects would result in many followers.

CDI Benefit #7: Easy to Learn

Learning is something that people either like or dislike. However, one thing is for certain—few people like to spend the time to learn new processes and methods and to learn how a new innovation works. Neither do people like reading instructions.

Learning is right at the heart of getting ahead in life. Innovations that make an innovation or process easy to learn—whether for adults or children, whether software or hardware—will have a much better reception than those that don't.

Computer software that shows users how to intuitively use a computer is one of Microsoft's newest product lines. Recently it has launched the Microsoft® Agent software that website developers can use to enhance user interface on a web page with interactive personalities in the form of animated characters. The software enables user interaction that can make using and learning to use a computer easier and more natural. The characters move freely within the computer display, speak aloud (or display text onscreen), and even listen for spoken voice commands. This approach is also entertaining (CDI benefit #6).

Learning new software or how to use a product or a new production process can become not only entertaining but also can, and should, be a dynamic learning experience. A user should be able to quickly learn the right way to use a new product.

ENTERTAINING AND EDUCATIONAL

This is the focus of the Leapfrog toy offering. We developed the first Leapfrog globe. It was interactive, upgradeable, and sound based. In other words, it was both entertaining and educational. The first one cost $399, had too many functions, and was too expensive to capture a large portion of the market. The second version was $199; it reduced the functionality but captured a greater market share because the price was more in line with the consumer's cost threshold. The device had a conductive sphere, game cartridges, and a speaker. It asked you questions like where is the Nile River? You would touch the globe in the location or continent where you thought it was, and it would answer you with "No, it's miles south" or something like that. It could even tell you the size of the country you pointed to and its weather patterns and population, and play native songs.
–Phil Bourgeois

In a work environment, a user should also quickly check his/her performance against the established standards, if applicable. With statistical process controls (SPC), much could also be learned from other operators' performances. This type of methodology could be a part of a computer-controlled machine or system. The immediate and continual feedback from the computer should quickly enable the user to identify what can be done to improve performance.

There are other opportunities as well—for instance, the resetting of clocks and other electronic devices. When you have a power outage, which clocks and devices are easy to reset and which ones aren't? At the least, they should either be programmed to reset themselves or at least talk to you when you press the clock button. Why not? There's no reason functions can't be programmed to go from one step to another by simple directions or by voice.

Another tough area is the assembly and re-assembly of rarely used apparatuses. Can't the parts be sequentially coded or sequentially numbered to show the user how to assemble them quickly? It would give the user comfort in knowing that the product will be easy to assemble next time it's used. Camping tents provide good examples. They are usually fairly difficult to set up.

For children, there are many untapped learning opportunities. Toys can and should provide learning experiences. Books do not have to be only pages with words and pictures. Why not jazz them up with bells, whistles, snaps, fragrances, and textures? For that matter, why can't books listen and then talk back to teach children how to read?

CDI Benefit #8: Appeal to the Senses

One way you can give your inventions character it to appeal to the senses of your target market. Appealing to the senses of your target market can be an important advantage and can represent a central marketing theme as well. We appeal to the senses of children in toys, videos, tapes, games and books, but there are many more opportunities with adults.

In homes and businesses, do machines, computers, and other devices have to be so noisy? Newer technologies tend to be more quiet, but frankly that should be an objective of every company in almost every type of business. Can't we use motors and methods that are quiet? Why can't dentists' drills make a soothing hum instead of the pain-eliciting "whir"? Can you imagine a drill that has a euphoric effect? Patients would be knocking down the dentists' doors to get their teeth drilled!

In industrial applications, can the mixing of two ingredients impart a certain telltale fragrance when they are thoroughly mixed? Do all industrial cleansers have to smell like ammonia or lemon?

Can you make the design of your inventions pleasing to the eye instead of square and mechanical? This is an important part of Apple Computer's success. Their monitors, laptops, and even their software, always appear futuristic. Or, like Celestial Seasonings, have teas with wholesome, natural packaging and unique fragrant ingredients. New innovations cannot look like old clones of established products. They must look like they're new and, of course, they must perform better as well.

CDI Benefit #9: Appeal to Vanity

Does your innovation outwardly or subtly appeal to the buyer's vanity? For instance, if

---

A SIMPLE TEST

Does your innovation require a manual? A training video? Extensive drawings to show assembly? An individual to train? If it requires any of these, it's probably too difficult to know how to use.

PERPETUAL CLOCKS

Think about this. You have to learn how to set and reset every clock in your house. It's hard to remember how to do this and it can be time consuming if there is a power outage. Is there any reason that all clocks can't automatically reset themselves? If you think about it, isn't it a bit ridiculous that we have to set and reset them in the first place? (Do you hear us, Seiko?) The twenty-first century is here. We have the technology. When will it be applied?

IT'S A BEAUTIFUL WORLD ACCORDING TO WATERPIK

Appealing to vanity also relates to home and car products. People like their home to be the most beautiful representation of themselves. That was one of the drivers of the "One Touch" home control system for Waterpik. The typical customer for the devices was the new high-end home construction company. The device we designed had to fit into the customer's home décor. That was why we styled it to fit with the "Decora™" light switch plates. It needed to blend into the wall of a high-end home and almost always used Decora Plate covers.

–Phil Bourgeois

a new or improved product improves the softness of skin, it could be a hit. If women are the most common users of this product, it will have a strong following.

A good example of this is the introduction of plastic grocery sacks. Believe it or not, the use of paper sacks by baggers and checkers frequently resulted in dry hands and paper cuts. Because plastic sacks cause none of this, many baggers and checkers prefer to use plastic sacks instead of paper bags. Could this concept be applied to reams of paper? Could a paper manufacturer develop a cutting process that eliminates or decreases the risk of paper cuts? That quality should be worth an extra twenty-five cents per ream or perhaps more.

Appealing to vanity doesn't stop there. In general, a vain person goes out of his/her way to purchase high-quality overpriced goods. Go into any department store and see for yourself the myriad of branded products targeted for the upscale market. Some cosmetics may cost three-fold that of an everyday brand, even though the quality is marginally better.

A particular brand label tends to appeal to certain customers because they just feel good having it or wearing it. Appealing to vanity can convey confidence in the purchaser.

CDI Benefit #10: Improved Productivity

One of the greatest industrial and commercial invention opportunities in the world today is founded in *improved productivity*. With high labor costs, virtually every manufacturer is striving to improve output or throughput. The basic economic principle that *productivity produces income* is a vital consideration in a manufacturing company's operations and purchases. Smart presidents, vice presidents, operations people, buyers, and managers know this economic principle well.

An example may help illustrate this economic principle. Consider the situation of a $25-million-a-year manufacturing company that can save $20,000 per year on a cheaper packaging film, thus lowering the per-unit cost. But the cheaper film slows productivity (output) by 3%. Is there really a savings? Of course not! The 3% loss in output equals $750,000 in lost sales. This far exceeds the meager $20,000 savings on the packaging film. Lower per-unit costs with even a small decline in productivity can be disastrous to many companies.

Look at the situation in reverse. A similar-sized manufacturing company presently using the cheaper film could switch to the new invention, the more expensive film, and expand its sales base by 3%. And, it will most likely not have to hire additional personnel to do it. Innovations that improve productivity are big hits today.

It is important to remember that:

**Companies pursuing niche markets
by providing improved productivity to their customers
should have great successes!**

If your inventions affect this market segment, you're right on track. If your invention is going to negatively affect productivity, or you don't know if it does or does not, you've got some homework to do. Reevaluate your invention's performance and make the necessary design changes to meet the important issue of improving productivity.

CDI Benefit #11: The Systems Approach

Instead of thinking of your inventions in terms of individual components, try thinking of them in terms of "systems." By thinking of your products as systems, you may find many excellent ways to improve productivity and make them easier to use. Some examples of individual components that have been effectively converted into dynamic systems are:

- A plastic bag—plastic grocery sacks in supermarkets that open automatically

- A cellular phone—a voice-activated dialing system

- Hand assembly—robotics

- A water sprinkler—an automatic sprinkler system

- A flashlight for blackouts—an emergency back-up lighting system

In the first example, trying to sell simple plastic bags to a supermarket would be fruitless. They would be too difficult and time consuming to use and would require hiring extra labor just so checkout lines didn't grow too long. But when the plastic grocery sack uses a bag holder, suddenly the combination becomes a viable system. When that bag system become self-opening, then it reduces labor and improves productivity. Plastic grocery sack systems became a smashing success and almost completely replaced paper sacks in a matter of a few years! Plastic sacks significantly improve productivity over old-fashioned paper bags and they sell for one-third the cost of paper bags as well! Lowering a supermarket's expenses allows them to sell their products at lower prices.

The systems approach automatically changes a supplier's image with the customer. For instance, no longer is the supplier "just another company selling plastic bags." The producer becomes a packaging-systems specialist with a focus on improving productivity and lowering overall packaging costs. This is music to the ears of economically minded corporations in the United States and other countries.

Systems also represent excellent patenting opportunities and an excellent way to expand a product line. Once your system is in place, it becomes a lot easier to add peripheral products and new components or to introduce new systems in other facets of your field.

CDI Benefit #12: People Friendly—The Greatest Innovation Opportunity of All

The computer and electronics industries have given consumers an incredible number of new engineering and science innovations in the last few decades. For many, these innovations were, and still are, too difficult to understand and operate. You can see this phenomenon in far too many products, machines and software in both the workplace and at home. To make matters worse, complicated manuals frequently make these products even more difficult to use.

To evaluate your invention, ask yourself, "Are they people friendly? Will my customers know exactly how to use them without complicated instructions? How can I improve their ease of use so it becomes a simple, operation, one that does not require instructions?" This may be a little ambitious, but it should be a chief objective for any new product.

While it may or may not require a feat of science or engineering to make your products easy to use, it is critically important to remember that people must know how to use them or they will not be used, let alone purchased. Customers and end users simply pre-

POOR PRODUCTIVITY?

If your invention does not improve productivity, but reduces it instead, what should you do? Do what every smart inventor does. Solve the problem and turn the product into one that does improve productivity. It'll never sell otherwise.

TOMORROW'S KITCHEN

"The ultimate objective for inventing tomorrow's kitchen would be to have 'no kitchen at all' . . . a fully automatic kitchen that can prepare food, cook it, set the table and clean up afterwards." This comment is from a home economist of the 1940s.

SYSTEMS REPRESENT GREAT PATENTING OPPORTUNITIES

As you will discover in Part Two—Patenting—if you combine two or more components into a new, unique system, you may be able to secure a patent. Systems patents such as these can afford tremendous patent protection, at times even greater than that for an actual product or apparatus itself.

fer products that are convenient, easy to use, and people friendly. Accomplishing this objective may require a significant amount of research and testing, but the end result will be a very profitable product launch.

If your inventions are not people friendly, you need to make some changes. It might be more appropriate to say that you have some great opportunity ahead for some innovative improvements. Making your products people friendly is also an excellent way to build brand loyalty.

Making Inventions People Friendly

The following five concepts relate to making products people friendly. The secret to doing this is to keep the end user in sharp focus. This is the most important customer and the most important buying influence. Other buying influences such as some middleman, like a distributor or supervisor, or an accountant or buyer are also customers of sorts, but are of secondary importance. Their needs must also be met, however.

Making products people friendly primarily refers to the end user. This is the individual who actually uses your innovation and will determine its acceptance. In focusing on the end user, you will want to be hyperaware and supersensitive as you determine the best ways to make your invention people friendly.

Your goal is to strive to make your new innovations a "no-brainer" to use. If you can do this, you will greatly improve your chances of success. When users ask, "Why didn't they make it like this in the first place?" or say, "You're not going to take my new ZYZYX away from me," you know you have made your invention very people friendly. Making products easy to use almost always causes a positive emotion with users and most frequently improves productivity in industrial and commercial applications.

Capitalize on Natural Tendencies

With a new or improved innovation or system, what are the natural tendencies of the end user? You can capitalize on making your new innovation people friendly by incorporating these natural tendencies into your design. That way, users will instinctively know what to do.

Step inside an automobile and you can see several good examples and perhaps recall a few bad or confusing ones. If you were to improve the design in one of the following automobile components, what changes would you consider?

1. **Steering.** Would you want to use anything other than a round steering wheel?

2. **Brake pedal.** Can you imagine pressing on anything other than a foot pedal to stop?

3. **Transmission gearshift.** Placed on the column or console, we know how to use it. In the early 1960s, Chrysler's "superior engineering" came up with a modern push-button version. Dodge and Plymouth had buttons in the dash and Chrysler's were placed where the horn button used to be. Can you see it now? You are driving along in your new Chrysler and someone cuts you off. You retaliate by blasting them with your horn. Instead, you screech to a halt as your transmission explodes and metal shrapnel is strewn all over the freeway! Or how about your new Plymouth? Your 5-year-old daughter reaches up to change radio channels, just like she used to. Whoops! Of course, Chrysler abandoned the push button system after just a couple of years.

4. **Headlight switch.** At first headlight switches were push-pull switches located on

the left side of the dashboard. Then some switches became push button, some moved to the right side, others were incorporated on the turn indicator, and yet others had a separate turn indicator-type switch on the steering column. Have you ever rented a car at night in an airport? After walking half-frozen out to the dark parking lot, you climb inside, close the door, and try to turn on the headlights so you can see what the heck you are doing and so you can put the key in the keyhole and start the engine. Not only can you not find the headlight switch, you cannot read the little names or tiny logos on them either. In frustration, you reach for the . . .

5. **Dome light.** No luck here either. The dome light switch is either part of the "mystery" headlight switch, button, or turn indicator post. Or, is it a separate switch somewhere on the dash or on the console? Or, can you reach up and find the switch on the dome light itself, or behind it? Who knows? Automakers are finally figuring this one out. The most obvious solutions are to either have the dome light be its own switch so that by pressing it in, you turn it off and on, or have the light stay on for a few seconds after the door is closed, so there is sufficient time to find the keyhole. Fortunately, many automakers are doing this now.

However, because of nonstandardization of headlight switches, it is still frequently a challenge to find them. The same goes for wipers. Recently, some auto manufacturers have been installing some interesting electric window buttons. Before it seemed that they were all vertically oriented buttons that flipped either up or down. Now, some are placed horizontally on the armrest and are pulled up or down instead; it's easy to flip those switches backwards.

One of the better examples of designing based on the natural tendencies of human beings is a computer keyboard. The original keyboard was developed so that the operator's most dexterous fingers would be used for the most frequently used letters. While there are new ergonomic designs being introduced, it is still best to use one that incorporates the natural typing motions so ingrained in the users. Most of the new, super-easy, superfast keyboard systems emulate the natural "QWERTY" typing motions. Could you imagine having to start all over with a different format?

In your invention-related matters, try to incorporate the natural tendencies of the users into the design, or else you may be forcing them to do something they consider unnatural or that will require a lot of retraining. Using natural tendencies in industrial and commercial applications can also contribute to improved productivity and the reduction of errors.

Aim for Instant Recognition

Make the functions of your innovations easy to recognize so the user will instinctively know how to use them. It is surprising how often new improvements are not improvements at all . . . after you try to use them.

The icons used in computer software programs give an instant explanation of their use. Some are better than others like the trash bin icon commonly known as the place to delete files, folders or entire programs. Other icons make program identification easy.

Some other good examples are fire alarms, knurled knobs for turning or indented knobs for pushing in and pulling out, a push-plate on an exit door, and color-coded computer jacks for peripherals (It's about time this was standardized!).

IS IT OUR PROBLEM OR A DESIGN ERROR?

We frequently point fingers at operators when production equipment is working poorly or products are defective. It is important to keep in mind that if operators are continually having the same problems with a machine, there is something inherently wrong with the machine's design or the production process. The most inexperienced operators should "naturally" understand how production equipment works. Don't blame the equipment—fix the design or process!

SUPERIOR ERGONOMICS

In striving to create step-saving innovations, you usually improve the ergonomics of an invention. While some view ergonomics as an engineering process, there is a much better way to look at it. Try looking at ergonomics from the viewpoint of what would be ideal. Attempt to visualize the most natural way for someone to use your product. By doing this, you are applying some basic ergonomics principles to your design. Then, have your invention engineered with these facets. By visualizing the ideal, you can really advance the look and performance of an innovation.

One bad example is the shower-bath-hot-water-cold-water, pressure-control knobs used in some hotels. You cannot be sure if you are turning the shower on or off, or the hot water on or off. Another one is the push-rod style bar on a door that makes it difficult to know whether the door opens to the left or the right, unless you can spot the wear on the handle.

Applying natural recognition techniques typically refers to those items or functions you don't use on a regular basis, as compared to those you do, such as a computer keyboard. Or, it could refer to those functions that are new to the user, which you want the user to quickly understand. For instance, color-coded peripheral jacks make it a cinch for anyone to set up a new computer.

Use Natural Mapping

Natural mapping refers to using or emulating existing natural shapes and forms. One of the best examples of this concept is the seat adjustment buttons used on Mercedes, Jeep, and several other cars. There are two buttons: one forms the seat and the other the backrest. The user moves each button to put the seat at the desired height and tilt and the back support at the desired angle.

In general, computer software companies strive to incorporate a form of natural mapping in their schemes, but they often fall short. In fact, the incredibly large "help folders" and user manuals are evidence of this shortcoming. One good example of natural mapping with computers is Hewlett Packard's touch screen, which makes it easy for the user to follow. There is no need to understand computers; simply having a finger is all that's required.

One of the best opportunities to use natural mapping still exists in remote control devices, but it is rarely used. Think about it. Remote control devices usually have a series of entirely too many little buttons, some with little words printed on them. It is assumed that everyone who uses them has 20/20 vision, can easily read them, and watches TV with the lights on. At least Sony has lighted buttons on their remote controls, but it's not enough.

OK, all jokes aside. Why can't at least some of these buttons be ones that the user can touch and intuitively recognize their uses? Here are some possibilities for a VCR:

- Opposing arrowheads, easy to identify by touch, for rewind and fast forward buttons

- The round play button located in between the two arrowheads

- A volume control button in a rocker switch with the "up" volume at the top and the "down" volume at the bottom (not sideways) and the top of the volume button raised and the bottom with a dimple

- A round channel tuner knob (It doesn't have to be numbered, just has to turn and click sequentially for each channel.)

DID MERCEDES LISTEN TO PETERS?

In a good example Tom Peters notes that he needs to move the seat back when he drives his wife's car. Since the seat does not move without the car on and he can't get in the car without the seat moved back, he has a great usability complaint. Benz solved this by allowing you to move the seat using a series of switches on the door, which look like the seat. The movement of the form directly relates to the seat movement. And, of course, you don't need to turn on the car to do it.

Figure 3.7 Natural Mapping for Ease of Recognition

Natural mapping techniques can be used in many types of control applications in industry. Natural mapping inherently reduces errors and improves performance and productivity.

Save Steps, Save Time, Reduce Confusion

Capitalize on the basic economic principle of "productivity produces income" by reducing the number of steps required to use a device or perform a function. It goes without saying that if you can reduce the number of steps required to use a product, machine, or device, productivity will improve in the workplace and profit potential will increase. Or, customers at home will simply prefer your innovations over the more time-consuming ones.

In business, operations managers are usually powerful decisionmakers, making their living trying to cut labor costs through timesavings. You will greatly appeal to these decision makers if you can eliminate entire labor-intensive steps in operations.

It's the same at home. Step-saving inventions actually became popular in late 1800s and into the early 1900s when the major thrust of invention was to make life easier for women in the home. It was the primary arena of invention for years, with the invent of General Electrics household appliances like toasters and irons, Maytag's washer, and so on. Everyone likes home innovations that reduce the number of complicated steps and time.

Another good example of step-saving innovations is software. Each time Microsoft introduces a new version of its software, the programs tend to require fewer steps to execute commands. Even macros are easier to make than ever before; the newest Microsoft software programs will actually track your keystrokes and automatically form, create, and execute the macro function.

Yet another example of eliminating the steps required to perform a function is in the supermarket. The customer checkout is faster than ever, due to laser scanners and plastic bagging systems. With laser scanners, the checker saves time in handling the purchases. In concert with scanning, a fast, step-saving, self-opening plastic bagging system keeps pace. Compared to paper sacks, plastic sacks save about 4–6 seconds to open and load each bag. The combined result is that you literally save minutes in the checkout line.

Here are some other step-saving innovation opportunities:

- The elimination of mixing two-part epoxy glue
- Preset settings to restart a machine after shutdown
- A machine's ability to track statistical process controls
- Automatic emergency shut-off or warning devices that detect human error (or presence)
- Totally automatic drip coffee makers that measure the amount of coffee and water according to your pre-determined settings
- Recipe packets for specialty foods combining the desired ingredients in a unique combination—for example, Top Ramen (this field is still wide-open for major exploitation in other areas)

A philosophical approach to step-saving is to strive to make your product or system totally automatic, with no potential for error, human or otherwise. This simple approach can result in a lifelong series of projects and innovations for a smart entrepreneurial company.

CASH REGISTERS

This generation of young Americans will not know what a cash register is unless their parents tell them about yesteryear. They will also not know that older folks learned to type on typewriters before there were computer keyboards. What's next? How about ink-jet printing presses for fabrics and printed packaging? Get ready, because these industries will be converting to the new technology soon. Yes, your clothes won't be dyed any longer in toxic bins . . . the patterns will be printed on instead!

The following example of a water temperature control unit, designed by Waterpik Corporation and called the Jandy Aqualink, is used with spas and swimming pool systems. The older version on the left can be programmed by the user, with various functions assigned to 23 pushbuttons (10 more buttons are in the flip down door).

Sound confusing? Waterpik thought so too. It was not only a programming nightmare, but it was also difficult to install and even harder for a purchaser to figure out how to use. The solution is the unit on the right, designed by Studio RED, that eliminated the confusion of the excessive pushbuttons and used a simple menu-driven approach instead. The new unit was far easier to use and was simple to program and simple to install. The result was a dramatic improvement in sales and a device that garnered top awards from both *Home Automation* and *Electronic House* magazines.

This is the kind of future thinking that is starting to show up in other areas of design as well, areas such as VCR, DVD, and TV and satellite programming. There's no reason that you cannot apply this type of future-thinking, chip-based technology to your innovations.

Monkey See, Monkey Do—Minimum Training Required

Invariably, some new products will require training. If you can substantially reduce the amount of training required, you have done your job of inventing quite well! Keep in mind that although your new product may be the eighth wonder of the world, if it requires extensive training, you have a tough, expensive time ahead.

Industrial and commercial innovations must be easy enough to use so that very little instruction and training is required. There must be a relatively seamless transition into a new product or system. If this is so, management will instantly give you its blessings.

In industry, training is expensive. So, if you can honestly characterize it in your sales presentations as monkey see, monkey do, the expense of training becomes a moot issue. In the home, training and learning a new process can kill any sale quickly. Who wants to spend time training to learn a new process that replaces an old one?

Training is important to learn new methods and to learn to be proficient in a new arena. But if substantial training is required in order to change from one methodology or one product to another, then the reception will be quite negative. Imagine the amount of training that would be required for you or your company's employees to switch away from the QWERTY keyboard to some other version.

The Best Kept Secrets of Great Inventions

We know that inventions are based on having some new, unique function that ultimately is patentable. We also know about focusing on CDI benefits so that the unique functions are ones customers want to buy. But perhaps the most important aspect of a buying decision is based upon a factor that may have nothing or little do with the physical merits of the invention and its performance.

Almost all of us have heard that "form follows function." This may have been dozens of years ago, but it is no longer true today. It is one of those old adages that never seems to die.

Great contemporary inventors and designers (artists too) know there is a higher truth to follow to success. It is "form follows emotion." Don't kid yourself otherwise.

People make their decision to purchase based on positive emotions. It may be related to the thrill of speed, the peacefulness of silence, the confidence of ease of operation, or the sense of power potential provided by horsepower. It could be the psychological appeal of a color or the softness of a texture that raises an emotion that elicits a favorable response from the buyer. It could even be the right trademarked name representing trust.

One of the best examples of a company using this concept is Apple Computer. Their designs started with the original Apple PCs, then the Macintosh, and more recently, the iMac and the iPod. An even modest design improvement such as the introduction of its 17-inch screen paper-thin notebook computer elicits positive emotions. Apple Computer has the act down, or perhaps their designers do. Apples designers are brilliant, which is a main reason they are able to survive in a highly competitive market against low-priced competitors.

Apple's computers have always been compact, futuristic, and exceptionally easy to operate, and they have had the software to back them up. In short, they were designed to elicit a positive emotion from their users. The result was that Apple created and quickly dominated a new niche market with weak competition from IBM and Digital, the industry giants in the 1980s. Today, Apple has differentiated less and has lost market share as a result, but nevertheless, it is still a powerful, influential player in the field.

Another example of form following emotion occured when Kawasaki Motorcycles took on the three motorcycle leaders—Honda, Yamaha, and Harley Davidson—with some dazzling new cycle designs. The emotion the designs elicited from buyers was a sense of speed and power. Kawasaki stole the power position from the others because of the emotions their cycles elicited in customers. Now, it is being emulated.

But Harley Davidson retaliated with some emotion-raising events of its own. Superior quality replaced old oil-leaking engines. The bikes become more comfortable, sounded a lot better than the little two-stroke imports (Harleys rumble instead of "wing-ding"), and offered incredible color selections. The new bikes could be customized in a myriad of possibilities, and suddenly began to maintain their resale value. Altogether, Harley Davidson re-established an image of superiority over all the other manufacturers.

The most recent release, the Harley VRSC, is an amazing combination of emotion-eliciting features, design, and performance that makes even the racy Kawasakis look archaic. Harley Davidson learned well and knows how to appeal to their target market with dazzling design that really makes you want to buy. And people do . . . just look at Harley's annual reports.

TRAINING?

In my invention efforts, I have at times trained employees in the manufacturer of a new innovation. Teaching various shifts about the quality the customer expects and showing them why they expect it is of great value. How else are they going to know? I have even posted quality standards in both English and Spanish. How was I able to do this? Simple. I asked the management, which is usually most willing, since training costs money and I did it for free!

–Bob DeMatteis

WHEN WAS THE LAST TIME YOU BOUGHT SOMETHING BECAUSE OF ITS FUNCTION?

When was the last time you bought something and afterward said, "Wow, I really like how this thing functions"? The reality is you bought it because something about the product and sale made you feel good. People select their automobiles not because they are the best functioning cars, but because they create a positive emotion within them. Think about why you bought your last car.

INDUSTRIAL DESIGN FIRMS

Another source to aid small businesses and independent inventors are industrial design firms. They are practiced innovators and creators of new products. Their knowledge of materials and processes, human factors and ergonomics, engineering and product development, and the consumer marketplace, as well as appropriate product and packaging design, can greatly expand the potential of an inventor's new product idea. Remember, ultimately a new product idea will need to be translated into a presentation form appropriate for testing, concept presentation to potential licensees, sourcing through original equipment manufacturers (OEM), and/or future presentations to potential retail buyers. Depending on whether you intend to license your new product idea or market it yourself, the form and depth of these presentation materials will vary greatly. Industrial designers are experts at preparing effective and appropriate product presentation materials.
–Paul Lapidus, CEO, NewFuntiers

Even more interesting is that the terrain in the world of motorcycles is blanketed with patents secured by Harley Davidson protecting their new prowess and innovations. More than 200 patents in the last ten years show they are bent on protecting every aspect of their re-emergence as a leader in the field. About 50% of their patents are design patents, an illustration of one of the wisest uses of design patents ever. They cover the design of gas tanks, radiator covers, fenders, and so on. Good luck trying to design around those!

Look for ways to elicit emotions that will drive your products to success. Then, design the products accordingly. Don't let function dictate your ultimate design; let the emotion that will be elicited drive the design. As Harley settled in on the design of the

VRSC, it had to invent many new technologies and created several new patentable inventions in order to accomplish its objective. If it had not taken this approach, then it would have been following the old-fashioned format of "form follows function." With such an approach, the VRSC would have looked like a motorcycle from the 1960s.

Following are examples of emotion-eliciting effects you can incorporate in your invention:

- High-speed performance: "I can feel (imagine?) the power, the exhilaration."
- Racy design: "It looks modern, sexy."
- Ease-of-use: "What a relief, it's works just like you'd expect."
- Long-lasting: "It's the last one I'll ever buy."
- Compact design: "It fits in my vest pocket, easy to carry around."
- Totally enclosed: "No exposed parts; it's safe to use."
- Sharp print copy: "Quality is exactly what our company is all about!"
- Easy to fix: "Even my wife or 10-year-old daughter can fix it quickly."
- Self-adjusting: "Anyone can do this. All I have to do is enjoy it."
- Great color choices: "It matches my boat . . . looks great and is a perfect choice."
- Patriotic: "It's made in the U.S.A. and I want to do my part."
- Comfortable: "Legroom at last! As good as flying first class."
- Classic curves: "It's beautiful, just like a Lexus."
- Free 24-hour service: "Now I don't have to worry about how it's used."
- Money-back guarantee: "I can't go wrong; these guys must really be confident of their product."

There are many other emotional responses you can look for in your innovations. How many different emotions does your present invention elicit? You'll want at least two or more that reflect your invention's central theme. Hopefully, your invention has several others that will satisfy other buying influences, too.

For certain, developing character in your innovations that creates positive emotional responses in the end user is at the heart of a customer-driven innovation.

Designing to Elicit Emotion

How do you design for emotion? It's not that difficult if you're aware of what types of emotions you want to elicit. When you discover something unique, usually it will have some emotional appeal. But you really should take it further and capitalize on all aspects that will give it sizzle, pizzazz, pop, sex appeal, or whatever fits.

One good way to identify attributes that will contribute to eliciting emotion is through visualization. Try this:

> **Visualize your invention in use and imagine the end user's emotions. What do you want the end user to feel? How can you incorporate emotion-eliciting attributes into your invention? What will they look like? How will you do this?**

Make a list of all the elements you can incorporate into your invention that will elicit the desired emotions. Then, design as many of them into the invention as possible and figure out how to manufacture it. Insist that your manufacturer go the extra mile to accommodate your design. Don't allow the manufacturer to dictate how the design should look based solely on traditional manufacturing processes.

An important facet of designs that elicit emotion is the need for manufacturing processes that can produce them. Part of the problem is that many typical, old-fashioned manufacturing companies tend to have square, functional-thinking managers and employees. Likewise, their equipment frequently follows suit and makes only square products.

Be determined to solve any manufacturing problems associated with design. Your manufacturer must be a part of the team that can get this done. Regardless of how your new invention is going to be produced, you cannot kowtow to the mediocrity of everyday mass-manufacturing processes. You must maintain your vision and insist that it be fulfilled.

Waterpik's Jandy Aqualink illustrates how to design for emotion. The old version had 26 complex pushbuttons you had to know how to program and subsequently use. The new version overcame those objections through the use of a simple menu-driven format.

GOT SIZZLE?

A commonly used concept in marketing is "don't sell the steak, sell the sizzle." You should think in terms of what "sizzle" your invention has.

PRINCESS LEADS WAY

Henry Dreyfus was one of the first creators of "beautiful, modern devices" that tried to appeal to "form" instead of "function." One of his first "beautiful creations" was the hand crank washing machine of the 1920s. A later creation was the Princess telephone of the 1960s that forever changed the way we used and purchased telephones and led the design revolution of modern products.

"I plan on selling the One Touch model exclusively. I have heard nothing but good feedback from all of my clients using the One Touch control system."
Joe Marcotte, Coral Pools, Lafayette, CA

Redesigned by Studio RED, the new Jandy Aqualink is a big hit. The redesign hit all the right emotional buttons. Not only did it hit the right emotional buttons regarding ease of use, installation, and programming, but it also made its appearance absolutely dazzling. No longer was it just a control box mounted on a wall. It became a work of art that matched any number of interior décors, switch plate combinations, and schemes.

The Jandy Aqualink buying experience by a customer was now complete. Every facet has been satisfied from installation, programming, and use right through to its exceptional aesthetics and attractiveness. Think about this. Every time the customer sees or uses the Aqualink, he/she is having a positive experience. Do you think that customer will buy another Waterpik product?

Just like with Waterpik, you want to strive to give your customers a complete positive experience with your innovations. It's not OK to be just "a little bit better." Do what Waterpik has done and make the experience complete.

Other Attributes and Secrets

The preceding CDI benefits are not all the possibilities. They represent many of the design attributes sought out in the early twenty-first century. They really serve to help get your imagination percolating so you can create innovations that will be well received.

Focusing on CDI benefits to the customer is also a key to evaluating the potential of an innovative idea. Determining exactly what benefits an end user may want is somewhat subjective. It may sound altruistic, but well conceived, creative, innovations that last long term are invariably founded in attributes that reflect the three driving forces of life: truth, beauty, and goodness. While this sounds idealistic, it is true and can be an underlying driving force behind any new creation. Design your innovations and products so they appeal to the deep-rooted ideals, desires, and emotions of the people who will buy them.

About CDI and QM

Customer-driven innovation and Quality Management principles and philosophies go hand in hand. In many respects, CDI is a QM system for monitoring and controlling products and innovations. In certain respects, where continuous improvement leaves off, customer-driven innovation takes over. The illustration in Figure 3.8 shows the continuity between the two concepts and where responsibilities lie for innovation, continuous improvement, and ongoing maintenance.

Figure 3.8 Total Quality Management and Innovation

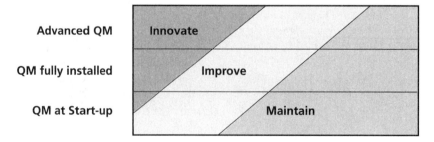

The book, *Essentials of Patents,* expands upon this concept and gives readers guidelines for developing products for their companies' futures. It discusses the responsibility every corporate CEO has toward his/her shareholders and society. Whether you are an individual designing or engineering for yourself or part of a company, you should develop PQM concepts in industry and put PQM policies in place. By designing your inventions based on CDI, you will be in a better position to partner your inventions with industry.

CDI and Patenting

Any invention that is new and unique and solves customer problems should be patentable. Creating highly desirable inventions almost always means you are creating highly desirable patent protection as well. In Part Two on patenting, you will learn that the owner of a patent has the right to exclude others from making, using, or selling his/her products. So, as you invent and improve your innovations, you can also secure more patent protection, which will further protect your interests.

You see, patent protection is directly related to how well you've developed and qualified your invention. Marketing prowess actually—literally—qualifies and confirms the scope of your patent protection. The claims in a well-written patent application directly reflect sales/marketing attributes. The stronger the sales potential, the more powerful the patent claims.

Follow this logic and you'll see how important product development and marketability are to your inventions:

- The more you focus on CDI, the more desirable an invention
- The more desirable an invention, the stronger the marketability
- The stronger the marketability, the better the patent protection potential

The conclusion here is:

**Making money on your inventions and
protecting them with bulletproof patents all
stem from proficiently pursuing CDI.**

TO FILE OR NOT

One thing is for certain . . . if a company does not file for patents on its new improvements, inventions, and discoveries, it may find itself always trying to stay ahead of copycats . . . or infringing on the patents of those companies that do file.

Patents also give you and perhaps others on your development team a chance to be a part of history. Whether in manufacturing, sales/marketing, or engineering, everyone on an invention-development team may receive recognition for his/her contributions.

Becoming patent-wise as you invent will also give you a sound basis to design around a competitor's patents. You will be better able to understand a competitor's present and future direction. As a result, you will be able to adapt your developments accordingly. Don't be surprised that your CDI knowledge allows you to outfox them in the marketplace altogether! It is quite possible that the nature of your patenting thrust will be quite different from theirs. Large established companies tend to develop an invention along the lines of engineering new improved processes and reducing manufacturing costs, whereas an entrepreneurial small company tends to focus on the creation of new markets.

CDI as a *Modus Operandi*

Applying CDI should be your focus in your invention activities. How often and how seriously you develop your invention is based on you or perhaps your company's philosophy. Obviously, it is based on the size of your pocketbook as well.

Rational-Emotional Design – How to
Design to Elicit Emotion

Founder and CEO Philip Bourgeois of Studio **RED** in Menlo Park, California, explains the importance of applying their unique Rational-Emotional Design (RED) process during development, which guides and ultimately determines the product's design and position in the marketplace.

According to Phil, product design is a mixture of physics and art, or in other words, "rational-emotional design." The rational side is the technical, functional, patentable side concerned with cost effectiveness and manufacturability. The emotional side is concerned with compelling design, human factors, and aesthetics. For a product to reflect its unique position in the marketplace, the blending of the rational and emotional sides must be balanced. However, both must be maximized.

The designer pushes the emotional. The engineer and manufacturer push the rational side. Both sides must come together to ultimately meet the objective. The designer provides his design and aesthetic input, while the engineer and manufacturer figure out how to accomplish the objective cost effectively. For instance, the manufacturing side seeking to lower costs may ask the designer, "How can we modify this design for cost reduction reasons without compromising the experience?"

The overall objective of applying the **RED** methodology is to clarify the innovation's unique position. Brand strategy becomes relative to the product experience. Each company is different and so is each product. For example, a corporate strategy may be that of the best and most innovative in its field. But a product strategy could be different. It has to convey the company's innovative image and must also reflect its own unique positioning in the field when put alongside other competitors. Studio **RED** accomplishes this through the use of questionnaires and market research that identify and qualify the product positioning quickly and cost effectively.

The process begins by first defining the competitors' positions and knowing what's going on in the field. There's no substitute for being an expert in your field to help accomplish this objective. Then, the company must define where its innovation fits in this landscape. This means where does it fit price-wise, quality-wise, and so on.

For instance, a company in the medical field may have a corporate brand statement of being an *innovative* producer of medical devices for *personal* use. It may already be producing disposable pregnancy testing devices. The company's new product offering might be a reusable pregnancy testing device that has a higher price point. The goal is to design the new device to reflect the corporate *brand* as well as bring forth a new product *position* at a higher cost option. The design language must represent both to the buyer/user and deliver on that promise in performance in order to be successful.

Once the unique position is clarified, the overall character of the product and its appearance must bring forth the message that conveys both this position and its functional uniqueness. Studio **RED** accomplishes this objective first, by identifying the words that convey the right product image (the unique position); second, by creating images around the words; and third, by using those images to explore and ultimately create the product's physical appearance, bringing forth those messages and emphasizing the special unique performance of the product. Last, statements are created to reflect the product's unique position.

For instance, the new high-end WaterPik One Touch control unit was designed to be the size of a quad electrical box positioned alongside a designer light switch plate. The goal, therefore, was to create a design that reflected a sleek, modern, and simple look that also matched existing high-end designer light switches, such as the Decora line. The product was intended for high-end homes, and the Studio **RED** design team researched a variety of home interiors and studied the other products people in this target market used and might have in the vicinity of the One Touch controller. Due to its higher cost and the custom installation required, the One Touch would remain in the home for quite a few years. The design needed to be timeless, as opposed to trendy, and it needed to exude quality in its functionality, user interface, and visual appearance. The added wrinkle to the design process was that the buyer of the product was the architect or pool installer, and the user was the homeowner. The design needed to appeal to both in order to be successful.

This image must then be reflected with its "at a glance impression" on the shelf. What is the impression, right from the onset, when a consumer sees the product on the shelf? For example, with the One Touch, the consumer must see the right corporate image reflecting WaterPik Corporation and the underlying fashionable style of the product's relationship to the user's home. You need to know who the buyer is, what the home interior might look like, where it goes in the environment, etc. Those words and goals are turned into iconic representations that ultimately direct and bring forth a favorable experience to the consumer. It must be compelling, engaging, and fashionable; tell what it does; and make the user feel good after purchasing the product.

This entire process goes back to the original concept of Rational-Emotional Design and fulfilling those product objectives.

When working with their clients, Studio **RED** uses a formal, technical matrix and facilitates meetings with the client at the onset of the project to achieve the correct direction in order to accomplish its **RED** objective. After defining the competitive position, it may produce several different concepts . . . some to the right and some to the left of the specific goal . . . for the client to review. It's all about finding the right words that can be turned into the right design, a design that reflects the right representation of the new innovation and its corporate owner. Personal design opinions are not a part of the process. "I like pink" is not a justification for using pink, but finding out that pink is a popular color among teenage girls, the target market, is justification. Even a shade of green used in a medical application should be a certain shade and shouldn't look like the green of a child's toy or the green of money.

Throughout development the image is consistently reflected in every aspect and throughout the matrix. Upon purchasing the innovation, then when it's in use, and afterward, a consumer should have the same experience, and delight, in having made the purchase.

Thomas Edison, perhaps the most famous inventor in America, had goals for himself and his team. That goal was "a minor invention every ten days and a big thing every six months or so." However, Thomas Edison was in the business of inventing and had assembled a full-time staff of engineers and scientists to assist him. Their aggressive pace would most likely be unreasonable for an independent inventor or small business based on current costs to invent, test, and patent a product line. Larger companies, like IBM, could conceivably keep up such a pace. Then again, it earns over $3 billion annually in licensing income from its patent portfolio.

Apply the CDI principles based on the nature of your invention and its field and continually improve it. Just one of those improvements might become a breakthrough opportunity and a "big thing." Edison was hoping that only one out of ten inventions might be a winner. So, if you hit one out of ten, you, too, may be in for windfall profits you never thought possible. It happens.

How often you and your development team (or partners) release innovations will depend on several factors. The primary ones being the complexity of the product and its related processes and systems, how broad the product line may become, and, of course, your budget. Depending on their philosophy and aggressiveness and their product's life, small businesses and product developers typically use one of two approaches:

- **Product centric.** Companies that have an established product line may elect to file patents as new discoveries are made. The patenting may focus on improvements to existing products, new spin-offs, or new product introductions altogether. The release of new innovative concepts for any given product is based on market demand and need for change or accidental or purposeful discovery. Product-centric companies tend not to focus on patenting as an objective but as protection. Over time, one of the new patented improvements may become a breakthrough opportunity.

- **Patent centric.** A company that focuses on patents and only developing patented products will more aggressively develop improvements at a faster rate. Depending on marketability evaluation and budget, such a company pursues patents. In a patent-centric mode, it will typically focus on a smaller number of innovations, perhaps only one or two.

Both philosophies are good ones. Both are far superior to companies that are not innovative and have no patent protection for their product lines. Regardless of your approach, filing patents on the important inventions that have commercial value is key to success in business in the twenty-first century.

Inventions with limited marketability and start-up companies with limited funds will need to carefully determine the merits of filing patents on any one invention. Once you're established, it becomes much easier to budget for patenting.

CDI Driving Your Future

It will become apparent that after a first innovation is launched or a product-centric company embraces patent protection as an important part of its future, there is only one thing to do. Create, invent, and do it again. Each time a new customer-driven innovation is launched, it gets easier. With manufacturing and marketing partners in place, you will find inventing the future and introducing new improvements a simple, natural process. Exciting too!

A PROLIFIC INVENTOR

Jerome Lemelson averaged a patent a month for more than 40 years, all of which he accomplished independently. An optimistic and vibrant man, he amassed the fourth largest portfolio in American history.

Staying ahead of the competition with innovative individuality is like building a retirement fund for you and your company and having fun doing it. You'll find that no one will be able to challenge your position once you become the expert in your niche. As the all-knowing expert, your position is strengthened each and every year.

Being creative is a gift only a few people have. There is nothing better than living your life today as you imagine and create the future. Applying the CDI tenets only gives a clearer vision of how to create that future. Never have truer words been said:

Your destiny is determined, moment by moment, by the achievements of your day-to-day creative efforts.
Your actions today determine tomorrow's destiny!

TOP OF THE LIST

The life of a struggling inventor can be a hard one, but really no more difficult than the lives of their families waiting at home with warmed-over dinners and missed soccer games. When taking stock of their most valued assets, inventors should place their enduring loved ones at the very top of the list.
–Don Kelly, Intellectual Asset Management Associates, LLC

ASSESSMENT, ASSISTANCE, AND PROBLEM SOLVING

*Until you have assessed and qualified your invention,
you should not be patenting it. To pursue an unqualified
concept is a waste of time and money.*

Over time your invention will undoubtedly go through a metamorphosis. How can you be certain this transformation is for the better? There are several factors that will ultimately determine your invention's marketability. You want to make absolutely certain you make the right decisions that will have a positive impact on your innovation's future.

For instance, you may make a design change so that your invention can be more cost-effectively produced. However, that change may also make it more generic and less unique. Or, you may incorporate a raw material that shortens shelf life or opens up the vulnerability of supply shortages. Or, you may get caught in a position where the prototypes or test models are not going to be as durable as the customer would like. So what do you do?

First of all, you can't walk blindly forward, hoping and wishing. If you do not have a marketing or manufacturing partner on your team at this point, you really should not proceed without properly assessing and verifying your invention's attributes. Ask yourself the question:

> **Are you prepared to answer all questions a
> marketer, distributor, retailer, or end user
> may ask about your innovation?**

If you hedge your answer, then you have not sufficiently assessed or qualified the merits of your invention. You should not be pursuing licensees or partners until you do.

Frankly there's really only one way to determine an invention's potential 100%—and that is by having your internal marketing department or external marketing partner or licensee give it its stamp of approval. That's your ultimate objective. But what can you do if you don't yet have your marketing partners or if your internal marketing department is not sure? While in search of marketing experts or gathering data for your internal marketing department, are there steps you can take to assess your invention? To make sure you're on the right track? Yes, there are.

If you do not have your marketing partner yet, or you're pursuing licensing, in most cases you will want to pre-qualify your invention's potential in depth before pursuing them. This is particularly important if you do not have an internal marketing department or established contacts to consult. Thus, assessing and pre-qualifying your innovation is very valuable before you spend a lot of money and time. The assessment research you compile will also be valuable when you prepare your New Product Summary or pro-forma as outlined in Chapter 10.

Perhaps more important than assessing an invention is knowing how to fix it and where you can find help, if needed. There are some simple proven methods you can use to identify and solve problems and to even create the future. You'll want to be ready to employ these concepts when the time is right.

If assistance is needed, there are many experts waiting in the wings to help out. Many of them work for government agencies—federal, state, and local—and do not charge for their consultations. You can find numerous experts at your fingertips if you know where to look. Regardless of where you live, every community has these resources to tap.

The Assessment Process

Assessing marketability to make sure there's enough potential earning power behind the project is one of the essential laws of invention development. When your project and prototypes are well developed, you'll want to make one more thorough review at assessing and perhaps fixing your invention again, just when you thought your innovation was ready to go

Hopefully any changes or modifications you make now won't be major ones, but if you do not address these issues, a product launch will probably flounder or your innovation will fall on deaf ears when you are searching for marketing partners. You've heard the axiom "you have to tend to the details." Truer words may never have been spoken when dealing with launching an innovation. Assessing your present invention will not only tell you whether it's ready for market but it also will help you plan for future opportunities and the all-important breakthrough opportunity.

Your assessment objectives include:

- How to anticipate and fix problems
- How to ensure that you're satisfying **all** buying influences
- How to use the Strategic Guide to keep you ahead of competition
- How to spot and exploit breakthrough opportunities
- How to create the next generation

Your objective at this stage is to apply all the knowledge and research you've compiled. Hopefully you've applied most tenets for a successful innovation already, which should include some preliminary feedback from manufacturers and marketers. Now, you've got to fix all the last little problems you might find and make things happen. At this stage, small differences resulting from seemingly small modifications can make a huge impact on marketability.

Invention assessment actually begins with the laundry list of CDI benefits listed in Chapter 3. These are the first elements to apply in order to have a new, unique, highly marketable innovation. But there are other factors to consider that will affect an invention as well. Miss one of them and you may be completely shut out of a market.

If some red flags pop up, as you test and evaluate your invention, take action now before you try to convince manufacturers and marketers to partner with you. If you repeatedly hear the same or similar negative comments, it's time to act. All of these warning signs are blessings in disguise.

But what should your research prove? What should you look for? Can you list your invention's strong points? Its deficiencies? If you can't, you need to assess it.

There are five key areas you may assess to qualify your invention. If you've done sufficient research and substantial testing, you'll find this fairly easy to do. If not, you'll want to uncover your invention's strengths and weaknesses.

The five areas to assess are:

- Marketability and competitive response
- Societal and environmental impact
- Engineering and manufacturability
- Intellectual property analysis
- Legal and safety analysis

In order to be prepared to talk with prospective partners, you really must have answers to the myriad of questions they may raise in the above categories. You're asking for problems if you've not researched and assessed these points. You may apply some of these points in a brief summary in your New Product Summary or in greater details in your prospectus as outlined in Chapter 10.

Assessing Marketability and Competitive Response

Assessing marketability is the most vital assessment you'll do. Since marketing new products is all about making money, and money can only be made if the product gets sold, marketability potential is at the top of the list. Here's what you have to assess:

- **Market size.** Have you substantiated the potential size of the market in actual units and dollar volume? Potential partners will want to know this. If your innovation is a very new concept, a marketing partner will want to get some feel for profits and return on investment (ROI) so they'll know if it's worth the investment.

- **Ease of entry.** What are the barriers to entry into the market? How many competitors are there in the field of your invention? What is their relative strength? Will you be able to establish a position in the market? Will the existing suppliers be willing to license? For example, it is a lot easier to enter into the field of backpack accessories (sporting goods), than it would be to compete in the field of manufacturing a new type of automobile. Competing against the major world-renown automakers would be a formidable task! Just ask DeLorean.

COMMERCIAL ASSESSMENTS

There are some commercial means available to help in the assessment process. Some colleges and universities provide limited services in various fields. The biggest problem with these assessment methods is that there are too many assumptions, which may not apply to your or your company's market objective. Overall, the most significant problem with assessment programs is that they tend to focus on telling you what's wrong, but they rarely tell you what you need to do to change or correct a deficiency. So, you'll most likely need to read between the lines, make various changes, and then retest them.

PATENT CAFÉ ASSESSMENT

The assessment points we cover are those used by the Patent Café's Invention and Product Evaluation. They cover development principles covered in the book. If you've done your research and product development well, you'll find your assessment will be an easy process to complete and evaluate. You may learn more about this invention assessment system on our website.

- **Market durability.** Is your invention a fad or a trend? If you're licensing or partnering the sales of a fad, you better have a marketing partner in place, or the opportunity will pass you by. Most established marketers want solid long-term trends. Also, is your innovation in a dying market or an emerging and growing market? How can you illustrate this with your invention?

- **Competitive response.** What will competitors do once you've entered the marketplace? It is doubtful they'll not respond should you be successful. You must have some idea if they'll try to low-ball pricing, introduce competitive products, or whatever. Can you answer these questions? Your marketing partners and licensees will want to know.

These elements are the basic consideration of marketability. Know where you stand before you proceed. You should read Chapters 9, 10, and 12 to better understand what marketers will be looking for and how you can position your innovation when interviewing them.

Assessing Societal and Environmental Impact

There are two primary concerns to consider to fulfill your obligation in this category. Obviously, it goes without saying that if you cannot provide an invention that has a positive impact on some aspect of society, it will fail. For instance, you don't want to create a new innovation that has a toxic impact in the home.

Your concerns are:

1. **Environmental impact.** Have you assessed the environmental impact of the raw materials your invention uses? Is a "cradle to grave" study going to be required in your field? This type of study illustrates where the raw material is derived, how it is processed and converted into your product, and what happens afterward. It details the overall environment impact from raw material formation through to the product being a part of the trash stream. It's not always required, but you should be ready if it is needed. At the very least, you'll want to know the answers to any environmental questions and issues before you proceed.

2. **Societal impact.** Will your innovation have a positive impact on society? Does it improve health, cure or treat disease? These are big pluses. The opposite can kill a new product. If an innovation makes a negative impact on children or affects those living in a certain region, you're going to have problems. Is your innovation best made overseas? If so, make sure it is not using child labor to produce it.

While your innovations don't necessarily have to have a positive effect on society and the environment, they almost always cannot have a negative effect or one that worsens present conditions. Review Chapter 3 if you're uncertain of these elements.

Ensuring Engineering and Manufacturability

We know inventions must be made cost-effectively in order to compete. But there are numerous other manufacturing considerations as well. Here are the basics you should know about your invention:

1. **Product structure and engineering.** What will your invention be made of? What is the best raw material to use? Are the raw materials easily attainable? Or will a new

formulation be required? Is your product's structure and quality adequate for the target market? Why? Does it consistently perform as expected? Or will your innovation require substantial engineering and technology development?

2. **Manufacturing process.** How will it be made? Will dies or molds or equipment modifications be required? What are the associated start-up costs? How long will it take to gear up production? Most important, what is the per-unit cost of the product?

3. **Product line.** What is the potential for extending and expanding the product line? The more it can be improved and expanded upon later on, the better the sales potential.

You have to be able to answer all these questions. You have to know how your innovation will be manufactured and what manufacturing partners will expect from you. You'll read more about this in Chapters 9–11.

Analyzing Intellectual Property

Nothing is more discouraging than developing a concept and finding out later that your patent protection is insufficient to protect it. You must have strong patent protection and hopefully other intellectual property to protect and support your sales marketing effort.

If you are a business and plan to sell the new product as an adjunct to your existing product line, this may not be as important. But if you're seeking marketing partners and licensees, then it's probably going to be essential.

Here are the key assessments to protecting your innovations:

1. **Strength of patent protection.** Will it have one patent with a narrow scope or several with broad-scope protection? Did you file yourself or did you use a competent attorney? Have you protected international patent rights or only rights in the United States?

2. **Trademark protection.** Are you using a new mark that is highly distinctive or one that is an extension of an existing product line? Are you not able to provide any trademark protection, which reduces your overall protection? Can you apply for trade dress protection as well?

3. **Other intellectual property.** Do you have copyrights, trade secrets and know-how that may also improve your intellectual property (IP) position? Can you protect these assets?

Throughout Part Two you will read about how to build strong patent and intellectual property protection. You can always proceed without strong patent protection, but it really is essential when competition is expected to respond. This begs the question, "When is the last time competition did NOT respond to a new product launch?"

Analyzing Legal and Safety Issues

Last, there are several issues from the legal and safety side you must address should questions arise. In order to be taken seriously, you must be aware of these issues before pursuing partners. Certainly you can't imagine introducing a new drug product without Food and Drug Administration (FDA) approval. But are you also aware that FDA approval is required for a wide spectrum of other products? Fortunately, it is usually easier to attain approval for these products, but if you are unaware of the specifications relevant to your invention and its field, you may be pursuing the development of a potentially banned product.

SHHHH!

Do not talk to anyone about your invention without a confidentiality agreement signed beforehand. The only protection inventors have until a patent is granted is 'trade secrets' which require the technology to be maintained confidential.
–Michael S. Neustel, U.S. Patent Attorney

There are two primary areas you should know:

1. **Legal considerations.** Are there federal or state laws you need to know about? For instance, is FDA, UL or Good Housekeeping® approval required? Exactly what is that and how long does it take? Does your innovation require some form of legal marking such as law labels or warning labels? Do you know what the requirements are?

2. **Safety considerations.** What is the liability exposure of your innovation? Do you know that some innovations may require exorbitant insurance rates? Is there a potential medical problem, such as carpel tunnel syndrome, associated with the innovation? Are there legal requirements in your state or county, such as with the department of health? Is your innovation a toy for small kids that must not contain small parts? Do you know what the requirements are?

This information is usually not difficult to gather, but you must have answers to all these issues. Someone has to insure the product, but if it's extremely expensive to do so, you better find out up front. Review Chapter 3 if you're not sure about these safety issues and make sure you have all the answers.

After making a thorough assessment and gaining a broad understanding of your innovation and how it stacks up, your job is still not done. An assessment will point you in the right direction, but there will still be many additional relevant buying influences you'll need to address.

Buying Influences

One of the most important challenges you will face is to address all the various influences on buying. The term "buying influences" usually relates to one of two areas. They either relate to the products themselves and how they influence how end users perceive them. Or, they pertain to influences from a marketing perspective and how potential customers throughout the entire purchasing chain will perceive the innovation. Sometimes satisfying just a single buying influence in your innovation's structure can substantially increase sales. Other times, overcoming a single negative buying influence that affects marketability and how the product gets to market can also substantially improve sales. For instance, the WaterPik controllers didn't sell well initially as they were too generic, over engineered, and frankly, very hard for people like us to understand. But once they employed Studio RED to design a unit that fits in with home décor and is intuitively easy to use, sales soared.

The question to you is "Have you truly and honestly looked at your innovation in the same light?" The WaterPik controller is dazzling, isn't it? Is your innovation? Or is it just . . . maybe . . . a little bit better? If you cannot answer this question with conviction, you're not ready to go to market. Go through Chapter 3 and see what may be missing. Ask yourself what you can do to make your invention truly dazzling, just like WaterPik did with its controller.

From the marketing side, can you make a comprehensive list of all the buying influences that may be posed by retailers? Distributors? And everyone else in between who will be handling the product, billing it, servicing it, installing it, and so on? Can you respond to every question an individual may have in order to assess those all-important buying influences, those key aspects compelling distributors to distribute, retailers to retail, and consumers to purchase your innovation. Here are some questions you should ask yourself. If you don't have answers, you're not ready for market or ready to seek out partners:

- Does it satisfy the end user? What's so great about it?

- Does it satisfy the licensee? Where's the money? Where's the profit opportunity?

- Does it satisfy the manufacturing partner? Why? How? What's it going to want out of this project?

- Does it satisfy the distributor? How will it make money?

- Does it satisfy the retailer? How does it profit? How are you going to obtain exposure on a retail shelf? What advertising will you do?

- Who else does it need to satisfy? Warehouse personnel? Trucking companies?

If you say that's for my licensee or my marketing partners to do, you're wrong. You should be prepared ahead of time or work with your licensees or manufacturing and marketing partners to address these issues. After all, how else will you become the expert in your field if you don't have the basic answers to what other experts and customers, retailers, and distributors in the field want or expect?

You'll read more about marketing-related buying influences in Chapters 9 through 12. In Chapter 9, you'll be able to prepare a pricing structure for your innovation. If you've not done this yet, you're just in the beginning steps of product development.

Frankly you'll be reading about satisfying buying influences throughout the book. Your challenge is to turn them into an endless stream of benefits for everyone involved and to effectively convey that message to them.

Measuring Your Assessments

Assessing an invention means measuring and comparing it in some form to the products consumers now use. So, how do you know how your assessments stack up? You may approach this scientifically and use measurable quantities to illustrate a quality or performance, or you may use simple qualifiers, such as "not as good," "same," or "better."

However you accomplish the task of assessing your innovation—written assessment, team assessed, or at times surveys—you must be able to convert it into communicable language, so partners, retailers, distributors, and consumers will understand how your innovation stacks up and how it is positioned in the marketplace.

You also need to know how to use this information to determine your innovation's marketability. For example, what if your invention is better in four categories, but inferior in one? Will that be acceptable to end users, retailers, distributors, and partners? The answer is "It depends." It goes without saying that if the one difference is severe, such as being twice the price of competitive products or having a negative environmental impact, it would probably kill the product's future. On the other hand, a small difference, such as it costs 2% more or takes up just a little more space than competitive products, may be acceptable, providing the offsetting benefits are sufficiently superior.

What you want to strive for is simple. You want to be as good or better in every category you measure, so that you may overcome any objections. Your challenge is to make an assessment so your innovation can be favorably compared to competitive

COST IS AN ESSENTIAL BUYING INFLUENCE

In the field of plastic packaging, there are more than 30 patents for square bottom plastic bags. But none of them has ever been commercialized. The reason is that they add from 20% to 300% to the cost of making a plastic bag. Thus, paper continues to be used in fast food applications. When we launched the M2K bag, we targeted our cost to be 20% less than the cost of paper; we achieved this and it resulted in several process patents to accomplish the objective.
–*Bob DeMatteis*

products. But you have to do this without bias and you have to be able to support the claims when required.

Experience tells us that if there is any one notable deficiency in any one area in which a new innovation is assessed, it can kill the product. Whether this is related to performance or any other of the critical factors or buying influences, you'll want to overcome these problems prior to launch.

<table>
<tr><td>

THE ROOT OF PROBLEMS

Many times the real problem and its root cause are obscured by the problem itself. The expert problem solver always strives to ensure that the root causes and not merely the superficial symptoms are resolved. TQM methods identify a questioning sequence you can use in order to identify the root cause. Start by asking, "Why did this problem happen?" Next, ask "Why?" again to go deeper into the underlying root cause. Then ask "why" several more times . . . possibly five times or more . . . in order to clearly understand the real underlying cause, or causes, of a problem so it may be permanently solved.

</td></tr>
</table>

Problem-Solving Methods

Problems with performance, processes, structure, or composition of a new invention usually arise quickly. Most will surface during in-house testing. A few may surface later during test marketing. Problems with sales, delivery, and marketing may surface after product launch. Overcoming these problems will make or break your innovation's success. For this reason, you want to be extremely sensitive to problems during these start-up tests and trials, identifying the underlying causes of the problems and solving them. The solutions will become a key to success.

Problems such as those associated with production processes must be conveyed to your manufacturing expert; problems associated with packaging, delivery, arrival and so on, must be referred to marketing. Of course, basic design flaws will go right back to you and to the engineers who must decide how to correct them.

Starting with a Single Creative Spark

Finding innovative solutions to problems usually stems from the creative, inventive process of a single individual, like you. Who knows where the seeds of your creative ideas and solutions come from? All we know is that if invention were easy, it would be a simple, obvious process—easy for anyone to do. In such a case, "invention" in and of itself, wouldn't exist. But inventing is not easy; at least, it's not easy to invent dynamic, dazzling inventions that excel in every way.

To a creative individual, sometimes just identifying a problem reveals a solution. Other times the process is far more complex and time consuming. There may simply be too many problems and buying influences that will take time to overcome. Overcoming these problems may result in new multiple inventions and patents. Most likely, your innovation discovery process will fall somewhere in between.

The invention process usually consists of extensive trial-and-error testing. As the inventor, you want to be hyperaware and supersensitive to all that is happening. Indeed, inventing is an ongoing learning and discovery process. All the while, you become more and more aware and curious about what is taking place. Innovative solutions and opportunities are usually found as a result of this discovery process. As an inventor, product developer, you will try to manipulate the various phenomena to attain the consumer-driven innovation you seek. You may not know exactly how that outcome will occur, but if you have determination and persistence, it will occur. After your innovation has been successfully launched, you'll truly become an expert in the field . . . in your niche.

By using your creativity and imagination, following your intuition, and persistently pursuing all the right outcomes, you will invariably find solutions to problems and the best direction to pursue. Always keep in mind:

<table>
<tr><td>

TO THE POINT . . . AND TRUE

In the middle of difficulty lies opportunity.
—Albert Einstein

</td></tr>
</table>

Successful innovators are successful not because of what they do,
but because of what they *strive* to do.

It is in the process of striving to design superior products that new discoveries are made and the emotional appeal of your innovation is improved. You will also find that solving a problem or discovering something unique frequently reveals a principle in physics you may not understand. You do not need to know why an invention works the way it does, but you do need to know that it does. It's not necessary to understand the physics or the engineering processes of your innovations as long as they perform well. Even when securing patent protection, you do not have to know why the invention works the way it does . . . as long as you can explain how it works and show the novelty.

Many times, after you have made an innovative discovery, engineers can take it, understand the physics involved, and engineer excellent products from it. If you are an innovative engineer or a member on a PQM Innovation team, it is best to let go of your preconceived notions and see what happens in the developmental process. There are always surprises along the way that seem to defy the physics that you were taught.

Using Teamwork in Problem Solving

Sooner or later, it will require a team effort to get your invention perfected and launched. Your team provides valuable input into the CDI attributes you desire and will quickly solve problems. Nothing will propel your project forward faster than this team effort. With a committed innovation team, it is amazing what can be accomplished and how dramatically you can reduce development time and effect a product launch.

If you are in a position to use a team in your problem-solving effort, you will also find you'll be able to dramatically reduce turn-around time for problem solving and quickly validate new solutions. Within a company environment, you may have several individuals who can help expedite this process.

The ideal invention development team will have at least one creative inventor type (most likely you), plus your manufacturing expert (the plant's General Manager, etc.) who usually has access to engineers who may help solve fundamental production problems. Also included is the all-important sales expert who will act as a liaison with customers and will be sensitive to their needs. Here is a synopsis of how team members may work together during product development and subsequent testing and problem solving:

- **Inventor.** You are free to imagine, create, and invent. To better understand problems and turn them into opportunities, you visit a customer's facility (if commercial or industrial) and observe end users. You learn the manufacturing processes associated with your invention so you'll understand various restrictions and limitations. Through your observations, you discover, explore, and question as much as possible and formulate and invent needed solutions.

- **Manufacturing expert.** With a commitment from the general manager of the manufacturing department (or subcontractor), problems with making the new product will be solved and new processes explored. Dedicated to the project, the manufacturing expert will ramp up the production quickly and effectively. The participation of this key member in the invention process solves production problems fast, keeps production on schedule, and makes sure orders are shipped as required.

EXPONENTIAL INCREASES

Adding one person to your team does not double a team's potential performance. It is more like a four-fold improvement. Adding the third team member is more like a nine-fold advantage. For instance, if you try to do the entire project alone, it might take you four years. With one key team member, the two of you (2 times 2 = 4) could reduce the time to one year. With a third team member, it could be reduced to five or six months.

- **Engineers.** They are challenged to fine tune an invention and solve problems based upon fulfilling your CDI intentions. They make the necessary adaptations and modifications to the manufacturing processes so that the new product will be cost effective from the onset.

- **Marketing expert.** Your sales/marketing department, licensee, manufacturer's representatives, or JDA partners will validate the various CDI attributes and will act as the ongoing liaison with customers, distributors, retailers, and end users. They will keep them informed of new developments and progress leading towards ramp up and will provide the invention development team with timely market-related feedback.

In your team efforts, remember that knowledge and information must be shared promptly among all members. Otherwise, the process is worthless. One other important concept that should prevail in your invention development team effort is:

**Innovative progress demands individuality; mediocrity is
perpetuated by standardization.**

In other words, no one on your team should accept mediocrity in manufacturing, quality, or performance. It can't be "halfway OK" and there can't be any unverified solutions. You have created something special, so make sure it is and everyone else will know it!

Testing Inventions

Almost every invention must be thoroughly tested. Testing starts in-house (your facility) and then later expands to end users and eventually into broad customer trials. Every possible in-house test scenario should be tried before starting trials outside your plant. Your testing follows the natural prototyping stages discussed in Chapter 3.

TEST IN EXCESS?

It's not possible to "over test" during the prototype stage. How is the product responding under test? Push to the limit and go over the edge, make it fail, and then make it better.
*–Matt Gilliland, author
Microcontroller Application
Cookbook series*

A new product must be completely tested in-house and you and the marketing department must be thoroughly satisfied before expanding those tests outside. You really cannot overdo controlled in-house testing. There always seem to be a myriad of problems—mostly minor ones—to resolve. Even if the innovation performs well from the onset, there is still much to learn from continued and even redundant testing. Hopefully, after your innovation team signs off on the in-house test models, the innovation will be ready for your customer's trials.

During the in-house testing and subsequent customer trials, *be patient*. Rome was not built in a day. Invariably, the in-house testing, product refinement, and customer start-up trials will take more time than expected. If you are new at innovating and new product releases, it can be a frustrating. When you take the testing outside to real time users, remember to start small with a limited trial at only one test site and then expand it later. Retail consumers—end users—need to be tested in a different environment, maybe even in their homes. Later, you can expand the trials with more end-user customers and into the retailers themselves.

With the initial testing with end users, it is best to not send the test models to the customer, but to deliver them personally. Ideally, you and your sales/marketing expert will personally set up, prepare, and perhaps even train at the initial testing grounds. During the initial customer trials, it is critical to observe the end users as closely as possible. Trials invariably uncover new problems, which should lead to new solutions. Keep everyone on the innovation team involved in his/her support role, until changes have been made and you know for certain the modifications work. You'll read more about how product testing ties in to test marketing in Chapter 10.

Regardless of which team members work to resolve a problem, there is a certain methodology that is generally followed. Here is a basic step-by-step team approach to innovation problem solving and an example of how it may be used to solve a particular problem:

1. **Plan the team meeting.** On the prepublished agenda describe the problem and your objective. Everyone should be asked to bring his/her ideas for resolving it. Allow sufficient time on the agenda to thoroughly address the problem.

2. **Break the innovation down into parts.** At the meeting begin by identifying the parts described in the Fishbone Chart in Figure 4.1 and discuss how they influence the invention's function and performance. The discussion should lead to a thorough understanding of: A) **materials**, their characteristics and how they may affect performance; B) the **machinery** used in the manufacturing process and how it may affect the product's quality and performance; C) the **people** using the products, thus qualifying they are the right target customer and they know how it's used (If they don't, this is a huge issue that needs to be corrected.); D) the product's **design** and whether it gives misleading mapping or promotes dysfunction; E) the **manufacturing process** and various aspects that may be causing a problem; F) the **environment** in which the product is used (or misused?) and; G) the **means of measuring** the performance and or the problem. To use this fishbone chart, first write in as many elements as you can on the horizontal lines. Next, discuss them thoroughly and last, circle the potentially problematic areas.

TESTS VS. TRIALS

Your in-house testing activities are best called "tests." However, when the product goes out to the customer, it is best to call your activities "trials." Your customers (or end users) should not think they are guinea pigs, testing unproven products. It is best to let them know that the "tests" have been concluded and that now you are taking it to the consumers to see if there is any fine-tuning needed. This approach is more positive.

Figure 4.1 Fishbone Chart

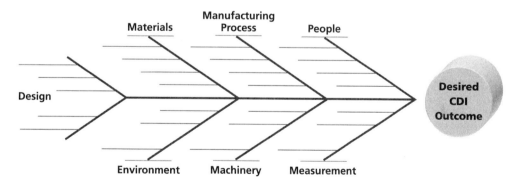

3. **Examine the manufacturing process.** If a problem has the potential to be solved by changing the manufacturing process, draw a flow chart with the various components (parts) in order to understand and identify what is happening or not happening. Can you verify that each component is performing the way it is expected to perform? If not, why? At your meeting, use a flow chart to help identify modifications and alternative manufacturing means. The flow chart in Figure 4.2 represents manufacturing stackable storage cubes for an industrial application.

WHY?

Ask yourself, "Why did they do that the way they did?" There is clearly a better way.

–Master inventor, George Margolin

Figure 4.2 Sample Manufacturing Flow Chart

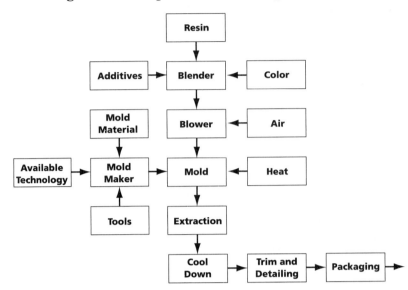

4. **Identify the dominant idea and crucial factor.** Dominant ideas are the ideas that dominate the issue and cannot be challenged. Crucial factors are those factors that are holding you up and keeping you to the old approach. You must identify the dominant idea and crucial factor in order to weigh viable alternatives. For example, you want to manufacture strong plastic storage cubes that can be stacked and won't easily slide off. Identify a *dominant idea* as using inexpensive, strong, high-density plastic material in a standard mold. The corresponding *crucial factor* is that this high-density material is naturally quite slippery. The task is to find a means to mold the cubes out of high-density plastic so they stack well.

5. **Brainstorm.** With all this knowledge, the team can now brainstorm about the many possible ways to give the blocks a non-slip outer surface. Solutions to this problem may include:

- Applying a special tackifier coating to the surface
- Applying heat to roughen up, or remove the "slip" from, the surface
- Modifying the die's surface in the desirable areas to include non-skid ridges
- Adding a no-slip additive to the resin
- Incorporating interlocking male-female components in the corners of the blocks

Figure 4.3 shows the previous flow chart with alternative possibilities in the shaded boxes. If desirable, any one of the components in the flow chart can be broken down into another, more detailed flow chart.

If your team is stumped and cannot find acceptable solutions to the problems you encounter, you can seek outside experts. You may also consider more creative, drastic approaches. This may require a major change in the product's design and its processes. For instance, the stackable blocks may be best suited to a different molding process that uses a generally stickier, low-density resin.

Be determined to not give up until every viable solution has been tested. Before shelving a project, retest all aspects to check for errors or to uncover some factor that you may have overlooked. The worse-case scenario is to put the innovation on the back burner for a while. But if you do this, don't forget about it. Frequently, a solution will surface later, when you are not expecting it.

Figure 4.3 Alternatives to Sample Manufacturing Flow Chart

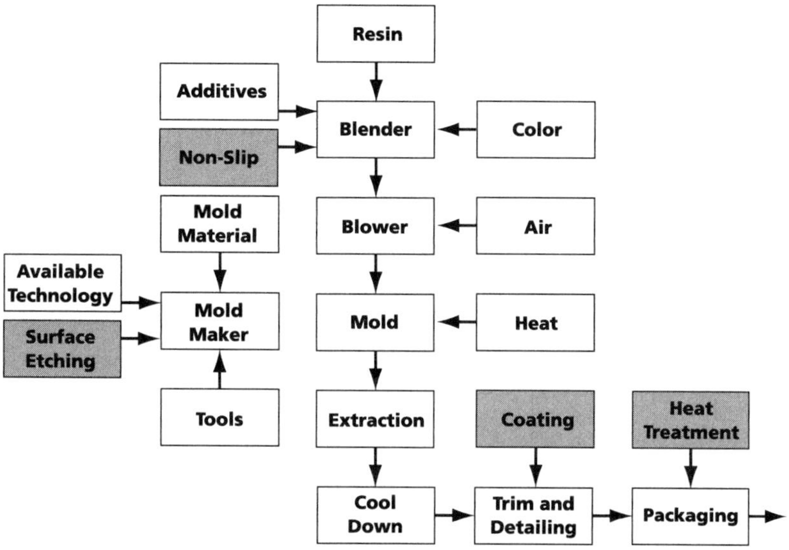

Being Sensitive to Red Flags

Red flags are defined as warning signals that something is wrong. If you're not familiar with this term, you should be. If you're not being hypersensitive to your innovation's performance and highly analytical so you can spot red flags when they occur, you'll want to develop this valuable insight in inventing.

Red flags may appear during manufacturing, in-house product testing, customer trials, at a plant, or in the field. As you stay close to your customers, turn on your intuition and learn to spot these red flags. When you spot them, you will realize that they are usually masked opportunities. Spotting a red flag is really the uncovering of a problem that will enable you to creatively improve your invention. Always keep in mind that. . .

Red flags invariably represent good news!

Red flags can be outstanding opportunities, if you have the awareness and insight to take advantage of them. Make it a habit to seek them out, because the resolution of them is crucial to your success.

Continually Improving

Continuous improvement in a person's life is the single most important key to success. Continuous improvement in a product's life is also the single most important key to its success. Through continual improvement, you'll maintain and improve market position.

What improvements have you considered for your innovation? How will they be assessed and tested? Can they be applied easily enough? Fortunately, with your team in place, it's a fairly easy process to apply.

OPPORTUNITY

Quit looking at all the problems as obstacles and start looking at all of them as opportunities. Problems when converted become your best opportunities.

—Stephen Paul Gnass

Simply put, with an improvement concept you've identified, get some preliminary feedback from your team members to make an initial qualification. This may include your manufacturing partner (or department head), marketing partner (or department head), engineers, and at times, outside subcontractors. If marketing agrees the concept could improve sales and manufacturing agrees it can be made cost effectively, then you can conduct an appropriate test market.

With your team in place, the process is simple. Follow the chart in Figure 4.4.

Figure 4.4 CDI Continuous Improvement

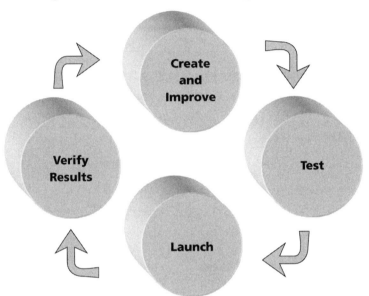

Aiming for Breakthrough Opportunities

The first invention you create or even the first improvements you make to it rarely provide breakthrough opportunities. Usually breakthrough opportunities occur sometime later and many times they will occur unexpectedly. At times, it is hard to pinpoint what innovation or improvement may turn into a breakthrough opportunity, but you'll certainly know after the fact. The term "breakthrough opportunity" really is nothing more or less than some inventive discovery or marketing event that significantly boosts sales.

One reason that first inventions rarely become breakthrough opportunities is that the inventor simply doesn't know enough about the field. In fact, frequently the marketers don't either. Another reason is that the first products developed from a new technology tend to be more basic, whereas future models will have many more bells and whistles, any one of which may become a breakthrough opportunity.

The invention behind a breakthrough opportunity usually represents excellent patent protection too. The mere fact that the invention or improvement has significantly improved marketability should also help support strong patentability. Even if the scope is narrower, the importance may be greater. Of course, you'll want to pursue this protection, and it may play an important part in extending your product's life cycle.

Minor breakthroughs may occur several times during development, with some being more significant than others. What's important to know is that most frequently breakthrough concepts are not a result of planning. They are usually the result of your attention to detail and your persistent pursuit of improving the innovation. And when you spot them, they may represent giant steps forward.

They do, in fact, serve as proof of the validity of your initial invention, which may have begun years earlier. Just be analytical and apply a continuous improvement approach and you'll discover several good opportunities.

Creating the Next Generation

Once established, you'll want to stay on top by continually improving and creating the next generation. If you don't do this, your competitors will!

There are a few proven approaches you can use to accomplish this objective. They are:

1. **Ask your customers.** Ask your customers what they'd like to see in the future. Of course, they'll want to say "lower costs," but that's only one of their concerns. They may tend to talk in broader terminology, such as "easier to use" or "more convenient to assemble." This just identifies which development opportunities you may pursue. The more you hear a certain request, the more important it may be to pursue.

2. **Listen to complaints.** A treasure trove of new improvements and inventive opportunities may be found by listening to or reading customer's complaints. If nothing else, you'll want to solve any lingering problems or complaints, but why not take it one step further and create the next generation?

3. **Watch legislation.** Frequently legislation may be enacted that may affect your innovation. For example, you may need to change ingredients or raw materials to conform to new environmental laws. Why not be the leader and make these changes early on? If you anticipate even further legislative changes, what else can you do to improve your products before the legislature mandates it?

4. **Dream, imagine, envision the future.** Close your eyes and let go . . . imagine the possibilities and then assess them with your team. Innovators are usually pretty good at doing this, so it's a welcome concept and fun to do in your spare time.

Once you've got some ideas that you believe may represent the next generation of the product, go back to the basics of development. At this point, they're a lot easier to employ since your team is already established and the development time frame will be relatively short.

PQM and Design

Patent Quality Management in a company setting means you have a mindset that is always striving to stay ahead of the competition, create the next generation, and protect it with patents.

To a company, applying PQM is essential to its existence. It's a way of doing business. Input for innovation may come from any department, from any individual. This substantially increases the ability to innovate and create the next generation.

With PQM, you encourage manufacturing and marketing to pursue innovation and you turn your engineers loose to think and create. Marketers will be attentive to spotting future trends and opportunities. Even your maintenance and accounting departments

PQM OR TQM?

Patent Quality Management is a new and emerging paradigm, a concept that new innovative businesses would be wise to adopt. Just like TQM replaced old-fashioned sales-driven management systems, PQM will replace regular customer-driven systems. One of the budding concepts in business today is that "patents are the new money." We agree. Its time has come and managing the new money to direct a company's future is visionary today and will be called "common sense" tomorrow.

may contribute. Maintenance may lower costs through new processes, and accounting may discover and even employ new business methodologies.

All told, one of the best outcomes from a PQM approach is the long-term establishment of standardizing a product line based on patented technology. Preferably, this is your technology, not a competitor's, or it may be a cross-licensed technology. From this perspective it is essential that companies build a war chest of patents over time so they'll have some patents to cross-license.

Independent inventors and product developers are at somewhat of a disadvantage from this perspective because a company with several employees has a lot more brain power and creative potential. However, if you've become an expert in your field, you'll have the advantage of having more knowledge and information to apply to future concepts.

If you are an independent inventor who has developed working relationships with companies and licensees in your field, there's no reason you cannot try to incorporate yourself in the PQM strategy of your partnering companies. After all, PQM promotes the use of outside technologies and partnering with IP owners, regardless of whether or not they are independent inventors or competitors. In fact, you might be surprised at the welcome you receive when offering your creative services to one of these companies as a development or JDA partner.

During the developmental stage of an innovation, it is best to work quietly and confidentially. Until all testing and trials have been completed, you do not want anyone telegraphing your innovation activities to competition. This means that your team of experts should not be talking to outsiders. Even your first customers, who may be helping with product testing before public disclosure occurs, should sign confidentiality agreements. Maintaining confidentiality also serves to protect your rights to file patent applications in a timely manner.

Once your invention has been launched, you can tell it on the mountain and get all the publicity you want, but only if you think it will help sales. If you are competing against some sleeping giants, it may be best to keep working as quietly as possible and round up as many new customers in niche regions you can protect. In other words, stay "under the radar." It is frequently best to keep competitors in the dark for as long as possible and let them be surprised later, after you have captured the new business and have solidly established your niche.

Assistance

During development, testing, assessing, and problem solving, you may come up against problems that are not getting resolved or barriers that seem insurmountable. When this happens—regardless of whether you are a small company or independent—there is a wealth of assistance available to you, most of which is free.

Assistance may come in any number of forms. It may include assistance with the following:

- Manufacturing
- Marketing and packaging
- Environmental issues and inventions
- Business plans and prospectuses

STEALTH

When inventing and developing with your team members, think in terms of "stealth" and remember all those innovations that were just sprung onto the market and captured a niche. Examples include, Stealth bombers, the personal computer, direct TV, Celestial Seasonings Teas and even the Internet! The element of surprise can be a powerful ally in marketing warfare within a new market segment. That's why it is called "marketing warfare."

FOLLOW THE ADVICE

Inventors are always pressing their resources for free advice, and that's not a bad thing. But, if only they worked as hard *following* that advice as they did in *chasing it,* they'd be better off by far.
–Don Kelly, *Intellectual Asset Management Associates, LLC*

- Business operations
- Loans programs
- Research programs
- Local training assistance and incentives

The various government and private agencies that may be able to help you are numerous, too many to mention here for every state, county and city in the United States. Some agencies are federal, some are state affiliated, and some may be county- or city-sponsored. Should you need assistance, look them up and see how they may help. The more common ones are:

- The Small Business Development Centers (SBDC)
- The Small Business Administration (SBA)
- State universities and colleges
- State environmental agencies
- State agencies assisting in manufacturing and training
- Federal, state, and local business loan agencies
- Federal and state international commerce development
- Local business relocation incentives
- Local inventors' groups

Almost regardless of the type of assistance you seek, there's a government agency to help you. Most of them are free, but some may charge nominal fees. In addition, you may find a large number of private entities to assist you in product testing and surveys, and in engineering and business consultation.

Federal Agencies

The most prolific of all innovation assistance agencies are the Small Business Development Centers. They are a part of the SBA, but they focus on new innovations, start-up companies, and independent developers. In other words, they provide assistance before you have formed a company (or new division or expansion) and usually before product launch. They can literally assist you in every category from A to Z.

In the United States there are more than 1,000 agencies to serve you. Most SBDCs have field offices, many of which are in the smallest communities in your state. Generally speaking, you can arrange a consultation with a phone call.

SBDC services include not only consultation but also courses and seminars on invention, product development, manufacturing, marketing, and small business start-up. Many are tied into the state university and college systems and have some amazing resources available—for instance, incubators, prototyping, package design, and much, much more.

SBDCs are also an excellent source for locating angels, initial financing, and other forms of loans. You can read more about financing in Chapter 10.

If you're just getting started, a visit to your SBDC is a must when you find yourself stuck and do not know what to do or where to go next. To locate an SBDC near you, go to our website at http://frompatenttoprofit.com/helpfullinks.htm.

SBDC—ASSISTANCE IN EVERY IMAGINABLE WAY

According to the SBDC, "We're the government, and we're here to help." You've probably heard that before, but this time it's true! Their offices are nationwide and offer existing and prospective small business owners free, confidential consulting, educational seminars and workshops, databases, and more. Whether you're new to business or looking to expand, the SBDC is the first contact you should make. Here's an A–Z list of their specialties:

Advertising
Business feasibility and planning
Cash flow management
Dealing with customers
Export/import assistance
Financial management
Growing your business
Home-based business
Internet access/training
January tax preparations
Keeping books and records
Legal business structure
Marketing plans
New technology
Office automation
Procurement assistance
Qualifying for a loan
Retail marketing
Strategic planning
Trademarks and patents
Understanding entrepreneurship
Venture capital access
Women business owners
X-Pert training
Your family business succession
Zeroing in on success

The SBA may serve your purposes better if you are an established business or when your start-up is growing. It focuses its consulting and assistance on established small business, whether they have 4–5 employees or 50. They can provide help with expansion of manufacturing, sales, and so on.

One of the most common means of assistance the SBA provides is helping to arrange loans.

State Agencies

Along with the SBA and SBDC, there are several state agencies that can help you ramp up and launch your innovations. Most SBAs and SBDCs are familiar with all the local agencies and can advise you of their areas of expertise. Three areas that they should know well are manufacturing and environmental concerns, education and training considerations, and financing.

Most states have agencies that assist in manufacturing, whether that be manufacturing processes, seeking out locations for plants, or simply locating a company for your product. Contact your SBDC to locate the agencies in your state that may help you.

Local colleges and universities also have the ability to train and educate. Most have night or weekend courses and many can even tailor a certain course to your needs should it be advantageous to your company and its employees. Contact schools near you and ask about community education or extended education courses for business.

Many colleges and universities also have the ability to assist in prototyping, and package design as well as conduct product testing and customer surveys. If you're not sure which ones to contact, contact your SBDC and ask it to find out for you.

Local Agencies

Aside from state agencies, most counties have great interest in promoting commerce in their region. It goes without saying they want to increase the labor force and tax base.

The primary means of assistance a county can provide small companies is usually loans, training assistance, tax credits, and at times, relocation assistance and lease information. There may also be some education assistance available through a night school program at a local high school.

Visit your county offices and see what they have available. Or again, visit your SBDC and they'll be able to help you.

Private Agencies

After contacting the various free government agencies, you may still find a need for specialized assistance in order to solve a specific problem.

There are a myriad of experts available for consulting in manufacturing, innovation and design development, prototyping, engineering, importing, and of course, patenting. To find these companies, you may search the Internet and local phone book and, of course, your local SBDC.

If you're just starting out with a new product development, you'd be wise to contact your SBDC before hiring private consultants. They may be able to help you locate a state agency or college or university to fulfill your needs.

PART 2

PATENTING

PATENT BASICS

Patents can provide security and a sense of accomplishment,
and they are a whole lot of fun.

Patents Are Assets

We frequently think of assets in the form of real estate, cars, and investments, such as stocks and bonds. But a well-written, carefully prosecuted patent application can result in another form of valuable asset—a United States patent. A single patent with a broad scope or several patents covering multiple aspects of your invention may protect or create your or your company's financial security for years to come.

Patents are commonly referred to as a form of intellectual property and have more respect today than ever before. In the past twenty years, record judgments have been awarded to patent holders against patent infringers. Because of the vast number of favorable court cases in favor of patent holders, it is well known that patent values have increased 20-fold to 100-fold in just the past several years.

More and more new products and new high-performance variations on old products are being introduced every year. Yesterday's high-volume generic product is being splintered into many new innovative niches. Almost all of these new product introductions are wisely protected by some form of patent protection. After all, it's only common sense to use the laws of the United States to protect an investment.

For this reason, many domestic and foreign companies want to develop new innovative niches and secure their futures with patents. Many of these companies are small-to medium-sized companies that do not have built-in research and development departments. Some may try to develop new products by simply copying existing products, but

FACT

Don Kelly, former Director of the U.S. Patent Office, has met thousands of inventors and says that only 3% of all patents make money for an inventor. Follow the teachings of this book and 100% of your patents can make you money! Remember, patents cost money and sales make money. Patents protect those sales. *From Patent to Profit* shows you how to develop only those ideas and inventions that *can* make you money!

more and more companies are looking for an advantage over their competition by developing or licensing patented concepts and technologies.

Simply put, licensing and developing new patented products can be a fast, economical way for companies to launch new niche products, gain new profits, and protect their futures. This is good news for small businesses and independent inventors and product developers who plan to partner their patented creations.

The best way for you to ensure your future as an entrepreneurial small business, inventor, engineer, and so on is to have a sound patent strategy. This strategy includes applying three important processes simultaneously:

1. Carefully developing and testing prototypes

2. Confirming the value of your test results with your marketing team

3. Simultaneously writing your patent applications

By applying this simple strategy, you will be accomplishing the following:

*   Developing the most highly desirable product possible

*   Proving its worth with verified market potential

*   Protecting your innovation with superior, broad patent protection

Positive feedback from marketing validates superior design attributes. In turn, you want to protect these superior design attributes with patents. From this perspective, you can say:

Patent value is substantiated by the sales/marketing potential of the various attributes claimed in the patent.

What Is a Patent?

A patent is a document declaring that a novel, unobvious, useful discovery or invention is being credited to an individual or group of individuals. When granted, the patent covering a new discovery or invention is given a number, filed with the U.S. Patent Office and publicly disclosed. Once issued, a patent provides the inventor, or assignee, a legal monopoly to practice the manufacture of products and processes as described in the issued patent. It can be thought of as a grant of property rights by the U.S. government to the inventor(s).

In broad terms, there are three types of patents: utility patents, plant patents, and design patents.

Utility Patents

Utility patents refer to the usefulness or functionality of an invention. They are by far the most common types of patents. As of June 8, 1995, they are enforceable for a span of 20 years from the date of filing an application. Prior to this date they were enforceable for up to 17 years from the date they were granted. Patents on drugs, food additives and medical devices are considered utility patents, but under special circumstances, these patents may be extended for an additional 5 years (25 years total). An extension is generally granted due to the amount of additional time that is required to test and subsequently obtain FDA approval.

THE AMERICAN DREAM

It's important to remember that people from all around the world continue to endure great hardships to gain entry into our country, each of them hoping to take part in the unique opportunity we proudly refer to as the American Dream. The Declaration of Independence might just as well have listed *pursuit of inventions* among the inalienable rights. Inventors bringing ideas out of their garages and basements and up the proverbial hill to market epitomize our most basic of freedoms.

–Don Kelly, Intellectual Asset Management Associates, LLC

U.S. PATENT NO. 1

It was not until 1836 that the U.S. Patent Office began assigning numbers to patents. Interestingly enough, the first numbered patent, U.S. Patent No. 1, was for a relatively important invention by J. Ruggles entitled Traction Wheels.

Plant Patents

These refer to new plant varieties that have been asexually reproduced, which includes mutants, hybrids, and newly founded seedlings. These patents are also enforceable for 20 years from the date an inventor files.

Design Patents

These refer to any new, original ornamental design for an article to be manufactured. Thus, only the appearance of the article is protected. They are in force for 14 years from the date of actual issue, not from the date of filing.

After the expiration of a patent's term, either 14, 20, or 25 years, the patentee loses the exclusive rights to the invention. It becomes part of the public domain and anyone may practice its teachings.

What Rights Do Patents Grant?

Patents can represent powerful property rights, a legal monopoly. These rights can be powerful tools if used effectively. Patent rights may also bring substantial income through the manufacture and sales or licensing of the inventions as claimed in the patent. According to the U.S. Patent Office, patents are negative rights:

A U.S. patent "gives the owner the right to exclude others from making, using or selling the invention."

However, if your patent application is not well thought out and the corresponding invention is not thoroughly researched and tested, you may have absolutely no rights at all . . . or a weak patent position at best. Be thorough in your patent-writing activities and verify the marketing potential on an ongoing basis. To have a genuinely effective, broad-scope patent, you must wisely plan and develop your invention, verify its marketing potential, and file an application with advantageous timing.

Who Can File?

The person who is considered the original (first) inventor may file for a U.S. patent. Upon submitting a permanent patent application to the U.S. Patent Office, you will sign an "Oath of Inventorship." This oath is a declaration stating that you believe you are the first and sole (or joint) inventor(s).

If someone other than the original (or first) inventor received a patent, it would be invalid. If there is more than one inventor, they should file jointly. It is important to list everyone who contributed to the novelty of the invention. To clarify this point, it is not necessary to list any participant in the invention process who was only "following your instructions," such as may be the case with a parts or raw material supplier.

Confine your invention activities to a tight group of people. It can be particularly dangerous to include outsiders as part of the process. They may try to claim some ownership of the patent, even though you already knew what they contributed or suggested. The last thing you want is litigation from an outside entity or individual claiming you stole the invention. Look at the application's claims you will be filing. (Chapter 7 talks

FACT

It is generally understood that 97% of all patents don't earn the inventor any substantial income. The two most common reasons are the initial marketing was poorly researched or the product is too easy to design around. Inventors should initially consider utility patent protection, but they should also not overlook the value of a design patent for a simple invention.
–Michael S. Neustel, U.S. Patent Attorney

HISTORICAL FACT

Liquid Paper was invented by a 17-year-old girl who was a high school dropout! It quickly became a multimillion dollar product! Bette Nesmith Graham originally marketed the product as "Mistake Out" and made it in her kitchen. When demand skyrocketed, she changed the name to Liquid Paper and applied for a patent and a trademark. In 1979 she sold the company to Gillette Corporation for $47.5 million, plus a royalty on every bottle sold.

in detail about claims.) If a participant has not created inventive matter covered in a claim, or at least influenced one of them, he/she should not be listed as a co-inventor. Should you need to employ outside individuals or companies in the invention development process, make sure to have them sign a confidentiality agreement or an independent contractor's agreement transferring all rights to any new discoveries to you.

Only when a person dies or becomes insane may another file for him/her. If an inventor refuses to file a patent application (or cannot be found) on behalf of a company for which he/she works, a co-inventor or proprietor may file on his/her behalf.

If you have any doubts about legal inventorship (which is directly related to legal ownership), consult your patent attorney for clarification.

Requirements for Granting Patents

The U.S. Patent Office has rules and laws that apply to the granting of patents. When filing for a patent on your invention, you must meet three important criteria: it must be useful, operative, and new or novel.

The invention must be *useful*. Remember Eli Whitney's cotton gin and Thomas Edison's light bulb? Or the invention of the paper clip or the spring-loaded mousetrap? More recently, highly useful inventions might include Velcro™, ZipLoc™ bags, and the process of reading DNA in genes. If the product or process in a patent application is not considered useful, the Patent Office will reject it. For instance, square tires would probably not be considered useful, nor a vaccine that would cause a disease such as AIDS or SARS.

The patented invention must be *operative*. In other words it must work according to the claims. For instance, square tires would be considered neither useful nor operative. A patent on a plastic material in which claims are based on improved strength, but the material does not perform as indicated, is not valid.

The invention must be *new* or *novel*. An invention **cannot** be patented if:

- The invention was known in any part of the world at any given time before you came up with the idea.

- The invention was previously described in an article published anywhere in the world.

- It was previously patented anywhere.

- The difference between your invention and a previous patent (or publicly known product, process, etc.) is such that it would have been obvious to any person skilled in the art. For instance, simply changing the size or color of a product is not acceptable.

- Your invention was put into use more than one year prior to filing for a patent application in this country.

You can file a patent application up to one year after your first public disclosure. In Chapter 6 you'll learn how you may use this law to your advantage and make it part of your patent and product development strategy.

Difficulty of Obtaining a Patent

If you have a genuinely unique discovery or invention, you will most likely receive a patent. To give you an idea of patent activity at the U.S. Patent Office, we are presently filing more than 220,000 applications a year and more than 115,000 patents are being granted annually. Of these patents granted, about 25% will be from independent inventors. The chances of receiving a patent are better than 50% and probably closer to 80% or 90%, if you do your research up front and follow through.

The chief reasons the Patent Office will not grant a patent are the unforeseen discovery of prior art in a patent search or the invention is deemed "anticipated," which would make the invention "obvious" in the eyes of a patent examiner.

There is another factor that contributes to a fair percentage of applications not resulting in patents—the inventors decide to abandon them. This generally happens because they are unable to get broad enough patent protection, sales of the invention are inadequate during the patent pending stage, or they lose interest.

What Cannot Be Patented

There are two general areas that cannot be patented. They are:

- **Perpetual motion machines.** The U.S. Patent Office considers perpetual motion machines as not being possible; thus, they are not patentable. So if you think you have developed one, call it something else such as an "improved power system."

- **"I have an idea about a product that would be super . . . "** Merely having an idea cannot be patented. Nor can you get a patent on the "suggestion" of a "new, unique invention." A patent is granted based on a specific design for an apparatus, a product, process, etc., which works in at least one form, as stated in the patent application.

Business methods used to fall into this category, but now are mostly considered patentable. However, some methods of doing business and their related printed matter may not be acceptable. You may ask, "If a company's method of doing business includes a special packaging means, would this be considered a method of doing business?" The answer is no, provided that the special packaging has a specific utility or design value in and of itself. If so, it may qualify for utility patent protection based on its utilitarian merits or a design patent or copyright based on its unique appearance.

If you employ a patent attorney, he/she will almost always ask you if you have a prototype or sufficient drawings showing how the invention works. If you do not, you will need to show due diligence and in the near future prove your invention works. Reduction to practice and diligence are discussed in greater details in Chapter 6.

If you are filing for a process or systems patent, you will sooner or later need to have some illustrations or perhaps a video showing how it works. Pie-in-the-sky ideas cannot be patented. You must clearly illustrate how the idea can be implemented into an invention and how it functions.

To expand on this, if you suggested an idea for an invention to a friend, without indicating how it could be made to work, and your friend subsequently figured out (discovered) a novel way to make your concept function, you will probably not be considered the inventor. At best, you may be considered a co-inventor. If your friend receives a patent for his efforts in implementing your idea, you're probably out of luck, even though the original big picture may have been yours. You cannot claim rights to any invention or discovery by merely thinking it would be a good idea.

On the other hand, if you suggested an idea to someone and showed the person in a sufficiently detailed drawing how it could be made to work, and he/she made a working model ***based upon your drawing***, you, not that person, would qualify as the inventor. If

HISTORICAL FACT

Today we have learned that America's independent inventors and small businesses have arrived. They effectively hold a seat at the table with larger intellectual property policy makers. And, as they slide their chairs up to the table, they are finding that with their newly gained influence comes a heavy responsibility.
–Don Kelly, Intellectual Asset Management Associates, LLC

FACT

Companies license 'patents'—not ideas.
–Michael S. Neustel, U.S. Patent Attorney

PATENTING IDEAS OR INVENTIONS

A good example: Suppose over one hundred years ago, a person had the idea of making a vehicle that could fly. He would not get a patent on that idea, but the Wright Brothers, who figured out how to make a flying machine, would and did get a patent on their flying vehicle.
–Bill Pavitt, Esq.

you saved the drawing and could prove that the concept was yours, you would be considered the first, original inventor, unless you decided not to pursue it. Patents are serious business and stealing "original ideas" and claiming "original ownership" is fraud.

Related Forms of Intellectual Property

There are five forms of intellectual properties that are not considered patents that you should be aware of. Some are processed at the United States Patent and Trademark Office (also referred to as the "USPTO" or the "PTO") or its related branches. They are trademarks, service marks (the two are sometimes referred to solely as "marks"), trade dress, copyrights, and domain names.

Trademarks

Trademarks relate to any word, symbol, or device used to describe the origin of goods or services and to distinguish them from other competitive products. Examples include McDonald's golden arches, Shell Oil's shell, and the numerous insignias used by Coca-Cola. Trademarks legally begin the first time they are used in public or commerce. You may file an application to register a trademark with the U.S. Patent and Trademark Office. If the PTO considers it "unique," registration will be granted. The two key considerations that qualify a mark as unique are:

- It is not in **conflict** with any other marks
- It does not cause **confusion** with any other marks

Commonly used words cannot be trademarked; they must be unique by their very nature. However, there is an exception to this rule. If common language words have been used exclusively by a company to describe an article and have not been contested (or used) by competitors for a period of five years, the owner may file for and be granted a trademark registration. Dual-tab® and Big Red® are examples of this kind of common language registered trademark.

Service Marks

Service marks are similar to trademarks but refer to a manner in which goods are offered for sale. In other words, they refer to a service, not the product. Burger King's "Have it your way," Wal-Mart's "We sell for less . . . always," and Sears "You Can Count On Me" are examples of service marks.

Unlike patents and copyrights, trademarks and service marks last indefinitely, if the owner continues to use them. Between the fifth and sixth year after the initial registration of the mark, a registrant must file an affidavit verifying its use to keep the registration in force. Thereafter, the mark may be renewed every ten years into the indefinite future.

To secure a registered trademark, a user files either a "use" application, meaning that the mark has already been used in commerce, or an "intent-to-use" application. The use of a mark within one state does not constitute commercial use. In other words, marks may be registered only when they are used in interstate commerce.

The owner of a mark is usually an individual, a corporation, or a partnership. If the owner is an overseas entity, the owner must have a representative in the United States process all paperwork with the U.S. Patent Office. Either the owner of a mark or the representative may file a registered trademark or service mark.

VALUABLE: PATENTS OR TRADEMARK?

Initially, a patent is more valuable because it keeps others from making the same product. However, in time, brand loyalty to a trademark can cause its value to surpass that of a patent. Patents tend to decrease in value over the years, whereas trademarks increase in value the more they are used.

SEARCHING TRADEMARKS

There are a few good ways to do an initial trademark or service mark search on the Internet. By searching the words or word string on the many search engines, you should have a reasonable idea of its availability. Also, search it as a URL. You can do a trademark search at the PTO or thomasregister. com. Keep in mind that this type of search only works with simple trademark names and service mark strings, but not symbols. Be careful to search for similar words and phrases, which may be in conflict. For a list of search sites, visit our website @ from patenttoprofit.com

Before a trademark is submitted to the Patent Office, the owner should do a search for conflicting marks. However, the PTO does not require it. A search can be conducted at the U.S. Patent Office, online at the PTO website at *http://www.uspto.gov/main/trademarks.htm*, through a trademark search company (or attorney), or at any of the Patent and Trademark Depository Libraries.

From the first date of use, the owner may identify products by simply placing a "TM" for trademarks or an "SM" for service marks after or near the mark and then put it into use. This, in effect, alerts the public to your claim of ownership. Once a mark has been registered with the PTO, the owner may use the "circle r" registration symbol, ®.

Some examples of marks are:

Trademarks: ***BIG*mouth™**

Registered marks:

Service mark: **Have it Your Way**SM

Trade Dress

Trade dress is a concept that has received much attention in recent years. It primarily applies to product and package configurations. Trade dress protection typically falls under trademark or copyright laws. Trademark protection tends to encompass physical configurations, thus protecting against unfair competition. An excellent example would be the classic shape of the coke bottle used by Coca-Cola®.

In order to receive registration approval, a product must meet three basic criteria: (1) nonfunctionality, (2) proof of secondary meaning, and (3) likelihood of confusion. Common law tort of unfair competition gives substantial protection against copying nonfunctional aspects of consumer products.

A few examples that have been cited in common law actions include the configuration of a cereal biscuit, a loaf of bread, a medicinal tablet, a root beer bottle, a cologne bottle, a crescent wrench, a washing machine, a clock, and a padlock.

To register trade dress as a trademark, follow the same filing procedures as those for a trademark or service mark. The same forms are used.

Every company, every entrepreneur, every inventor should consider registering the trade dress associated with their new products, if at all possible. Frankly, in most cases, it is possible. Like a trademark, trade dress does not expire. After years of use, it can result in a form of intellectual property much more valuable than a patent. It can also result in an extended license term if you plan to license your innovations.

Copyrights

Copyrights protect the writings of an author or artist/designer. They protect the form of expression, not the subject matter or content. Neither names nor titles are copyrighted. Examples of copyrights are:

• Books, magazine or newspaper articles, maps

• Artwork, patterns and designs, sculptures

TRADEMARK FORMS

On the Internet, the U.S. Patent and Trademark Office has forms for filing trademarks that you can download (prinTEAS), fill in, and mail. Or, you can file electronically (e-TEAS). The Internet URL is: uspto.gov/teas/print/welcome.htm

TRADEMARK STRATEGY

Include trademarks in your patent strategy. Try to develop a catchy trademark, register it in due course and use it as part of your company's marketing strategy or in a license agreement. Licenses on patents usually expire on the last expiration date of the licensed patents, but trademarks last indefinitely. Upon the expiration of the last of the patents, a new license for the use of the trademark can be struck.

- Songs, sound recordings, motion pictures
- Notes written on a piece of paper

A copyright arises the moment the work is completed, and no copyright notice is required. However, when you publish a work, it is wise to include a notice of ownership. Publishing and circulating the work constitutes a public disclosure. If you include a notice, it substantially strengthens your protection against infringement by warning others. It is an important step if you want to sue for damages in the event of infringement.

Damages may be granted for two primary reasons: first, if others copy your mark and use it for personal profit and second, when the commercial value of your copyrighted material is harmed by the action of another party. This second reason is enforceable regardless of whether the other party tried to profit from the use of the copyrighted material. The correct form of copyright notice and examples are:

Copyright [date(s)] by [author/owner]

Copyright 2009 by IPT Co., Grass Valley, CA

The circle "c" (©) may be used instead of the word "copyright." The phrase "All Rights Reserved" was a requirement in some countries, but it is no longer needed. Materials that were copyrighted after January 1, 1978 last for the lifetime of the author, plus 50 years after death.

The posting of your copyrighted material to the Internet, Usenet, or a web page does not constitute putting it in the public domain. Placing in public domain means that anyone can use it without obligation to pay royalties and so on. Placing copyrightable material in the public domain can only occur when the owner/author has explicitly said so, using language such as, "I hereby grant this writing to the public domain."

"Fair use" of a copyright is a legal provision and not considered infringement. Fair use refers to the use of a copyright without the author's consent in commentary, news reports, research articles, education, etc. The intention of the use must not be to damage the commercial value of the copyrighted material.

Free ads promoting an owner's copyrighted material are not considered "fair use." It is up to the copyright owner to determine if it is acceptable. It is interesting to note that government publications and notices are not protected by copyright laws. They can be freely copied and disseminated by anyone. As you'll see later on, you may copy and incorporate language and drawings in existing patents into your patent applications.

Copyrights are registered with the Copyright Office of the Library of Congress in Washington, DC. A copy of your copyrighted work must be accompanied with an application form and a check for the appropriate registration fee. See the Appendix for more information.

Domain Names

Domain names constitute a field of intellectual property that is becoming an important part of product identification and related business. They can be thought of as the "trademarks of the Internet." A domain name represents your "address" on the Internet. A "top level" domain specifies the broad category and usually ends with **.com** (company),

---

THE VALUE OF COPYRIGHTS

Since copyrights include artwork, designs, and patterns, you may have heard that you do not need a patent because the design or the production blueprints (patterns) for manufacturing of the invention is copyrighted. In part, this might be true. But keep in mind that copyrights on a particular design are limited to only those aspects of its artistic "printed" appearance. Blueprints, or patterns, can be modified and changed and be easily designed around. In contrast, a utility patent tends to cover the entire functionality—utility—of the invention. The clever use of a copyright would be that of Celestial Seasoning Teas. Mo Siegel developed an entire product line identified by ingeniously copyrighted cartons. Over the years it also became the company's trade dress.

.net (network infrastructure), **.org** (non-profit organization) or **.edu** (4-year degree-granting institution). A second-level domain name represents an individual as in **yourname.com**.

Domain names are used much like trademarks, logos, and brand names. They are entered into the computer of your Internet service provider (ISP) and are referred to as domain name servers (DNS) The DNS (your domain name) represents your business identity. Once registered, others are prevented from using it. The use of a domain name must not infringe an existing trademark, service mark, or company/corporate name. Domain names are also portable. No matter what server or ISP you may use, your domain name goes with you.

Registering a domain name is relatively easy. You conduct an Internet search and if the name is available, you record it with InterNIC. InterNIC is the Network Information Center—the Internet's administrative registration agency. Your current ISP or other companies on the Internet can conduct this registration process for you. They typically charge a one-time fee of $7 to $10 to register and establish a DNS. InterNIC subsequently charges $100 for the first two years and $10 to 20 a year thereafter.

Patent Laws and Processes

There are several important laws and patent processes you should be familiar with. These laws and processes will help determine the best patent strategy for you and your invention.

Patent Application Types

There are two types of patent applications. One is referred to as the regular or nonprovisional patent application, which becomes the permanent, numbered patent entered into the annuls of U.S. history. The other is the temporary, provisional patent application, or more correctly, the "provisional application for patent."

The regular, nonprovisional patent application is a complete specification in compliance with 35 USC #112 in U.S. patent law. It includes the all-powerful legal claims covering the novel invention and should be written by a qualified patent attorney or patent agent. Once granted, your patent rights are defined by these claims. They are the ultimate determining factor upon which conclusions that others are making products similar to yours are infringing.

The permanent patent application should be as tightly written and professionally prosecuted as possible, in order to give you adequate patent protection. You should leave no margin for error. The next twenty years may depend upon the scope and integrity of this document.

In summary, the regular patent application includes:

- Complete specification of the invention
- Drawings representing the invention (if needed, which is almost always)
- Claims
- Abstract summary of the invention
- Oath of inventorship
- Power of attorney (Your attorney will act on your behalf to file and prosecute the application.)

FACT

When you're seeking a Sovereign Granted Monopoly (i.e. a patent) to market an invention, keep an open perspective on having several "picket fences" around your invention concept: e.g. utility patent, design patent, trademark, trade dress, copyright, trade secret, and even a license.
–*Doug English,*
Patent Attorney

FACT

Good domain names are getting harder and harder to find. If you can zero in on one that reflects your trademark, it'll make it difficult for others to get a mark using the same words with a different suffix. The reason is that it would infringe your trademark.

After an application is received by the USPTO, a confirmation of receipt will be mailed to your attorney or agent.

Declaring "Patent Pending" Status

Once the U.S. Patent Office has received a patent application covering an invention, you can post the words "patent pending" or "patent applied for." It is illegal to post patent pending notices on products or their related printed materials if the U.S. Patent Office has not officially received your application. It constitutes fraud and you may be subject to a fine.

The best method to establish the fact of receipt is to send your application via the U.S. Postal Service Express Mail. The U.S. Post Office is an agent of the federal government, which owns the U.S. Patent Office. Thus, once the Post Office has officially taken possession of your package, it is considered as "being delivered." You can legally post patent pending status the same day. This methodology has been tested in court and validated.

When you use this method, make sure you save your receipt. Also, it is wise to write the title of the application you've sent on the receipt.

Patent pending is a warning notice to others. During the patent pending time span, an inventor does not have any rights to claim infringement or cause a competitor to cease and desist. However, once a patent application is filed and subsequently issued, an infringer who has been knowingly infringing may be liable for past infringement going back to the date the patent application was published. In light of this law, it would be unwise for a company to pursue the manufacture of an invention, knowing it could be liable for patent infringement awards or damages once the patent is issued. For a public corporation, it could even represent a breach of shareholder trust.

The preferred way to give patent pending notice is to clearly mark your products and literature. This does not mean blatantly doing so, which can, at times, send the wrong message, as though the patent is more important than the product. You want to be tasteful about it. There are generally three criteria you want to apply. First, a patent pending notice should be in a visible location—for instance, the outside of a box, not inside the box lid. Typically you'll place the words "Patent pending" at the bottom of a carton or the product, where there may be other legal data posted, such as a trademark or copyright notice or the legal posting of ingredients or contents. Second, use a type size that is representative of the other fonts used on the product or packaging. Make sure it is large enough for a user to read—but once again, don't be blatant or distasteful. Third, use a type color that is easy to read in contrast to the background. For example, you wouldn't want to use yellow print on a white background. It would be hard to read. A red, blue, or black ink color would be much better. This approach should be used for both patent pending notices and for issued patent numbers.

Provisional Patent Applications

On June 8, 1995, Congress put into effect a new law regarding what is commonly referred to as a provisional patent application. This greatly simplified patent application allows an inventor to file at the U.S. Patent Office "a detailed specification" of the invention accompanied by the necessary drawings. Upon receipt of the provisional patent application at the USPTO, an invention officially has a "patent pending" status.

In summary, the provisional patent application has the following elements:

- Complete specifications of the invention
- Drawings representing the invention (if needed, which is almost always)
- Cover sheet (a simple form from the USPTO)
- Small entity status declaration is optional
- A filing fee (presently $110 for small entities)

That's all! The simplified format greatly reduces the time and expense required to file an application. The claims, abstract, and oath of inventorship are not required in provisional applications. The toughest and most costly part of filing a patent application is usually the claims. For this reason, provisional applications are a lot easier to write and far less expensive to file.

Here are a few other key elements of a provisional patent application:

- "Patent Pending" can be posted on your products.
- It is not reviewed by the Patent Office.
- A regular, nonprovisional application must be filed within one year and reference the provisional application.
- The PTO discards it after one year.
- It can preserve your worldwide filing rights for one year.
- It establishes the priority date.

Once filed, a provisional patent application establishes a legal filing date, referred to as the "priority date." To preserve the priority date, an inventor must file the regular, permanent patent application within one year of filing the provisional. When the regular patent application is filed, the provisional is referenced, hence preserving the original filing date. The permanent U.S. patent will have a 20-year life, beginning on its original filing date, not that of the referenced provisional.

The priority date may become important if two or more inventors have filed for the same inventive matter. This is discussed in more detail later in this chapter.

The real beauty of a provisional patent application is that it gives an inventor one entire year to develop and test the marketing of his/her invention without incurring the high cost of filing and prosecuting the permanent patent application. With the one year deferral, an inventor may use his/her finances to improve the invention and its marketability. The invention should be earning profit by year's end or showing strong indications that it will do so. With the income or the imminent promise of profit, it becomes easy to justify the higher cost of filing the regular patent application.

Provisional applications should not be construed as a substitute for an "invention disclosure." You will read about protecting your ideas with invention disclosures in Chapter 6. Provisional applications are well-written documents clearly depicting the invention. Thus, they can support the claims that will be made in the regular, nonprovisional application. If they do not, the subject matter in the provisional may be deemed irrelevant to the content of the regular application and the priority date may be lost on the new subject matter.

The best way to write a provisional application is to use the same guidelines as are

PROVISIONAL VALIDITY

Laws regarding the filing and use of provisional patent applications are well established and accepted worldwide. To ensure your provisional applications will not be scrutinized by the Patent Office, you'll want it to read as closely as possible to the permanent patent application.

PROVISIONAL PATENTS AND STRATEGY

With provisional patent applications costing a fraction of what the non-provisional application costs, you can file several applications and post the words "Patents pending" or even the number of patents, such as "Five patent pending." As long as you've not publicly disclosed the invention prior to filing, you have preserved your international rights to file PCTs later on and are sending a powerful message to industry that several patents are pending, not "just one."

used for a regular, nonprovisional patent application, but excluding the claims and abstract. How to write a provisional patent application is discussed in detail in Chapter 7. The book *Patent Writer* shows you in greater depth how to write dynamic, successful patent applications.

Provisional patent applications offer a great opportunity, but some caution is needed. Provisional applications do not give an inventor rights to exclude others from making, using, or selling the invention. A provisional application only serves as an interim step before filing a permanent, nonprovisional application. Only the granting of the permanent patent application implies any rights.

Another caution regarding the filing of a provisional patent application and the use of the one-year rule is that once your invention has been publicly disclosed, you have one year to file a patent application in the United States. The patent application you file may be a provisional application. However, when incorporating the one-year rule into the filing of a provisional (or permanent) application, you lose your rights to file internationally (except for Canada and Mexico). This is discussed in detail on the following pages.

Cost of Patents

The patenting process does cost money. However, most of that cost is incurred later on in the more advanced stages of development. Frequently this is after the initial product launch. At that stage of development, you should have sufficient cash flow to justify the cost, whether your income is generated from sales made by your company or income through a licensee.

You may reduce some of your patenting costs by writing your own application, preparing your initial drawings, and of course, beginning by filing an inexpensive provisional patent application. Doing so will defer the bulk of the cost for a year. Provisional patent applications prepared by inexperienced engineers or inventors should be reviewed by your patent attorney prior to filing.

The Patent Office charges a patent application filing fee and, when the patent is granted, a patent issuing fee. There are also maintenance fees at $3\frac{1}{2}$ years, $7\frac{1}{2}$ years, and $11\frac{1}{2}$ years after the patent issue date. See Table 5.1.

Filing fees for small entities are half those for large entities. A small entity is one with a maximum of 500 employees; a large entity has more than 500 employees. Nonprofit organizations are considered small entities. When counting the number of employees, include all of the company's divisions, plus *all* those employed by the owner's affiliates, sister companies, and subsidiaries. Both full- and part-time employees are counted. If the size of your entity changes anytime in the future, you will need to adjust your payment status accordingly. See Table 5.1.

In addition to filing fees for provisional and regular, nonprovisional patent applications, you will incur legal expenses for an attorney or patent agent. Patent attorneys generally charge $250–$500 per hour and patent agents $125–$200 per hour. To write and file a relatively simple patent application, an attorney will charge about $1,500–$3,000 in addition to the filing fee. You can save a lot of money—as much as $1,000 to $2,000—by writing your own application and having an attorney prepare the final draft for you. This subject is covered in detail later.

If you hire a draftsperson to do the drawings, he/she will typically charge $40–$60

114

Table 5.1 Patent Cost Estimates

Low Cost	Description	Action	Description	High Cost
$0	You write	Write disclosure		
$0	You write	Write provisional	Attorney writes	$1500–$2500
$110	You file	File provisional (includes postage)	Attorney writes	$175–$250
$500	Attorney converts your provisional application	Write regular application	Same as provisional	$0
$1200	Attorney writes	Write claims	Attorney writes	$2500
$0	You do	Prepare drawings	Out-sourced	$300–$600
$500	Attorney files	File regular (includes postage)	Attorney files	$300
$1200	Attorney prosecutes	Prosecution	Attorney files	$2500
$750	Attorney submits	Issuing Fee	Attorney files	$900
$4160	Low estimate		High Estimate	$8175–$9550

per hour. The cost to have professional drawings prepared usually runs about $300–$600, but it could run more for sophisticated inventions. You can save money by hiring college or university engineering students, who can produce excellent work for as little as $12–$20 per hour.

Patent costs can also vary depending on the number of claims. Claims refer to the legal description of the inventive matter. If you have more than 20 claims, there will be a surcharge per additional claim. Independent claims are those that "stand alone" and do not rely on any other claim in the application. The costs increase if you have more than three independent claims in an application. Multiple dependent claims also incur an additional cost, as they add to the complexity of reviewing and studying the inventive material. Multiple dependent claims can usually be avoided with a well-written application. In Chapter 7 independent and dependent claims are discussed in greater detail.

Next, once your application is reviewed by the Patent Office, you will incur patent prosecution costs with your patent attorney, which could run an additional $1,000–$2,000.

Last, you have an issue fee once the subject matter in the patent application has been approved. The Patent Office fees are published every year in October. Don't let all these patent costs and legal fees discourage you. As you can see, most of these costs come later on, after you're into product development and if you apply the *From Patent to Profit* Strategic Guide, the bulk will be incurred after product launch. Well-conceived patents developed during product development earn far more money than the cost of patenting and help preserve your long-term rights to make money, not spend it.

The cost estimate in Chart 5.2 is a high-low cost summary of the entire patent writing process right through prosecution and issuance. It is based on a simple invention and having an attorney write at least the final draft to a regular application, add the claims, and then file and prosecute it. It assumes no major prosecution problems and a small entity status.

Don't forget that these costs are spread out over a period of several months or even a few years. With the right development strategy and aggressive marketing in the early stages, you may have sufficient cash flow before incurring the bulk of the legal costs.

THROWING AWAY YOUR MONEY?

If you have not become expert in your field, and you have not proven your market opportunity and ability to sell your invention, *ANYTHING* you pay for a patent is too much money. If you have evaluated, refined, proven your marketability, and have become expert in your field, anything you pay for your patent will be small relative to your future success.
—Andy Gibbs, CEO, PatentCafe.com

SMALL VS. LARGE ENTITY STATUS

If you have been negotiating a license with a large entity, but have not made a commitment, you can still file as a small entity. The key word is "commitment." But be careful. An error in identifying the size of your "entity" may invalidate your patent. If you have any doubts use your attorney to accurately assess the situation.

FACT

First to invent laws favor the true inventor who pursues his/her ideas in a timely manner. A "first to file" patent system, like Japan's, is best for large companies with a lot of R&D money to spend on patents.

FACT

A prospect that every inventor must bear in mind is that she or he was probably not the first, nor the only inventor to recognize the problem being solved by her or his invention. Don't forget, after all, there are nearly 8 billion people across our planet struggling against the forces of nature, facing and overcoming barriers with the one tool that defines us from all other creatures: inventive thinking.

–Donald Grant Kelly, Director, U.S. Patent Office

Later you can read about how to hire a patent attorney and how to pursue an economical patent-writing strategy. See Table 5.2.

Table 5.2 Patent Office Fees

Patent Office Action	COST
Provisional patent application filing fee	$110
Basic filing fee for a utility patent application	$165
Basic filing fee for a plant patent application	$110
Design patent filing fee	$110
Claims in excess of twenty per claim	$26
Independent claims in excess of three per claim	$110
Utility patent issue fee	$775
Plant patent issue fee	$595
Design patent issue fee	$430
3½ year maintenance fee	$490
7½ year maintenance fee	$1,240
11½ year maintenance fee	$2,055

First to Invent

Because the United States is a "first-to-invent" country—not a "first-to-file" country, only the first, true inventor(s) will be acknowledged as the patent grantee(s). Any invention or discovery an inventor is working on that has not been abandoned has precedence over subsequent discoveries that are the same or similar in scope.

If two persons are granted patents on the same subject matter, the inventor who can prove his/her discovery has precedence over the other will have the valid patent with one provision. This is regardless of who filed first or which patent was granted first.

First-to-invent rights are established by validating your date of original conception and then following through and reducing to practice your invention. Typically, this is accomplished with an invention disclosure and then keeping good journal records during development. By diligently pursuing the development and finally the patenting of your invention, you are preserving your rights. This is referred to as "diligence." It, you could say, proves you have not abandoned the concept. How to preserve your first-to-invent rights by writing an invention disclosure and keeping journal records is discussed in detail in Chapter 6.

First-to-invent rights are very powerful rights and many an inventor has used them to his/her benefit. First-to-invent rights also tend to improve the quality of a patent application (and the subsequent patent) because they tend to allow the inventive matter to be more thoroughly developed and the value and potential marketability qualified.

Interference

Some say that a patent application should be filed as soon as possible after the initial conception to establish a priority date and protect first-to-invent status because the inventor who files first is more likely to be proven to be the one with the valid patent.

The first-to-file person is considered to be in the senior position. The inventor who files his application second is considered to be in a junior position. This is an uncommon occurrence and is referred to as "interference."

Since a valid patent can only be granted to the applicant who was first to invent, a proceeding referred to as interference is initiated by the PTO to determine who is the first inventor. An interference proceeding may also be initiated regarding an application and an already granted patent, providing the granted patent was not issued more than one year before the filing of the new application. This is one provision.

Interference is rare. It occurs in less than $1/10$ of 1 percent of the regular patent applications received by the U.S. Patent Office. In other words, fewer than one in every 5,000 applications is ever involved in a dispute over who was the first to invent. Ian Calvert, an administrative patent judge at the USPTO, declared there were only 147 interference proceedings in 1995, many of which had been initiated in the previous years. Of those, 52.5% of the decisions were in favor of the senior party, 31.7% were in favor of the junior party, about 9.4% resulted in no patents being issued, 5.8% involved no interference, and 0.7% resulted in a split reward. This indicates that the first-to-invent laws truly work.

With 200,000 plus applications being filed every year, it is easy to see that there is only a marginal advantage over being first-to-file, if any at all. Today, interference is less common, even though more patent applications are being filed.

Some pertinent statistics, according to the Association of Patent Law Firms, Chicago, Illinois, include:

- Presently pending interferences: 156

- Pending interferences declared before 1998: 27

- Pending interferences declared between 1998 and 2001: 67

- The average duration of the presently pending 156 interferences: 28.6 months

- The average duration of interferences declared after 1998: 18.1 months

- The greatest number of interferences originating from biotechnology

It's easy to see that interference in not something you should be concerned about. It's best to invent, create, and not worry about a rare event that most likely will never occur.

In interference proceedings, each party submits facts trying to prove two facts: One, when the invention was conceived and two, when it was reduced to practice. With no log or supporting facts, the filing date of the patent application is considered the date of reduction to practice.

A board of three Examiners-in-Chief then determines which inventor has the priority date based on the evidence received. The inventor who proves to be the first to conceive and the first to reduce the invention to practice will be seen as having the priority over the other inventor. More complicated situations are not so easy for the examiners to determine.

Maintain good journal records as pointed out in Chapter 6 and you'll have little to be concerned about.

The All-Important One-Year Rule

As much as the first-to-invent laws may help inventors, there is even more good news.

MIS-INFORMATION

There has been much misinformation about the importance of being in the "senior position." It is apparent that the "first to invent" laws in the U.S. work and that there is no significant advantage to being the first to file. More important is good documentation and reducing your invention to practice in a timely manner.

PATENT OFFICE FACTS

With interference being less and less of a problem, you can ensure your position by maintaining good records in a log book. The Scientific Journal (see appendix or visit out website @ frompatenttoprofit.com for a thorough overview) is an attorney-approved legal binder you may use to document your first-to-invent rights as well as invention development objectives.

In the United States, once an inventor **publicly discloses** an invention, he/she may still file a provisional patent application or a regular, nonprovisional patent application within one year of the public disclosure. A first public disclosure is usually the first time the invention is offered for sale, but it may be in a trade magazine, a test market, a trade show offering, and so on.

You can use the one-year rule to your advantage. If you are unsure of the marketability of your invention and want to extend your cash flow, you can try marketing the innovation for up to an entire year before filing a patent application. Prior to year's end, you can then further preserve your cash flow by filing the inexpensive, provisional patent application. That's almost two years of sales— and that should determine whether you want to proceed. If sales are marginal, you may decide not to file the permanent patent application or you may abandon your effort altogether.

There is another case for using the one-year rule. At times, you may unknowingly discover an improvement to your innovation and inadvertently disclose it publicly. This can happen in a test market mode and you make some on-site adjustments that are considered new inventive matter. Many times, this new inventive matter is important and may even represent a breakthrough opportunity. Thus, you'll want to protect it with patents . . . and you'll have the legal right to do so.

Carefully developing your innovation and timely filing of your patent applications are keys to having a successful and effective patent. By waiting until the end of the developmental period when sales have begun, you will have most of the development problems solved. You will be filing a patent application on the preferred embodiments of the invention and will be making your claims stronger, more accurate, and less vulnerable to scrutiny and potential invalidation. Using the one-year rule in this mix gives you another tool to improve your flexibility to invent.

However, there is one downfall to employing the one-year rule. You will have lost the right to file for international patents. The exceptions to this rule are Canada and Mexico; they respect our laws regarding the one-year rule and will allow you to file in those countries.

Protecting Worldwide Rights

You can preserve your ability to file worldwide by making sure that you file a provisional or the regular, non-provisional application before your invention is publicly disclosed.

Foreign filing must occur within one year of the filing of a regular or provisional application in the United States. Read more about developing your worldwide patent filing strategy later in this book.

The Prosecution Process

Once a utility patent application has been filed, it will be assigned to a designated patent examiner for his/her review. It generally takes about one year before the review begins. When the application is reviewed, you will receive an **office action** based on the examiner's review.

Almost all patents are rejected in the first office action! You should take some time to review the objections and rejections the patent examiner cites. You have 90 days to do this. There are usually two or three, sometimes more, office actions before a patent is granted, finally and formally rejected by the PTO, or abandoned by the inventor. Each office action typically takes another 2–4 months, sometimes longer.

Don't be discouraged during these office action delays. The examiner is trying to evaluate the claims based on the subject matter in the application. You will have the opportunity to file an amendment and revise the claims or correct other subject matter in the text to clarify your inventive matter. Following through with revisions and corrections only benefits you in the long run.

Sometimes the examiner may cite that what you are claiming is not related to the inventive material in the text. In other words, the examiner may bring up the objection of "unrelated subject matter." This is not the end of the world.

You may still file a Continuation In Part (CIP). A CIP is treated much like a new application and allows you to carefully rewrite parts of the original application to more specifically address and clarify the inventive matter. It is not an uncommon occurrence. Filing a CIP might prolong the application's issuance by as long as 6–9 months or more. Your original date of filing will not be lost.

Figure 5.1 Utility Patent Application Process

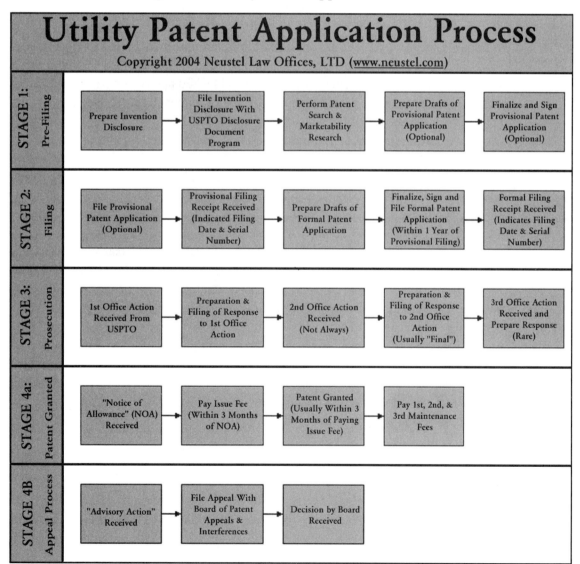

It can easily take 12 to 24 months or longer for a patent to issue. Some patent attorneys believe that the longer the wait, the better, because it indicates that the examiner is having a difficult time finding any prior art. If you wish, you can also include a petition with your application that will expedite the review of your application. This can be particularly important to a company that is relying on the patent's issuance to secure financing or combat a serious patent infringement problem. The petition costs about $150.

When a Patent Is Granted

When you have overcome the objections and rejections of the patent examiner, you will receive a notice of allowance. In other words, you are about to receive your patent! You then have 90 days to pay the issuing fee. When your check is received, your patent will be officially granted and issued about three or four months later.

Your legal rights begin on the date of issuance. From this point on you have the right to stop all others from making, selling, or using your patented material. Figure 5.1 is a flow chart of the patenting process from beginning to end.

When a Patent Application Is Finally Rejected

You may have taken your patent application through numerous office actions, perhaps even filed a Continuation in Part for clarification, but the examiner still refuses to accept your application's subject matter as being patentable. You have received a "Final Office Action," which contains a final rejection by the examiner. Upon receipt of this final notice, you still have two more alternatives.

First, you can prepare an immediate response and try to restate your position and clarify (argue) the differences. You will usually do this if you feel the examiner's rejections are in error. Hopefully, you can provide additional facts to support your position. If the rejection is promptly pursued, you can send this correspondence to the examiner and ask him/her to reconsider. You should have a response within 30 days. Your patent attorney will know what to do in such a case, so work closely with him.

Second, if you (and most likely your attorney) feel strongly that your material is patentable and the examiner's judgment in rejecting the subject matter is wrong, you can file an appeal with the Court of Appeals for the Federal Circuit. This is a special appeals court in Washington, D.C. that handles only patent-related matters. To appeal, you need a well-prepared legal argument to present your case. Your argument should include other related cases supporting your argument. Again, you patent attorney should write and present this appeal.

The appeal process can be costly, but it usually occurs after you have realized some success and profit from the manufacture and sale of your invention. Keep in mind that a patent granted after it has endured the appeal process in the U.S. court system will be more difficult to invalidate at a later date. If infringement occurs, the appeal process should help to strengthen your patent's integrity.

Your Patent as Property

When you receive your U.S. patent, you will have accomplished something that few people in history have ever done. Feel good about it and celebrate. Your patents are intellectual property and are personal assets, like a boat, a house, or a business. When someone infringes your patent, unfortunately, it feels the same as though someone stole your boat or robbed your house. For twenty years, this intellectual property is exclusively yours.

As with real estate, you can sell your patent or license it, which is comparable to leasing real estate. You can earn substantial income from the sale or license of a patent.

Inventorship, Ownership, and Patent Assignment

Only true inventors can be named as patentees, but any individual, firm, corporation, or partnership can own a patent. An inventor automatically owns his/her patent when granted, unless it is assigned to another entity.

For someone other than the original inventor to own a patent, the inventor's rights must be transferred via a patent assignment. It is just like signing over a pink slip to an automobile or boat. The assignment of a patent may be registered at the U.S. Patent Office, although this is not a legal requirement.

If you are an inventor and intend to assign your patent to an individual or a company, make sure you are getting a fair deal. Remember that patents are now worth 20-100 times more than just twenty years years ago. Ask yourself if you are satisfied with the amount of compensation you are receiving in exchange for giving up your title and rights to the asset.

Inventing as an Employee

When you work for a company and invent something related to your company's business, on company time and for its use, you cannot ask for compensation. Your continued employment is considered your compensation for the invention. After all, your inventive discovery probably occurred in the job environment and you would not have been able to make it otherwise.

If your invention was conceived before you joined the company and you are patenting it for the company's use, you can ask your employer for additional compensation. However, making this request after your employment begins may put a strain on your relationship and end it. A company may also declare "shop rights," which means it can use the inventive matter without compensation. If this sounds like your situation, make sure you do two things. First, clarify your intentions before your employment begins, and second, consult with your patent attorney whether or not your invention has "shop rights."

Ending a relationship with an employer may not be bad news. If you provided the company with solid patents that were shaping its future and your efforts went unrecognized, don't worry. The competitors will find you and make you a better offer.

Last, federal laws state that if you are employed by a company and receive a patent on an unrelated idea in another field, and development was not on company time, your employer cannot claim any rights.

Misconceptions about Patenting

There are several misconceptions about patents. Let's dispel some of the more common ones:

- **You have to be a scientific genius to successfully invent and patent your ideas.** Patents are not necessarily of a scientific nature. There are many considerations for the granting of patents, most of which are not scientific. You will find that education is not necessarily a prerequisite to being a good inventor. White-out® was

ZIGZAG STITCH

The zigzag stitch was invented by Helen Augusta Blanchard of Portland, Maine. She earned at least 28 patents, many of which were for improvements to sewing machines, including the first zigzag stitch machine. She also patented related items such as hat-sewing machines and surgical needles.

JUST LIKE A DEED

Patents are similar to deeds in a way. If you have a patent and someone wants to buy it, but wants to pay on installments, never assign it until you have received the last payment. Just like a deed to a house, you would not assign it to a purchaser before it was paid in full.

49 U.S. PATENTS

Beulah Louise Henry of Tennessee was dubbed "Lady Edison" in the 1930s. She earned 49 patents, but her inventions numbered around 110. Her patents include a vacuum ice cream freezer, an umbrella with different colored snap-on cloth covers. Literally overnight she invented the first bobbinless sewing machine in 1940.

invented by a high school drop-out who needed to keep her job in order so support her child. The oil can was invented by an uneducated slave—McCoy. Even children have been issued patents for their inventions. What's more important is that you understand the field of your invention and have the knowledge and insight to invent, develop, or improve on an existing product, process, or system.

More frequently patents are granted because they are considered useful, not because of their scientific nature. For instance, an innovation that saves time, reduces scrap or makes a product easier to use may be patentable and quite valuable. Sometimes you are better off not having a degree in the sciences with preset rules and regulations to follow. The chances are that you are creative and intelligent enough without a scientific education.

- **A university or college can teach you how to create, patent, and develop your ideas.** Colleges and universities that teach how to invent and patent are hard to find. So don't count on signing up for some classes at the local college. However, there is a growing trend in educational institutions to specialize in—even encourage—dreaming up ideas that improve our lives.

- **Inventors are well trained in engineering.** While some inventors are engineers, most are not. Frequently, inventors "create" or "discover" new concepts from which engineers then engineer new products. Engineering deals with established, commonly used processes, formulae, and so on. On the other hand, inventing tends to focus on something yet to be discovered. Engineering schools train engineers based on what is already known. They rarely teach how to think creatively or outside given norms and standards.

- **It cannot be patentable because the idea is too simple or the change is obvious.** This is frequently not true. For instance, what if a product has been made of leather for 40 years and an inventor has just discovered how to make it from a new plastic material that has several benefits over leather. At first glance, you might say that the change is obvious because it is the same product with just a change of material. This usually means it would not be patentable.

But wait a minute. No one has introduced this product in plastic for 40 years! There must be a reason. The reason usually represents the bona fide patentable concept. For instance, it may be that the plastic ripped too easily, making it unmarketable. If you have discovered a means to prevent it from tearing, then you would have new inventive matter, but not necessarily the resultant product.

- **Why even get a patent? By changing a product 10%-15% someone can design around it anyway.** There is no such thing as changing a product by a certain percentage that validates a "design-around." We know that merely changing color or size will not change the scope of a patent. Changes must be "substantial" to not infringe upon an existing patent. This fact of law is substantiated by the Doctrine of Equivalence, which was established in a 1952 court case. In summary, the Doctrine of Equivalence states:

If it operates in substantially the same manner, by substantially the same means, and produces substantially the same result, it infringes.

On March 3, 1997, in the Hilton-Davis case, the Supreme Court upheld the Doctrine of Equivalence and further clarified its interpretation. In summary, the Court's decision states that:

A. We will continue to adhere to the Doctrine of Equivalence.

B. The determination of equivalence should be applied objectively on an element-by-element basis.

C. If a patent holder can demonstrate that an amendment required during prosecution had a purpose unrelated to patentability, then it should not affect the scope of the patent's claim.

More recently the FESTO case has introduced the concept of limiting the scope of a claim to its literal interpretation if the claim had to have been modified during prosecution. The final ruling on this may still be a few years away, however. The Doctrine of Equivalence works in your favor most of the time. For a more in-depth understanding of how to apply it in reading patents, see Chapter 7.

- **Once you have a patent, you can sit back and "cash in" on the royalties.** Unfortunately, this is not normally the case. But you certainly can realize significant financial gain. Developing an idea into a patentable concept and then into a product requires a lot of hard work. The amount of time it takes to dream up the idea, plus the time spent on patenting, is probably only 5% of your total time investment. The majority of your time involves product development and design, and working with your manufacturing and marketing partners. But don't be discouraged; the process is a lot of fun and very exciting. The result can be a piece of intellectual property with real value.

- **It costs a lot of money to get a patent.** While some companies may invest a lot of money up front, there are several strategies you can use to either cut costs or defer them until later. You should do much of the work yourself and are usually better off when you do. For instance, writing your own patent application gives you more control over the outcome of the inventing/patenting process and the quality of the resultant patent itself. If you follow the guidelines in this manual, you can at the very least defray a substantial amount of the cost of patenting . . . even for years, if you should choose.

In addition, this manual will teach you how to successfully bring in manufacturing partners and develop sales relationships or licensees in a timely manner. At the very least, you'll be able to qualify the marketability of your patentable concept with little up-front money. If changes are required, you can make them during the initial start-up period before you've incurred the higher cost of filing non-provisional patent applications. Once your invention receives the stamp of approval from your marketing arm, you should be able to generate a positive cash flow during this patent pending phase.

Learning More about Patenting

It is important you learn all you can about patenting and become an expert in your field. There are a lot of sources of information about patents and patenting, but much of it is repetitive and much of it may not apply to you, a small company or independent inventor. Generally speaking, you will find it somewhat difficult to find patent tactics and strategies for small companies due to the following reasons:

A LOT OF HARD WORK

Obtaining a patent is like anything else in life . . . it requires a lot of hard work, hours of frustration solving problems and always hoping for new discoveries that will make your job easier. If anyone thinks that Edison, Lemelson, Land, Yamamoto and the many brilliant inventors who have contributed so much to our American way of life, had a gravy train waiting for them, think again. The monetary rewards to inventing and patenting inherently come from the process of problem solving and enjoying what it is you are doing. It almost never comes from casually inventing something and selling the patent for millions of dollars.

THE IDEA IS THE MONEY

People often tell me they don't have the money to pursue their ideas. I tell them that if the idea is good enough, the money is in the ideas.

—Stephen Paul Gnass

FROM PATENT TO PROFIT WORKSHOPS

From Patent to Profit workshops are given in many parts of the U.S., in colleges and universities and in cooperation with SBDCs. If you have a chance to attend one, it can be a highly informative and valuable experience.

NCIIA

The National Collegiate Inventors and Innovators Alliance or NCIIA is a unique interdisciplinary educational program started at Hampshire College in November 1995 with the generous support of the Jerome and Dorothy Lemelson Foundation. The mission of the Alliance is to foster and promote the teaching of invention, innovation, and entrepreneurship at colleges and universities around the country. The Alliance provides support in the form of grants to faculty and students, services and information for members, and meetings, workshops and so on.

FRAUD INDICTMENTS

The FTC has indicted 7 owners of dozens of invention development companies in recent years for fraud. They create an image that this is a booming business by seemly competing for your business. The truth of the matter is that they are frauds. For more information visit the FTC website at: www.ftc.gov

- **Classes and seminars on patenting for small business and independent inventors are rare.** Other than a few short courses that focus on certain legal aspects of patenting, it is very difficult to find classes or seminars in which you can thoroughly learn about the patenting process. The workshop series *From Patent to Profit,* offered by IPT, Inventions, and Patents and Trademarks Company, includes all the basic laws about patents and information regarding how to develop your patenting strategy. *Patents in Commerce* is another seminar series; it uses the book *From Patent to Profit* as its textbook. You can gain an immense amount of knowledge on patenting and product development from either of these workshops. Frankly, we cannot recommend any others since our focus is on making money, not just "getting a patent."

These seminars teach, in reasonable detail, information about all types of intellectual property, how to write a patent application, and what to do when your patent is infringed. More importantly, the workshops focus on patenting as it relates to invention development, manufacturing, marketing, and licensing. This approach ensures that an inventor will file for claims on the invention's very best attributes, all the while developing and designing products that will make money.

You've seen it before and will continue to see it throughout this manual: It is unwise to write a patent application or pay a lot to have it written, without a clear idea of the value, merit, and marketability of your invention. Why would you want to spend thousands of dollars without knowing whether you can earn a single cent?

Learn all you can about patenting, inventing, and marketing and make some sound decisions before you spend a lot of money.

- **Patent attorneys offer limited help.** There are many competent patent attorneys willing to provide information and guidance, but they generally charge between $250–$500 per hour. A phone call or office visit to learn the basics of patenting can cost you hundreds of dollars. This is a high price to pay.

In addition, keep in mind that while patent attorneys can help you with patenting, they are generally not able to guide you creatively or offer manufacturing and marketing expertise. Even if they could, they would not be cheap. Most patent attorneys don't even ask about whether you think you can market your invention and earn income from it. Don't get lofty ideas that having a patent automatically means that you will automatically make a lot of money. Remember, only about 3% of all granted patents ever earn an inventor any substantial income.

Patent attorneys are generally a pleasure to work with and usually bring you good news, but you should use them in sparingly. Many attorneys have the gift to gab . . . at your expense. Keep this in mind as you learn about patenting in your field.

- **You cannot always use another inventor to help.** There are successful inventors who could help, but there are many reasons why they may not. To begin with, you probably don't want a co-inventor on your patent. Also, most successful inventors are busy with their own inventions and don't have much time to spare. If they are good at what they do, and have the time to devote, they may charge as much as a patent attorney. Finally, keep in mind that an inventor creating in a different field may not understand your sphere of knowledge. Can an inventor working on jet propulsion systems help someone invent a machine to mass manufacture bagels? Probably not.

- **There are few books and handbooks to help.** There are some specialized handbooks on patenting, but again very few focus on patenting from the perspective of small businesses and first-time inventors.

From Patent to Profit is written for the layperson and provides a complete overview of the patenting and inventing process. It is the only manual available that combines patenting and patent strategy with an invention development strategy and the all-important manufacturing, marketing, and licensing processes. It is based on the experience of successful inventors. *From Patent to Profit* is intended to teach you how to plan your strategy, protect your idea, and write a patent application and get it filed in a timely manner. It also explains how to hire a patent attorney and defer major patenting costs while you are developing your invention into a new profit center.

Some additional books you may read that focus on individual elements of patenting and product development are being marketed by Patents in Commerce. These books can be of significant help in your development activities. They include:

- *Patent Writer* shows you in step-by-step details how to write a successful patent application. It dovetails perfectly with what is taught in *From Patent to Profit*.

- *Anatomy of a Patent* teaches all about how to read patents and how to research them through patent searching.

- *Essentials of Patents* talks about patenting in the corporate environment. From a licensing viewpoint, it encourages companies to license in technologies from outsiders . . . small product development companies and independent inventors. You can gain substantial knowledge of what the various departments in a corporation look for when licensing your inventions.

- **The Internet** is one of the best information sources on invention and patent-related matters in the world, but it can provide misinformation if you're not careful. We encourage you to keep in touch with our website http://frompatenttoprofit.com for up-to-date information. We maintain only qualified links on our website (click on Helpful Links) so you won't be misled.

Invention Promotion Companies

The fact is invention promotion companies promise help to inventors, but provide little. Several invention assistance companies advertise on radio and television and in magazines, promising to help inventors develop their ideas and/or find licensees. It is doubtful that any of these companies can provide you with help that has any real value. In fact, they may even do considerable harm.

Invention promotion companies want you to think they are well-connected with multimillion dollar corporations and that these corporations are ready—right now—to license your idea and pay you hundreds of thousands, even millions, of dollars. Don't believe them. The best you will get for your money is a book containing boilerplate copy providing information you already know or could readily find out. It will be a well-written dissertation using lots of positive phrases intended to make you feel good and to con-

LOW RATE OF SUCCESS

You can always ask for the success rate of an invention promotion company. It'll be extremely low. For instance, I don't know of a single success story related to any promotion company ever. A few years ago, I called one of the largest companies and interviewed with their West Coast rep. When I told her I made money from my inventions, she gasped and replied, "I have never met an inventor who actually earned any money from them."

AVOID INVENTION ASSISTANCE COMPANY ATTORNEYS

You have to remember that invention assistance companies only want your up-front money and rarely have your best long-term interests in mind. (Frankly if they did, they would immediately shut their doors in good faith.) Similarly, attorneys to these companies can only survive with this business and are not looking out after your interests, but those of their true client—the invention assistance company!

TAKE CONTROL

Don't abdicate your power. Be accountable and responsible for your own success and you'll never have to worry about unscrupulous scam operators. Education is the key. Be a smart consumer and don't become an invention scam statistic.

–Stephen Paul Gnass

vince you that you are going to become a millionaire if you spend thousands of dollars with them. Your odds of winning the lottery are much better!

Read the company's advertising and see for yourself. Check their references and you will be shocked at the extremely low rate of success. The reality is you have to pilot your inventions yourself. You can't rely on someone else to bring success to you on a silver platter. Don't waste your time or money.

If you have any questions about a company you are considering employing with your invention's development or patenting, visit our website at www.frompatent toprofit.com/inventorsalert.html or the National Inventor Fraud Center at www.inventor fraud.com.

PATENT STRATEGY

There are many more ways to gain
patent protection than you may think!

The Big Picture First

The first prerequisite to your patent strategy is to make sure you have an idea that will make money. It must be worth the investment. Chapters 2 and 3 discuss the basic tools to help you evaluate your invention's design and its money-earning potential.

Once you are confident that your invention is worth pursuing, you should protect it. Based on U.S. law, your protection begins by establishing the **date of original conception**, which is, fortunately, inexpensive and only takes a an hour or so to disclose. Then, follow through, reduce it to practice, and subsequently file your patent applications. As you recall, the United States is a first-to-invent country, which means that only the first inventor will have a valid patent. Your first action to establish your first-to-invent rights is writing an invention disclosure. This establishes your date of original conception. However, before you do so, you should know about the various types of patent protection you can seek. There is probably far greater patent protection for you to pursue than you may think, and it doesn't really cost much more, if anything.

Then write your disclosure to cover these various inventive aspects. Knowing this from the onset will also influence your patent development strategy, which should dovetail with your invention development strategy. What you discover as you invent is what you will want to include in your patent applications. If you have already begun your invention development and have written a disclosure covering it, you'll want to pay special attention to the following pages. In most cases you will find additional patent protection you can pursue without adding to the cost of patenting.

THE STRATEGIC GUIDE

Remember to follow the Strategic Guide in Chapter 2 and apply its tenets, so you do not spend a lot of money on patenting before you confirm marketability. There is a copy in the appendix your should use to track your progress. By following the Strategic Guide tenets, you will also find that you will not make the serious error of filing prematurely before you understand the inventive attributes of the invention.

Patent Protection Scope

Patents are usually thought of solely as products—something physical. But patents are commonly grouped into other categories that can have a greater sphere of influence in an industry. For instance, cost-cutting systems and processes of manufacturing can make an otherwise costly, hard-to-use product an extremely valuable one.

Study and understand the following patenting approaches. The eight patent categories that follow refer to utility patents. The names and definitions are well-known jargon among patent attorneys, agents, examiners, and experienced inventors.

Product, Device, or Apparatus Patents

Product, device, or apparatus patents refer to an actual physical product or group of products. Examples are Edison's light bulb, a paper clip, the Macintosh® mouse, a plastic grocery sack, and The Pump® basketball shoes.

Product patents are usually fairly straightforward and easy to understand. They are the most common and the easiest to conceptualize. Since they are easy to envision, they are also usually the easiest for which to draft a patent application. It is also fairly easy to determine if they are being infringed. However, product patents may not necessarily give you the best and broadest patent coverage, as you will see.

Process Patents

Process patents generally refer to processes of manufacturing. Examples are a process for extracting cotton seeds (Eli Whitney's cotton gin), cutting and cooking large quantities of tortillas, mixing large batches of ice cream (without a churn), or making automated batches of bagels. A process patent may also include the intangible process of reading DNA in genes.

Process patents are commonly secured and applied in industry and can be powerful assets. Can you imagine if all tortillas were handmade or ice cream was made in churns? Their retail prices would be exorbitant. Can you imagine the cost of a pair of jeans if cottonseeds were manually extracted? How about the tremendous benefits of reading DNA and how it is affecting medicine, law, and justice?

Process patents frequently offer significant improvements to our quality of life. The timesaving afforded from high-speed manufacturing processes means we have more time to do other things. The reading of DNA for disease prevention holds the promise of eliminating many diseases and birth defects.

If an industry is to make a significant advancement in product improvement, it must also have a cost-effective means to do it. Hence, it is easy to see that a process patent at times may be more valuable than the (patented) product that is being produced.

Systems (or Method) Patent

This describes a unique system, sometimes referred to as a method of use, in which a product is used. While patents cannot be obtained on a commonly known product, a patent may be obtained when it is used in a certain novel and unique way.

A unique method of use typically consists of a combination of components used together that result in a "system." One or all of these components may be prior art. For instance, systems patents have been granted on the methods in which a clerk uses a common, prior art, plastic grocery sack on a rack-style holder to facilitate loading of the sack. Patents have also been granted on the systems used to scan groceries in supermarkets, using prior art laser technology. These two examples have saved millions of hours of labor and saved customers a lot of time waiting in checkout lines.

Systems patents tend to make products people friendly. Scientific discoveries like the polyethylene resin used to make thin-gauged plastic film like that used in grocery sacks frequently are precursors to new product developments. But without new inventions like the plastic grocery sack and laser scanning systems, the supermarket checkout revolution would never have taken place. It was the systems approach that made the difference.

Similarly, it was the electronic age that brought us into the era of the computer, but it was all the people-friendly discoveries that made them sell so well. Among them are people-friendly computer software applications, email systems, the cordless keyboard, the mouse, and all the other systems that have made computers far more popular than televisions.

Systems patents frequently provide broader patent coverage than a product or device patent. For instance, would it be better to have the first patent on the handled plastic grocery sack or the system patent that uses those sacks in an efficacious manner—hence dramatically increasing usage? Take the system. When plastic grocery sacks were hard to use, they didn't get used in large quantities. But when they became easy to use in the systems developed in the late 1980s, they began to replace paper sacks.

Systems patents represent a great opportunity for businesses and product developers. In your patenting activities, when you think of writing an application for a product patent, try to include systems coverage as well. Of course, you can file for both.

Business Systems

Prior to 1998, business-related patents were not considered acceptable patentable subject matter. However, in 1998, court rulings on the State Street and Priceline patents changed the course of the U.S. Patent Office and patent statute. These court cases decided that Internet-related methods, such as the online reverse auction and the one-click shopping cart, should be protected.

Business systems patents may refer to Internet-related applications or how employees are used in a manufacturing process and may even include training methodologies. The Priceline patent and the amazon one-click patents were forbearers of a revolution that never before existed in U.S. patent law.

Frequently business-related systems are accompanied by software-related technologies such as an Internet patent, but this is not a requirement. If your product may be used in combination with a business operation, you may be able to include a business patent as well.

Machine Patents

When several components and elements are assembled to make a machine that does something useful, it is referred to as a machine patent. Examples may be an automated machine that forms the hole in a bagel, a high-speed plastic bag manufacturing machine, or a machine to sort plastic waste.

Frequently, machine patents have accompanying process patents. Many times, when new machine technology is being developed, several other subinventions and discover-

LEMELSON'S FIRST

The first patent issued to famous inventor Jerome Lemelson was on a "Flexible Manufacturing System." He later developed hundreds of other systems-related patents. Systems patents tend to be very common among prolific inventors.

PRIOR ART?

We know prior art articles cannot be patented, but some systems patents use prior art articles. If a company supplies prior art articles that are used "in your system" without your authorization, they may be liable for "inducement of infringement."

BILLIONAIRES CLUB

Countless numbers of Internet patents have changed businesses and made several billionaires, including Jeff Bezos of Amazon.com and Jerry Yang and his partner David Filo at Yahoo.com, who become billionaires in just a few years.

THE FIRST GAS MOTOR ENGINE PATENT

German inventor Nicolaus Otto was granted Patent No. 365,701 in 1887 for the first four-stroke gasoline engine.

ies are made. These individual aspects of the machine may also be qualified as patentable matter and receive a patent.

Composition of Matter (Chemical Compositions) Patents

Chemical compositions are usually scientific by nature. Excellent examples of patents granted in this field are the various types of plastics. A newer area of patenting that falls under this category includes genetic and biological engineering discoveries. If you are developing patents in this field, you may possibly be a scientist working for Lawrence Livermore Labs or perhaps Stanford, Cal Berkeley, or MIT.

There is, however, one application of chemical compositions those of us who are not so scientific-oriented may apply for in our invention activities. For example, if you have an invention that is made of molded plastic and your design uses a certain combination of plastic raw materials to create a certain effect, you may be able to patent it.

Let's say you have a new hairclip that must be rigid, but at the same time, must have some flexibility so the closure is durable and will snap closed consistently. High-density polyethylene plastic will provide rigidity and low-density polyethylene will provide the flexibility. Your challenge is to find the best blend of the two to get the desired outcome. In this example, you may find that a combination of 70% high-density and 30% low-density polyethylene provides an optimum performance. Thus, you can claim this chemical formulation in your patent application. Typically you'd claim a range, for instance 60% to 80% high-density to 20% to 40% low-density polyethylene.

Blending raw materials is also common for metal products, chemical formulas for cleansers, paints and so on, and of course, foods. However, for most food recipes, ingredients and their sources and processes are not always patented and are maintained as trade secrets instead.

Software Patents

This category refers to software as it is written and also the system or methodology in which it works—for example, the one-touch computer screen and its associated software system. Other examples include the sophisticated means of garbling Internet credit card transactions to protect the buyer or any number of the software programs you currently use.

Some of the more famous software patents are those that originally differentiated Macintosh with its "window" and "mouse" applications. Today, this concept has become the standard mode of operation for all computers. Another interesting computer patent is that of the "blinking cursor," which was developed by an independent inventor and licensed worldwide.

Software patents are more common today than ever before. When considering patenting in this field or making this part of your inventive matter, keep in mind that it is a rapidly changing field. Patents in this realm tend to become obsolete quickly, unless they are very broad and yet specific in scope like the one-touch screen patents.

It is a particularly tough field for patenting right now because the PTO is severely backlogged in its review of software patent applications. It may take three or four years to get a patent issued. The downside is that with technology changing so rapidly, that by the time the patent issues, it may be obsolete technology. Nevertheless, software patents are exceeding 20,000 grants per year and the stakes and patent values are incredibly high.

BAKELITE INVENTED

The forerunner to modern plastics was patented on December 7, 1909, in U.S. Patent No. 942,699. It was granted to a scientist L.H. Baekeland of Yonkers, New York.

20,000 NEW SOFTWARE PATENTS A YEAR!

According to Greg Aharonian of PATNEWS, the USPTO has been and will continue to issue more than 20,000 software patents per year . . . double that of just 15 years ago. Just a few years ago some believed that the field of software patents was so heavily covered that few new discoveries could be made! Now, the PTO expects to grant 15,000 a year for the next several years!

If your inventions require some form of software application, you'd be wise to include a claim on the use of the software relevant to the performance it produces. It's just one more form of patent protection to add to the portfolio.

Improvement Patents

Improvements to existing products or processes can be extremely valuable and frequently represent breakthrough opportunities. *Bona fide* improvements frequently make a big impact on our lives, just like the pneumatic tire did when it replaced hard rubber wheels, or like the impact Apple's mouse made on computer use.

There is an endless stream of improvement patents in our history—power steering systems, power brakes, convection ovens, automated banking systems, and even the self-opening plastic grocery sack.

Improvement patents rely on the existence of some other product or product attribute. If the other product or attribute was patented prior to yours, there could be the potential danger of the use of your patented invention inducing the infringement of the underlying dominant patent. This can be overcome when the holder of the new patent purchases the other product from an authorized manufacturer, thus implying license. The other way to avoid a problem is to license the other technology and then incorporate it in your product's design. Once the dominant patent's term has expired or if the maintenance fees go unpaid, there is no further obligation.

> **IMPROVEMENT PATENTS**
>
> Most patents today are for 'improvements' on existing technology.
> *–Michael S. Neustel,*
> *U.S. Patent Attorney*

More About Other Patents

Design Patents

Keep in mind that design patents refer only to ornamental design and not to functional design. They tend to be easy to design around. For example, assume that you have a design patent on a plastic kitchen hook that is used for hanging pots and pans. If it has certain smooth, rounded lines, it can be easily designed around by eliminating the smooth, rounded lines. Unfortunately, design patents generally do not afford much patent protection, but there can be exceptions.

The clever use of design patents is illustrated by Nike, the sports shoe company that has used design patents to its benefit over the years. Their purpose for using design patents is to keep others from cloning their attractive shoe designs. While not utilitarian, design patents certainly help preserve their original, proprietary look and have a positive effect on marketability. An example is the shoe in Figure 6.1.

Another example of the excellent use of design patents is Harley Davidson. Its new V-rod is an exceptionally good looking, futuristic motorcycle. Harley Davidson has protected the new V-rod with dozens of

Figure 6.1 Design Patent on Shoe

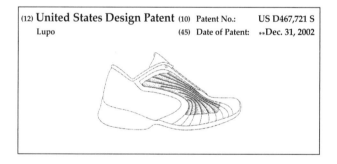

Figure 6.2 Design Patent for Harley Davidson

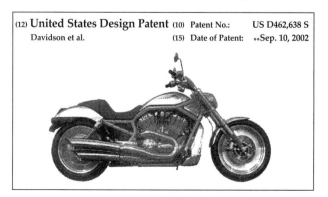

design patents covering the motorcycle itself and many of the individual components. The design patents are complemented by a slew of utility patents blanketing many new features, including a new form of radiator and muffler system. See Figure 6.2.

Utility patent protection is generally considered the best overall patent protection, but there are times when a simple design patent may serve the purpose. With a start-up company, or for those who plan to license their products, utility patent protection will most likely be essential. However, if you are an established company that plans to protect new designs like Nike and Harley Davidson have done, then design patent protection may give the desired outcome of protecting the product for several years to come.

After all, designs tend to change fairly quickly so a fourteen-year term on a design patent may outlast the product line. In the case of Nike, the product's design will probably not last 14 years. However, with Harley Davidson, you can probably bet that in 14 years the V-rod will be selling as well or better than it does today.

Plant Patents

Plant patents are granted to "botanical inventors" who create new plant variations asexually. Plant patents are commonly granted for award-winning roses such as Queen Anne, New Day and Pascali roses, or the many varieties of patented tulips. If you are creating patentable roses, tulips, or other flower types, you might consider working for one of the nurseries.

Plant patents are perhaps more common in the commercial trade—for instance, patented blueberries, blackberries, cranberries, walnut, almond, and avocado trees. All the various types have different traits that may have commercial value. For instance, some may grow to maturity faster and bear fruit sooner. Others may simply live longer. Yet others may bear fruit in colder weather, while even others may bear much larger fruit or seedless fruit.

Although photographs are not normally supplied with patent applications, the applications for plant patents are an exception.

Multiple Inventions

Filing One Application for Multiple Inventions

You can file one patent application that covers multiple inventions and inventive matter. By filing a single application, you can save a lot of money. You just want to make sure that the subject matter is reasonably related. Don't try to combine a new florescent light bulb with an ice cream manufacturing process, unless the light bulb's function is specifically a part of the unique process.

An important part of your strategy should include filing patent applications on all the inventive discoveries you've made. By doing so, you'll have a lot more success in your endeavors and you'll have a much greater deterrent to keep out competition. For example, if you invented a remote control device for finding lost keys, you may be able to file a single patent application that covers the following:

1. **A product patent application:** The remote control device itself

2. **A systems patent application:** The method in which it is used

---

BUDGETING

Filing multiple inventions in a single patent application is common practice. If you really want to string it out for a very long time, include a software application. The PTO is so backlogged in that field that it could take up to four years to review! Couple that with using the one-year rule, filing a provisional application and next filing the permanent application and you may postpone review for 6 years or more!

132

3. **A software application:** That sends out the signal and records the whereabouts

4. **A composition of matter application:** It just so happens that the remote control device must be of a nonmetallic substance that does not affect the microwave tracking transmission

5. **A second product patent application:** The stick on tracking device that attaches to a key

When combining related attributes such as these in one patent application, you will pay a single application fee. After the examiner reviews your application, he/she will respond by sending you a rejection based on your application containing "more than one invention" or "unrelated inventive material." The examiner will cite the different inventions and you can then select which one to prosecute first.

This action is referred to as a "divisional" patent application. In this type of application, the contents are "split-out" into two or more patent applications. When you file in this manner, the time required to review all the inventive aspects is lengthened.

Filing Multiple Applications for Multiple Inventions

Your alternative strategy would be to file up to five patent applications on the individual inventions. There are a few advantages to doing this.

First, patent applications will be easier to write and easier to organize. Second, they will serve as an excellent means to ward off competition. Reading the marking on a product stating "5 Patents Pending" is a lot more powerful than saying you only have one pending.

The downside is that it will cost a bit more, but it is still affordable, if you submit provisional patent applications first. The multiple patent applications will most likely be prosecuted faster, although it is uncommon that they are all examined simultaneously. They tend to be submitted to the same examiner or split between just a couple and examined in order, as opposed to all at once.

So it may result in a somewhat greater expenditure over the course of a slightly shorter time frame. But it's a powerful statement to make if you're entering a marketplace.

What You Do Not Want to Patent

There are several concepts and inventions you shouldn't patent, all based on either the ability to earn money (or the lack thereof) or the ability to have extended intellectual property protection. Here is a list of items you do not want to patent:

Trade Secrets

During the course of product development, some concepts and processes may be discovered that might be better kept as trade secrets. However, they must be tightly held or they will become public knowledge and anyone can use them. Those concepts that can be tightly held within a closed, discreet group or perhaps even in locked boxes can be more powerful than patents.

The chief advantage of a trade secret is that it may be used for a much longer time period than a patent. After 20 years, patents expire and anyone can practice the invention. On the other hand, trade secrets can be held for years, even centuries.

ROUND CHESSBOARD OR CHECKERBOARD

Design Patent No. D251826 was granted in 1977 to Norman Betros for a round game board used for playing Checkers or Chess. This obscure, yet novel, patent obviously did not take off in the marketplace. Upon looking at it, you would need to put your brain in a different perspective altogether. While the patent covers the design of the round board, an underlying utilitarian function is not covered under the scope of the patent. Whatever the solution may be, it would fall under utility patent protection, which is not covered by the design patent.

HISTORICAL FACT

A product that was originally developed as a cleaning compound has become a mainstay in almost every child's toy chest—Play Doh®. Today, its exact formula and process of manufacture is a trade secret. An independent inventor created it.

Examples of valuable trade secrets are Coca-Cola's formula(s) and Lea and Perrin's Worcestershire Sauce formula. It has been reported that a few individuals hold Coca-Cola's secret formulas in confidence. Each individual holds the key to only a portion of the formula and process.

Trade secrets are also common in manufacturing processes. These trade secrets may be more difficult to guard, since employees within an industry tend to move around and may divulge some of the trade secrets of prior employers accidentally or otherwise. Valuable manufacturing processes can easily make the difference between profit and loss. It is a common practice to guard them

Another area of substantial trade secret usage is in recipes and food formulas. There are two key elements that go into a recipe. First, there are the ingredients and the percentages of each used. Even the locale in which they were purchased will have an effect on taste. Second, there are the processes that are used to make the food product. This may include baking temperatures and times, aging processes, means and sequence of combining ingredients, and much more. Altogether, it's easy to see that there would be thousands of combinations for making a loaf of regular white bread.

One additional definition of the term trade secret applies to your products before and during the patent pending phase. While the patents are pending, they are effectively protected by trade secret. You can keep them as such by having those involved in the manufacturing and engineering processes sign confidentiality agreements. It's not absolute protection, but it certainly can serve as a deterrent, with the potential of some severe consequences if the secrets are revealed.

Fads

Pursuing a patent on a fad is a waste of time and money. By the time you receive your patent, the fad will have passed. The more important issue with fad-type product launches is to go out there and sell as much as you can as fast as you can. If it is a real gimmicky product, try to come up with an attention-getting trademark.

The only option you might want to consider that makes any financial sense would be to file a provisional patent application so you can post a patent pending notice. Then abandon it when the year is up. Otherwise, save your money and look for something that is worthy of long-term patent protection.

Low-Sales Volume

Products that will not generate significant sales will most likely not merit the costs of patenting. Even if you see factors that will improve the moneymaking ability of your invention, be careful not to invest a lot of time, money, and effort in something that will provide marginal income. It will also be extremely difficult to find licensees that will be willing to license or partner development. A small start-up company may be satisfied with $100,000 a year in sales, but small established suppliers to the local chain store trade would want a minimum of $1 million in annual sales.

Your medium- to large-sized licensees and sales agencies will not consider anything unless it has at least $3–$5 million potential annually and some won't even touch it unless it's $10 million potential or a whole lot more. In perspective, megagiant corporations like new product launches that are a minimum $50 million in annual sales and up.

Inventions of Insufficient Improvement

With an invention that insufficiently improves upon an existing product, there will be either low-sales volume because customers will not have enough incentive to switch, or the patent coverage will be so narrow that the invention will be easy to design around.

You'll be in a losing situation should you pursue the development of a new product that is of insufficient improvement. You probably will not earn much income, let alone true profit. And, even if you are successful and sales are strong, you'll end up opening up the market for a much larger competitor to step in and take over. Or, competitors may just roll back prices and run you out of business with low-priced sales promotions. They'll be able to withstand the lower profit margins for awhile, but will you?

Invention Disclosures

When you are ready to start the protection process on an invention, you'll want to begin with an invention disclosure. This is the best way to begin the establishment of your first-to-invent rights. An invention disclosure at least establishes your date of **original conception** and may establish **reduction to practice** if well written and verifiable.

An invention disclosure typically consists of two to four pages of text and drawings. It can be handwritten or computer drafted and printed. Disclosures are brief summaries and not nearly as long as patent applications. You don't need a functioning model at this time. All that is required is a reasonably good idea of how the invention will work.

Writing an Invention Disclosure

Try to write your disclosure as soon as possible after the date your discovery was made and validated. Here are the guidelines for filling out an invention disclosure:

- **Title of invention.** Descriptive titles are best—for instance, "Sanitary Pot Scrubbing Device." Do not use a title that might be a trademark and is not descriptive. For instance, "Potzella" may be the trademark of your new scrubbing device, but it describes nothing at all.

- **Background of the invention.** In a paragraph or two, discuss the products that are currently in use. For instance, in the genre of pot scrubbing, you may say that sponges, Brillo® soap pads, and assorted scrub brushes are most commonly used. You'll also want to talk about the problems associated with these prior art products. For instance, Brillo might clean well, but may scratch the surface, and brushes are cumbersome. Sponges don't clean as well as the other products, but at least they don't scratch the surfaces. Last, you can say that all of these cleaners tend to harbor bacteria after use because they retain water.

- **Brief description of the invention.** For clarity, try to describe the invention from two different perspectives. For instance, your pot scrubber may be described as such: "The present invention is a plastic cleaning device made from a netted fabric that dries out quickly and tends not to harbor bacteria." Then you may add, "It is a nylon mesh pad shaped like a ball and made from a continuous length of nylon webbing about 30 feet long that is folded back upon itself and bunched up." The first definition is more or less conceptual, whereas the second one is more concrete, physically describing how the product is made.

THE RIGHT INVENTION DISCLOSURE

The invention disclosure in the *From Patent to Profit* Resource Guide is the perfect tool to help you think about your invention in the right light. It follows the same context in the book, and its content is the foundation for future patent applications, both provisional and nonprovisional.

DISCLOSURES ARE IMPORTANT FIRST STEPS

Many lawsuits have been settled in favor of inventors who have been able to prove their original date of conception, along with their proof of reduction to practice. The only other factor is to verify that you, the inventor, did not abandon the project. Prepare a legal verifiable disclosure and keep it in a safe place. The Scientific Journal is perfect for tracking your day-to-day activities.

AN INTERESTING ALTERNATE USE

Play Doh was originally developed as a wallpaper cleaner. But when the wife of the inventor noticed her daughter playing with it—modeling it like clay, but without the mess—a light bulb went on. The cleaning compound was made in colors to make it more attractive to children and marketed as a safe, reusable alternative to clay. The inventor became a multimillionaire before he reached 27.

- **Detailed description of the invention.** This is the most time-consuming section. You want to explain in sufficient detail how your invention works. If your concept is just an idea right now, try to elaborate on how it would work as an actual invention. Remember, this is where you want to assess and validate all the inventive matter, including the product, the method of use, how it is manufactured (if applicable), and the composition of the materials (again if applicable). In this section, do the following: 1) describe the basic structure of the invention in a simple plan view or perspective view; 2) describe the inventive matter very clearly so there is no misunderstanding what you are claiming; 3) describe how it is used (a systems patent?); and 4) describe how it might be made or any other variations you can think of. Include numbers and lead lines on the drawings and reference the physical attributes that form the inventive matter and all those elements that are important to its function. It's important to know that if you have not adequately or accurately depicted how the invention works, your disclosure may not have validity.

- **Drawings.** No matter how crude, drawings are almost always necessary. It does not matter how well you can draw, as long as the sketches can be understood and the invention is adequately depicted. If you have any doubts, have someone assist you with some basic sketches. This person may also serve as a witness who validates your disclosure. If you're unsure of how to do the drawings, try to emulate the drawings on an existing patent. Chapter 7 also gives you the basic instructions for preparing patent drawings. If you are using the Scientific Journal (see appendix), just follow the instructions. In our Sanitary Pot Scrubbing example, your first sketch may show the following sequence: 1) a perspective view of the pot scrubber bunched up in a round ball; 2) a blown-up view of the netted mesh fabric showing its porosity; 3) a perspective view showing how it self-cleans; and 4) a view showing how it's bunched up and made. You may also show some variations on shape (a square one?), means of manufacture, and so on.

- **List the unique attributes.** Try to describe the unique attributes throughout your disclosure. After all, this is the basis for the granting of a patent. You may also list them in summary form after the disclosure is written. Doing this forces you to think about and define the important inventive matter that you'll be patenting. When writing comments in the disclosure, use words such as "unique," "faster," "easier to use," "more efficient," "convenient," etc. and then describe exactly why it is!

- **Other possible uses and variations.** Try to list as many possible uses and variations as you can imagine. Include alternate materials, ways of making it and different designs. If there is an overlap into other fields, include them in the disclosure. During development you may find some alternative uses that may prove to have more potential than the original intended use. If so, include these new uses in your log book as you reduce your invention to practice.

- **Name, address, date, and signature.** Print your name and address and those of all other co-inventors, if applicable. Include your phone number and e-mail address, which may come in handy if it is subsequently used as a disclosure to conduct a patent search. If all the inventors are not available to sign the disclosure, make reference to their participation in the disclosure and at least verify your signature. The others can sign later.

Validating an Invention Disclosure Document

There are several ways to validate a disclosure's date and content. Use one of the following methods. You need only one.

1. **Have the disclosure signed and dated by two reliable witnesses.** A witness must not be a party who may have an interest in the invention. A reliable witness may be anyone, such as an engineer, a college professor, a community businessperson, and so on. Anyone who stands to gain from the invention, like your wife, brother-in-law, or business partner, would be not be considered a reliable witness.

    When validating, show the witnesses the description of the invention so they understand it. Have them sign a confidentiality agreement before you disclose details. (See Chapter 10.) If you ever have to use them later on as witnesses in testimony, you want to make sure they can say that they understood how the invention works. This method of witnessing is usually the easiest way to verify the date of original conception.

2. **Have the disclosure notarized.** Notary laws are pretty tough and well-respected. In most states, notary laws are fairly strict and penalties against fraud severe. A notary's seal and signature on your document will probably cost $10–$15, but it serves as a reliable verification of your disclosure date. This method may be the quickest and easiest way of getting the date and invention validated, if you prefer to use a disinterested third party as a witness.

    The only problem associated with using a notary is that he/she may not understand your invention. It may also be difficult to prove that the notarized disclosure was not subsequently altered. To avoid this problem, make a copy of the original disclosure and have the copy notarized instead. It would then be obvious if alterations were made to the original.

3. **Mail the disclosure to yourself.** This is the least preferred method. If you absolutely have no other choice, then seal it securely in an envelope, take it to the post office, request a U.S. Postal Service date-time stamp on the envelope, and mail it to yourself. *Do not open* the envelope when you receive it. Use a copy of the disclosure for your ongoing development activities. Keep the date-time stamped copy in case it is ever needed. *IMPORTANT:* This method has been defeated in court. However, if you have a subsequent, immediate, and substantial paper trail that follows, it becomes just one component of many that can prove your position. The combination of the two has withstood legal scrutiny.

4. **Fax the disclosure to your patent attorney.** The date on his fax receipt will be a valid date of disclosure. You should ask him/her to let you do this at no charge, until you take action on the invention. The attorney can establish a file for your disclosures. Your patent attorney should welcome this approach as it's an indication that you are going to use him/her to file and prosecute the patent application.

    The U.S. Patent Office provides a service for registering a disclosure and charges a small fee (currently $10). The paper size of your disclosure must not exceed 8.5 inches x 13 inches, and you must include a self-addressed stamped envelope. You have two years to act on your disclosure before it is discarded. It is best not to file your disclo-

HISTORICAL FACT

Model airplanes began as a means for airline manufacturers to sell their real airplanes. Actually, patent models also become a means for marketers to sell their products as well. However, patent models are no longer required.

THE SYSTEM WORKS

Early in my invention career, I avoided a litigation incident with a company that was going to develop one of my inventions after I had disclosed it to them, but without compensation. The central issue was, "Who's idea was it really . . . mine or theirs?" In order to prove that I was the original inventor, I produced a copy of a dated, sealed envelope along with nine other copies of supporting documents, including my log and prototypes from my archives. Upon receipt, they immediately entered into negotiations.
–Bob DeMatteis

ALTERNATE APPROACH

Using the U.S. Patent Office for filing disclosures is the preferred method of many inventors including George Margolin. He says, "What's better than having it on record with the U.S. Patent Office?" Even if they do throw it away after two years, you will still have a copy and a receipt from the PTO with the date it was originally received. It's a good argument.

sures with the U.S. Patent Office. Two years can be limiting. For about the same amount of money, you can have your invention notarized regardless of paper size.

There is no time limit on how long you can wait to file a patent application after you have written a disclosure. However, waiting too long to file may result in complete loss of rights. If you are busy finalizing the development of other inventions and patents and arranging their marketing, long delays can be justified—even delays of several years. The only danger is that in the interim someone else may begin the development and patenting of the same inventive material and challenge your first-to-invent position. See "Interference" in Chapter 5.

During your initial stages of disclosure it is most important to maintain a journal (or logbook) of your invention's development and all drawings, sketches, and prototypes. A well-maintained journal is the primary source of proving your first-to-invent position. You can read more about the various types of logs, drawings, prototypes, and records you should maintain in Chapter 2.

Using Disclosures to Clarify an Invention

Disclosing and summarizing your invention helps you to understand the invention and even helps establish the direction of your developmental process. Writing a disclosure may also help solve some basic problems associated with it. It almost always gives you a clearer picture of your objective and the outcome you seek. As you know, there are several forms of patent protection you can secure while preserving your cash flow. As you write your disclosure, consider as many of these as you can.

With experience you may consider reformatting the disclosure into a brief, but somewhat descriptive, summary that can be used in discussions with development partners. Within your company setting, confidentiality should already be established. If you're going to use it in discussions with outside engineers or manufacturers, then you'll want to use a confidentiality agreement to preserve secrecy. Read about using confidentiality agreements in Chapter 10.

Reducing to Practice and Diligence

Reduction to practice refers to showing (or proving) that your invention works the way that you say it does. Along with recording your date of original conception, this establishes your first-to-invent rights. The more important date is the date of reduction to practice because that's the date the invention is considered a proven, verifiable concept. The three most common methods of verifying reduction to practice are: 1) adequate descriptive drawings, such as in a journal, invention disclosure, or other engineering drawings; 2) the filing of a patent application that effectively cites the inventive matter; and 3) the building of a working prototype that proves the concept.

Using a log book, such as the Scientific Journal, is a more common method of verifying reduction to practice and establishing this important fact. In Chapter 2, the use of a journal to record valuable design and marketing concepts and to protect your first-to-invent rights is discussed. In other words, a log book protects your patent filing rights.

However, you can lose these rights if you are not careful. First, you want to make sure that you record your journal entries on a regular basis. A continuous stream of dated recordings verifies that you are working on the project and that it has not been abandoned. This is referred to as ***diligence***. If diligence has not been demonstrated in your product development and there have been long lapses of time and journal entries, you may be subject to ***abandonment***, which essentially means you gave up on the project.

The allowable time frame in between journal entries is not established in statute and is more or less arbitrary. At times, an inventor may have large breaks in development due to other work-related responsibilities or with a female inventor, due to an event such as pregnancy; these breaks are justifiable. You would most likely not lose your rights because of these situations, but you'll most likely need to prove that you intended on picking it up again.

Should you have any concerns about whether or not you have abandoned your invention's development, then you may be vulnerable. Should you have some excellent record keeping to verify your development and position, you should feel much more comfortable. Last, if you've effectively reduced you invention to practice and verified the date, you should be yet more comfortable with your position.

Patent Searching

Sometime soon after your disclosure is written, you will want to conduct a patent search. Nothing is more discouraging than to be well-advanced in your product development or to have even filed a patent application only to find out later that there is already an existing patent, or your patent will have narrow, ineffective coverage. A patent search saves you time, money, and a lot of anguish.

A patent search also helps support the validity of your patent. If you are uncertain about the novelty of your invention, or inexperienced in your field, a thorough patent search will be an important next step so you don't spend too much time and money on an innovation that's not feasible to produce or may have weak or no patent protection.

A patent search can indicate if there is prior art that will affect your patentability and can help to determine if your invention may infringe on existing patents. You must know this early on, as this can be extremely embarrassing later. Can you imagine going into production or licensing a manufacturer and after a year or so discovering that your products are infringing another patent? This is not only illegal but it could also destroy everything you have been working on, including your credibility.

A patent search will determine if related art exists that may cause the patentability of your invention to be narrow in scope. In other words, you may be unable to get the broad coverage you want and may have to settle for a narrow scope. But what if that makes your product easy to design around? Is this going affect your marketing effort? It will almost certainly affect your ability to license your patents and related know-how. If it is easy to design around, it simply may not be worthwhile to patent. You can read more about narrow and broad patent claims and coverage in Chapter 7.

Patent searching can be done in one of several ways. There are benefits and drawbacks to them all. The various approaches are:

1. **Internet searches at the patent offices.** You can connect directly to the U.S. Patent and Trademark Office website, the European Patent Office, and any number of others. The benefits are that you'll have fairly instant access and it's free. The drawbacks are that you may not know what you're doing! A misguided attempt could cost you dearly in time and money.

WARNING

If you have made the unfortunate mistake of having a patent search done by one of the invention assistance companies, you would be wise to have another one done and reviewed by a qualified patent searching expert or patent attorney. Unfortunately, many of the searches done by these companies are inadequate, which is detrimental to the inventor's best interests. Many of them even overlook the utility patent potential of the invention!

PATENT SEARCH STARTING POINT

If you're not sure where to start on your patent search, start at the home page on our website frompatenttoprofit.com. The main searching sites are just a click away.

2. **Internet searches using search engines.** IPsearch, Delphion, and IPsurf may be much more thorough and may include other forms of prior art searching outside the patent offices. The downside is that they can be somewhat costly and you'll need to have some experience to know what you're looking for.

3. **Internet searches using software.** This is another alternative to consider. The easiest to use software is PatentHunter®, which connects to the Internet and goes through the "back door" directly to the patent office data bases. It then allows you to quickly download patents (at no cost), organize them in folders, and even mark them up if you'd like. The PatentHunter format shows you how to search and makes searching fairly effortless. The downside is the upfront cost, but it is a small price to pay if you'll be doing a lot of patent searching. Most sites charge up to $3 or more for patent copies; with PatentHunter, you can download as many as you wish for free. Check our website for more information.

4. **U.S. Patent and Trademark Depository Libraries (PTDL's).** Throughout the United States there are U.S. Patent Office libraries with experienced librarians who can assist you in patent searching. However, their actual connection to the patent office data base they use is the same you'd use if patent searching on your home computer. And you'll be doing the searching yourself. This may not be a negative, since once you have an understanding of the process, you'll be more proficient at home. If you're inexperienced, it's advantageous to have some help.

5. **Employ a search professional.** You can hire an attorney or a patent search professional to do the search for you. This generally costs about $300 and up. Patent attorneys generally cost substantially more since they have to outsource the patent search to begin with and then will usually provide some of their time to give you a legal opinion. Any patent search that is done too inexpensively should not be considered. The most effective patent searches are probably those done at the U.S. Patent Office Library in Arlington. The USPTO's library is well known throughout the world as the largest single information source anywhere, period. It includes patents from every country in the world, making it one of the best sources for international searches. It also includes the largest collection of supplementary reference materials, including scientific documents and databases outside the realm of patents. In addition to the cost of a professional search, there is the time factor. It could take up to four to five weeks to complete.

What and Where to Search

The best starting point is your own computer and searching on the Internet or via PatentHunter. The best methodology to use is a word search. A Boolean search methodology is also helpful to know.

Internet searches cover word searches going back to the early 1970s, and images of the issued patents before that date. When searching you'll probably want to search all years. Many modern inventions will have some related art from decades, or even a century ago. Don't be discouraged if you discover age-old prior art since it probably won't pertain to your exact inventive matter, but it's best to know up front.

Many say that you should also search the various fields to assure that your discovery is being conducted in the right ones. But with word searches, they'll overlap into any given field anyway. There are about 125,000 product classification groups and frankly you don't need to spend your time learning which ones do and do not pertain to your invention.

You'll also want to consider worldwide searches if you believe your invention may have had some influence overseas in the past. Keep in mind that the U.S. is by far more progressive when it comes to inventions that save time, are more convenient, and so on. Improving productivity and saving time is just not as important in most other countries. If you're in doubt, search internationally. If nothing comes up, it confirms your beliefs.

Last, published patent applications should be searched throughout the world. Most countries publish them after eighteen months. The U.S. gives an option to an inventor to have his/her applications published at eighteen months if a patent has not yet been issued.

INFRINGEMENT SEARCH

When inventing, inventors also need to be concerned about infringing on existing patents. If a patent is close to your invention, have your patent attorney review it for potential infringement issues.
–*Michael S. Neustel, U.S. Patent Attorney*

How to Search

When doing your own search, here's a reliable straight-forward method used by most searchers and patent attorneys.

1. **Start out with a word search.** Use those words that identify the function and usefulness of your invention. However, it is critical you don't overlook similar, related words. For instance, the word "glue" could also be described as:

bonding means	binder
adhesive	stick
adherent	fastening material
tackifier	epoxy
paste	gum
sticking substance	mucilage
conjoining compound	means of adjoinment

    How many more can you think of? Can you see that the process may be tricky? There are two good methodologies you can use to find the various words to use in your patent search. First, you can use a thesaurus, either in book form itself or use the one on your word processor software. Second, start your search with the words you know and when you read some of the patents that are identified, search for additional words and terms. You must also search phases such as those listed in the example above. A phase such as "means of adjoinment" may be more commonly used than a single word such as glue or adhesive.

2. **Review previously referenced patents.** After reviewing the patents you've uncovered in your word search, you'll also want to review those patents that it has referenced. All patents have other patents that have been referenced on the cover page. If you're searching online at the U.S. Patent Office, it's very easy to just click on the related patent and then search it. If you are using a software program such as PatentHunter, you'll want to download the other patent and then read it. Downloading the patent may actually be faster than trying to click on a link at the PTO's search site, which requires clicking a second time to view the images.

PATENT SEARCHES AS
PART OF A PROPOSAL

When approaching
manufacturing and
marketing partners early
on in the development
process, a common
mistake made by many
first-time inventors is to
say, "I have a patent
pending," which they
don't. We know this is not
a wise thing to say (in
fact, it's fraud). A simple
alternative that carries as
much—perhaps even more
weight—is to say, "Our
patent search has revealed
that we should be able to
get broad patent protection
for the new innovation.
My attorneys and I are
incorporating this broad
scope into the patent
application." It is honest
and a more powerful
position as well.
Remember, we are a
first-to-invent country,
not first-to-file.

3. **Review more recent patents.** After reviewing the prior patents that your initial word search has divulged, then you'll want to also review any recent patents. These would typically be more patents that reference the ones you've just reviewed. In other words, you're now searching for the more recent patents, instead of older ones.

   This method is not a guarantee that you'll find all the related patents, but it is very effective. The crucial factor is identifying all the various words that may identify the inventive matter. A problem may arise with foreign patents that are in foreign languages. However, be aware that most foreign patents of any importance are also patented in the United States so they will be traceable in English.

4. **Review published patent applications.** You want to make sure your patent searches include the published applications that are pending. Most Internet websites and patent search engines offer this service. The software PatentHunter will do it automatically. At the U.S. Patent Office, you'll have to click on the link to search patent applications, instead of granted patents.

Searching for Prior Art

In addition to a patent search, a prior art search is also important. Regardless of whether a previous patent that covers your inventive matter exists, a product that is already on the market will invalidate your ability to get a patent. This product may be the same or similar, and it may not even be on the market any longer (for whatever reason).

The best way to search current and recent products is via the Internet. The three best search sites for prior art are www.hoovers.com, www.google.com, and www.yahoo.com. You can also search specifically on any major company website that specializes in the field. For instance, you can search Toys R Us and KB Toys for toy items. Trade associations in the field of your invention may also be good sources to contact. Perhaps one of the very best, however, is the marketing expert who will be heading up the sales of your invention. Invariably, the more experienced ones have a broad knowledge of all the various products in the field.

Previous publications revealing the inventive matter can also affect your patentability. This is more uncommon, but it can happen. Try searching trade associations and the archives of trade journals in the field of your invention or try amazon.com. Trade magazines may also be reviewed regularly by engineers and R&D personnel in your company. So you may ask them as well.

Advantages of Doing a Patent Search

All patent searches are preliminary by nature—that is, they are subject to being searched by the patent examiner who will review your application. While doing your own search is not a guarantee you'll find all of the prior art, it does significantly improve your potential for success.

By reading the patents related to your invention, you can use the information to your advantage in the following ways:

- **Prove that your invention is unobvious.** In other words, no prior art pointing toward your inventive material improves your odds that it will be considered unobvious.

- **Prove the novelty of your invention.** A search that reveals nothing similar to your invention helps to prove it is novel. Patents that point to the inventive matter in your application as being inferior or not possible (based on the prior technology) also improve your novel position. Revealing how your invention or new process or material is superior or more viable than those before it should provide solid claims.

- **Discover other fields of uses for your invention.** A patent search frequently reveals new opportunities in other areas and in other industries that you may not have previously considered. You may want to try to broaden your patent coverage.

- **Help guide the development of your invention.** You may learn what to do or not to do from reading about related patents. Reading patents in your field can be a tremendous help in learning about new or related technologies. You can literally learn about problems associated with the various processes and what to avoid and what methodologies are superior.

- **Counter attack similar patents.** This can be an important step when writing your patent applications. A patent examiner will most likely not object to prior art you cited in your application, if you reveal it in the body of the application and carefully describe why it is inferior to your invention. As you do, you will, of course, describe how your invention overcomes the problems associated with the cited patent. The examiner's acceptance infers that he/she agrees that your inventive matter is different and that it's not in conflict.

- **Understand the terminology in your field.** Reading several prior art patents will help you obtain a broad knowledge of the technical vocabulary that is used in the field of your invention. This will be important when discussing your product with prospective partners, engineers, financiers, and licensees.

- **Use a prior art patent as a model for your application.** Patents are not copyrighted documents and anyone can copy and use them. When reading the patents you've found in your patent search, select one or two you'd like to use as a model to write your own patent application. By copycatting an existing patent you can save several hours writing time and a whole lot of money.

How to Read a Patent

Arrange your schedule so that you can read the patent search without being interrupted. If you need a search expert, send a letter with an analysis of the related patents with your search. Read it carefully since you are probably more experienced in the field of your invention than the searcher. In addition, you are the one who must make decisions about how to proceed.

Follow the sequence below when reading.

1. **Read the title and look at the picture on the cover page.** If it looks like it might be related, then,

2. **Read the abstract.** If the invention appears to be related to your invention, then,

3. **View the drawings in the patent.** Look at the referenced drawings and see if they appear to be related. Next,

4. **Read the field of the invention.** This is on the first page of the written subject matter. If it still looks like there might be a conflict,

5. **Read the summary of the invention.** This should tell you specifically how the present invention solved the problem it previously described. If there's a conflict now,

BECOME AN EXPERT

Patent searches help you become an expert in your field. You'll learn more about the state of the art, who's who, and the terminology used in the industry through searching than by any other means and in a fraction of the time.

BE CAREFUL

Short claims with just a few elements tend to be more powerful with a broader scope than long ones, but don't be misled. A long claim may list ten elements, nine of which are commonly used elements in every product in the field!

6. **Go to the back of the patent and read the claims.** You'll want to read every element in a claim very carefully to understand the patent's scope and its legal clout. Claims are not necessarily easy to read. However, if you dedicate some time to reading them, you will soon understand their unique way of describing inventive matter.

Reading Claims in a Patent

When you read claims in a published patent, you're most likely doing this to determine if any of them "read on" your invention. In other words, is the inventive matter you want to pursue already claimed in one of the patents in the patent search?

Here is a three step approach to understanding how to read a claim:

1. **Claims contain several elements.** In order for you to infringe, you have to be using all of the elements. Therefore, if there are four elements in a claim, and you're using only three of them, you do not infringe. For example, a claim in a manufacturing process might include the following elements: 1) one or more lanes of continuous plastic film; 2) said lanes of film gusseted at a predetermined point; 3) said gusseted film printed on at least one side, and; 4) said printed film passing under a sealing means, sealed at one end, and converted into a bag. If your invention uses all of these elements but is not gusseting the film, it doesn't infringe. Supplying pre-gusseted film would also obviate the infringement.

2. **If you think a patent's claim may read on your invention, go back to the body of the patent and read it more carefully.** Look for the exact language that you are concerned about and make sure you are interpreting it correctly. For instance, in our last example, the patent's claim might describe three lanes of film side by side, whereas you might have interpreted it as three lanes one on top of the other. Further research on your behalf reveals that three lanes one on top of the other has been considered impractical in the industry. Thus, you've got two important things to consider. Either your technology may not be well-received because it impractical, or you may have come across a very interesting solution that could dramatically increase output.

3. **If you're uncertain or have any doubt, consult your attorney.** Don't assume anything. If you have any red flags, you must consult your attorney now to make sure there are no problems later.

There is a bit of art to reading patents and claims. Reading claims in business-related patents is quite different from reading claims about machinery or toys or consumer products. The key to reading claims is understanding the operative words and knowing their true meaning.

Should you require more specific information on how to read a patent in a specialized field or perhaps how to read patents for other purposes, such as to determine validity, state of the art, and so on, read *Anatomy of a Patent*, which explains in great detail how to read the various parts of a patent and how to decipher them. Reading claims is an important part of your research and may directly affect your ability to obtain the patent coverage you want. Understanding claim language in the field of your invention will also make you an expert.

Dealing with Prior Art or an Infringing Patent

If you find prior art or believe you may be infringing an underlying, dominant patent, you have four choices: 1) You can improve upon the existing patent, which may not necessarily mean you'll still not be infringing; 2) You can design around it and make sure you don't infringe; 3) You can purchase the products falling under the infringing patent or license them from the owner; or 4) You can drop the project altogether.

Before you decide to take action, analyze the conflicting patent(s) to determine:

- What's the value of the dominant patent? How is it different from your invention? Does your invention have something that makes the product more marketable?

- What are the patent's strengths? This analysis may also reveal its weaknesses.

- What words and phrases specifically included in the claims have narrowed the scope of the patent? For instance, if a claim is limited to some form of the use of adhesives, how can you avoid this altogether? Would it make more sense if the elements were prefabricated so adhesives weren't necessary?

Every claim has certain operative words that it must have in order to have been granted. These words are key to your understanding of potential infringement and how you can avoid it. For example, let's say we have identified the operative words and phrases used in a claim for a plastic grocery sack with the following four elements:

1. Handle straps extending upwards from, and adjacent to, the bag mouth
2. Apertures intermediately located on strap handles
3. Upwardly extended, centrally located tab
4. Tab defined by a severable line between it and the bag body

As you scrutinize the claim, you find that certain operative words have specifically defined the scope of a patent. If so, these words can indicate how you may discover improvements or a means to design around the patent.

You now have four or more possible alternatives to design around this bag style. Hopefully, you can design around at least one attribute that will give you a competitive advantage. Four potential design-arounds are:

- Handle straps that do not extend upwards from the bag
- Apertures that are not located in the handle
- A tab that does not extend upwards (or perhaps two tabs laterally spaced)
- A tab that does not sever itself from the bag

When reading claims and attempting to design around other patents, apply the teachings of the Doctrine of Equivalence discussed in detail in the next section. If you decide to design around an existing patent, review Chapter 3 to help you decide on the new inventive attributes you may want to apply.

Doctrine of Equivalence . . . Literal vs. Figurative Interpretation

The Doctrine of Equivalence as recently clarified by the U.S. Supreme Court in the Hilton-Davis case states that:

If the elements of a product do not literally infringe a patent's claims, but essentially serve the same function, then it infringes anyway.

OH OH!

Unbeknownst to prolific inventor Jerome Lemelson, his first invention for a "universal robot" that could rivet, weld, measure, drill, and pick up objects and reposition them was found to infringe a patent application filed just two weeks before his. Lemelson immediately set out to work on a second application of the robot, his "flexible manufacturing system," which became today's automated machine shop.

A DRIVING FORCE

Lillian Moller Gilbreth (1878–1972), the mother of 12 children, had good reason to improve the efficiency and convenience of household items. She patented many devices, including an electric food mixer, the trash can with the step-on lid opener, and an ergonomically efficient kitchen that benefited homemakers and disabled persons.

Patents can be interpreted literally or figuratively. If you plan to design around an existing patent, you must thoroughly understand the scope of the infringed claims.

We now know that relying on the literal interpretation of claims can be dangerous. Generally speaking a broader interpretation of their meanings should be used. Nevertheless, some patent claims might be interpreted more literally than other claims, in particular if they were subjected to the examiner's demand that they be restricted as such in order to grant the patent. Usually this is revealed by reviewing the file wrapper, or the file folder that contains the patent's documents during prosecution.

In applying the Doctrine of Equivalence to the preceding potential design-around attributes, a question arises about serving the same function. You would be wise to get an attorney's opinion. Your objective is to create a new, improved functionality from your design-around. Hence, it should not infringe the prior art patent.

Here are a few examples of literal versus figurative interpretation:

Literal interpretation	Operative word	Figurative interpretation
a round hole	hole	any cut slit, shape, or aperture configuration
dial with a readable means	gauge	any means to measure—an idiot light, etc.
pneumatic tire	tire	any wheel—hard rubber, metal, or otherwise
treated to 250°	treated to 250°	treated to a temperature near 250°
round	circular	oval, elliptical, oblong

Commonly used phrases that reflect the Doctrine of Equivalence are "under the scope of," or "in the spirit of." The most important point to remember is that you can also use the Doctrine of Equivalence to your advantage. If your patent has well-written claims, you will be in a superior position to defend it.

When is a Search Not Necessary?

You may not need to do a patent search if your invention is so unique that its novelty is apparent. Some examples of unique inventions would be Eli Whitney's cotton gin, the first laser, the television picture tube, and many Internet-related inventions.

If you have been developing patents related to a certain invention or field for a few years, you may have accumulated a substantial library of related patents (or a folder on your computer if you use a program like PatentHunter) and an understanding of the field. If you pay for a search, you would be paying for copies of documents you already have, in a field in which you are most likely already the leading expert.

Anyone inventing on the leading edge of a technology knows it. They know the chances of finding prior art when they've been the one setting the state of the art is pretty slim. If this is you and you have not conducted a search recently or want to get some current knowledge about your competitors, some searching might be in order.

Writing and Filing Strategies

It's rare that an individual will come up with an invention and will immediately file a patent application. Frankly, this is almost always a poor strategy to take. There are several factors you should consider before you pursue patenting, including when to write, when to file, who should write, and who should file. Do you file a provisional patent application first or the permanent regular application? You'll also want to consider protecting your worldwide filing rights.

When to Write a Patent Application

After you do a patent search and are confident you want to proceed with development, it's time to start thinking about writing a provisional application. It's best to begin writing during the early stages of invention development because you probably won't be completing your invention in the next few days or perhaps even weeks. Sometimes it takes months, depending upon your strategy and if there is any urgency to get a patent issued.

There are two good reasons why writing a patent application early is helpful. First, it just makes sense to dovetail your patent writing with your product design and development. It makes your ability to describe the inventive material much easier.

Second, it allows you to say to prospective partners and licensees, "We are in the process of writing patent applications." In some ways, this is more powerful than saying "We are patent pending" because there's no telling what might be going into the patent application you're writing, but if one has already been submitted, it can't be added onto.

Who Writes the Patent Application?

There are three choices: you, someone associated with you, or a patent attorney. Most experienced inventors agree that it is best for the inventor to write most of the patent application. In the case of a provisional application, an experienced inventor can write the entire document. It is best to begin the writing process during development and complete it later when the inventive matter is best understood.

There are significant benefits to writing your own patent application. Four important benefits are:

- You can learn a lot about patents and patenting.

- By writing it during development, you learn more about your invention and potential design-arounds.

- During the writing process, you make new discoveries.

- Writing your own application saves you thousands of dollars.

Most patent attorneys and experienced inventors agree with the patent-writing approach described herein. It really enhances an inventor's ability to include the best, most inventive material relevant to the invention's success and to the patent's coverage and integrity.

If you feel uncomfortable about writing an application, at least consider writing a draft using the standard writing format covered in Chapter 7 or have someone you trust help you. If writing is just not your thing, you have three other choices:

- Make a videotape of the invention and narrate it. An associate or attorney can draft the application from the video.

- Describe the invention in an audiocassette. An associate or attorney can use it to draft the application.

- Schedule an interview with an attorney and he/she can either record the meeting or take notes.

ALTERNATIVE APPROACH

At times, you may elect to file a provisional patent application, instead of trying to get a confidentiality agreement signed by a big corporation. Most giant corporations will not sign confidentiality agreements. With a provisional filed, you are in a patent pending status; hence, you have the ability to reveal your invention to corporations with less fear of theft.

"WHEN?" . . . "BEFORE!"

All the patents I own have been qualified and licensed before or during the patent pending stage. In this manner, there is a return on my investment before any major out-of-pocket patenting expenses. The key is qualifying marketability before you file. Aggressive licensees like it, too. This way, they'll be ahead of the competition.
–Bob DeMatteis

USE AN EXPERT PATENT ATTORNEY

It is unimaginable to apply for the permanent patent application without using a qualified patent attorney, or in other words, to patent it yourself. You must rely on your legal experts to proofread your final drafts, write the claims, and file and prosecute the permanent patent applications. Relying on anyone else, including the attorney of an invention promotion company is a huge mistake. You might be a great inventor, but you're NOT a patent attorney or an expert in patent claim writing.

TAKE A LEAP

Most good patent attorneys will tell you that inventors seldom know what the INVENTION is. They may know what they meant to accomplish—what problem it was meant to solve—but the INVENTION itself usually remains a mystery to them. A good patent application provides the framework on which to hang together those aspects of what the invention does and how to show its novelty and usefulness. The application illustrates the inventive leap that goes beyond what persons skilled in the same art had done and had been able to do before the creation of his invention.

–George Margolin

Before a patent application, provisional or regular, is filed, it must be edited and scrutinized carefully for errors. A simple overlooked detail may change an entire meaning that could affect the integrity of the patent. A misplaced period or improper punctuation could change an entire chemical formula or give a sentence an entirely different meaning. It could invalidate the patent. Two examples of sentences conveying entirely different meanings are:

> **The formula uses a half-processed protein.**
> **The formula uses a half processed protein.**
>
> **Do not touch the part or roll in the solution.**
> **Do not touch the part, or roll in the solution.**

Even if you have drafted the provisional or permanent patent application, your attorney still plays an important part. He must review your draft, edit it, and make the necessary corrections to ensure you'll have a broad scope. After your draft has been edited by your attorney, you must still proofread it to make sure the content is correct.

Experienced inventors agree that your attorney or patent agent must prepare and file the final draft of the regular, nonprovisional application. It is particularly important to have an attorney write the claims on the regular, nonprovisional patent application.

The claims are the most important part of the patent application. They must be legally written and must properly and adequately reflect the inventive matter you want protected. It is best to have an expert write them. Most patent attorneys concur; this is where they earn their money . . . by writing bulletproof claims and getting you the broadest coverage possible.

If you are going to file the patent application yourself and rely on your own ability to write the claims, you are taking a risk. If this is your choice, you should first read *Patent Writer* (see appendix) so you'll at least have a fighting chance. However, to forego an attorney in your permanent patent writing and filing activities will substantially, if not entirely, destroy your chance to succeed.

Besides, if you develop your invention according to the guidelines in this book, you should have a cash flow before any major costs of patenting are incurred. The cost of hiring a patent attorney is therefore moot. When it is time to write the regular, nonprovisional application with the legal claims that will give you twenty years of protection, make sure your attorney does it.

About Filing *"Pro Per"*

Experienced inventors and businesses invariably use patent attorneys to file and prosecute their patent applications. Another good reason to hire an attorney to prosecute your patent application is that when infringement occurs, you will have an attorney on your team who intimately knows your patent. During the writing and filing process, your attorney will become an expert on your patent's subject matter and its field. If you have to hire another legal firm to defend you, their review and evaluation of your position could be costly. The patent attorney who knows your patent is the best individual to help you when infringement occurs.

There is a law that says that an inventor can file his/her own patent application, referred to as filing *"pro per."* The law further states that a Patent Office examiner must help write at least one claim. The problem with this law is that the claim the examiner may help you write is usually very narrow and thus ineffective. You already know that you can most likely include several claims and inventive matter in a single patent application. To limit yourself to one narrow claim is a huge mistake and a waste of time and money.

By filing yourself, you also take the huge risk of not knowing how to respond to the PTO examiner's office actions. While filing *pro per* may sound appealing, it is foolish to pursue this avenue. Do not patent it yourself. Your invention's success, your patent scope, and your future depend on it. Do it right and hire a competent patent attorney. It'll save you time and money in the long run.

Filing a Provisional Application in a Timely Manner

There is some disagreement about the best time to file a provisional patent application. For the most part, your timing depends on: 1) how fast the invention development is proceeding; 2) whether or not the invention's marketability is confirmed; 3) when a product launch is anticipated; 4) your manufacturing and marketing strategy, and 5) your finances.

A few say that since provisional applications are so cheap to file, they should be filed right away, regardless of the content. Some even say they can be filed instead of recording an invention disclosure. Frankly, this makes little sense. Why waste the time, let alone the money, filing patent applications on unqualified inventive material? As an inventor/product developer, you can spend your time much more wisely pursuing your inventive efforts and zeroing in on only those that prove themselves out. When new inventive matters are discovered, they too can be qualified and new provisional applications can then be filed.

A few attorneys claim that early filing of the provisional application is an administrative affair to convert it to a permanent, nonprovisional application and get the claims they want. This not only makes no sense but is also misleading and obviously the words of someone who doesn't understand the inventive process. The facts are: 1) a provisional application that describes an invention that has been subsequently designed around can't be modified to include new subject matter; 2) designing around your original concept happens all the time—in fact, more frequently than not; 3) the time lost in pursuing the writing of a patent application that is going to have to be abandoned leaves the inventor vulnerable to others who may be passing him/her up in the inventive process, and 4) the interference issue, which is the only real argument for the fast filing of any patent application, is weaker today than ever before.

You have to remember that in the event of an interference proceeding, the more important criteria is the date of original conception and the date of reduction to practice, both of which should have been recorded and verified prior to filing a patent application.

It makes no sense to rush to file patent applications on phantom inventions.

The only reason to consider filing provisional applications in a shotgun manner is if this is your strategy and you have a budget to support it. If filing a slew of provisional applications is your strategy, and you're deathly afraid of the minute odds of encountering an interference, then be prepared to spend a lot of money chasing and administering your patent applications and a lot less time on doing what you do best: inventing!

Hopefully you'll not feel that you have to rush to file on inventive matter you haven't

GOOD LUCK!

I always tell inventors: "Oh, yes. You certainly can file your own patent application. The laws and rules clearly permit this. Laws and rules also, by the way, allow you to perform your own neurosurgery." Patent prosecution is said to be the most complex area of the law. The pass-rate for agents and attorneys attempting the Patent Bar Exam can be as low as 30%, and it takes a professional patent examiner around six or seven years to become proficient enough to make unsupervised judgments on the patentability of patent claims. "Sure, write your own claims. Who says you can't?"

–Don Kelly, Intellectual Asset Management Associates, LLC

THE MODEL

The U.S. patent system has worked well for more than two centuries and is often referred to in the context of the American Dream. I've had the great honor of getting to know many of the world's greatest and most revered contemporary inventors. I've found they have at least three things in common, apart from their membership in the National Inventors Hall of Fame: 1) As kids they were lucky to have a dedicated teacher who somehow inspired in them a lifelong interest in science and technology; 2) As patent applicants, they never, ever dreamed of filing their own applications; and (3) They always did, and still do, count family more important than work.
–Don Kelly, Intellectual Asset Management Associates, LLC

validated. Almost all experienced inventors (and almost all patent attorneys) agree that you should wait until the invention is sufficiently developed to qualify the basics—manufacturability, marketability and performance.

There are three excellent reasons for this approach. First, during development, an inventor tends to better understand the inventive material after it is somewhat developed. That's not always true at the beginning. The outcome is more accurate and broadly describes attributes of the invention itself. Second, by waiting until sometime before product launch, before the invention is publicly disclosed (for instance, offering it for sale), an inexpensive, provisional patent application buys an entire year's worth of protection before incurring the high cost of filing the permanent, nonprovisional application. Thus, your finances can be spent towards product development and marketing instead. Third, filing one or two qualified provisional applications, instead of several shotgun applications, costs less and saves time.

In addition, if you've been writing your patent application since the early stages and modifying and adding content as changes and improvements to the product are made, then you've effectively established as early a priority date as possible. It's verifiable too. Just keep your patent application drafts and print them out on a regular basis.

Who Files the Provisional Patent Application?

Either you or your attorney files this document. It's best to spend a little more and have your attorney do it as it will then be on the firm's docket and you will automatically be apprised of the one-year expiration date. Usually your attorney will notify you on or about the ninth month so that the legal claims may be drafted. If you've filed more than one provisional application, there may be the additional consideration of combining them into a single permanent patent application as well. This will take time to sort through the inventive matter and to blend it all together into one easy-to-read document.

If you are experienced, you can file the provisional patent application yourself. If you do, it would be wise to send a copy to your patent attorney and have the expiration date put on the docket. Once again, you'll automatically be notified, in that ninth month or so, when it's time to prepare the formal, permanent application. If you rely on your calendar, make sure you provide sufficient time to notify your attorney of the upcoming conversion. Do this during the eighth month or so, as your attorney may have to schedule the claim writing and conversion into his/her schedule.

When to File the Regular Patent Application

After you file your provisional, you have up to one year to file a regular, nonprovisional application. Remember this one-year rule also applies to the time you have to file a provisional application after the first public disclosure. You must file either a provisional application or the permanent application before the end of one year after the first public disclosure. As previously discussed, filing a provisional application can give you an extra year to run up sales before incurring the more expensive costs of filing the permanent application.

At times, you may want to file your regular application long before the provisional patent application expires. There are two compelling situations that may instigate this.

A potential licensee may have some doubt as to the patentability of your invention and thus you'll want to get it issued with the claims then being a fact. Another consideration is potential infringement. To enforce your rights, you must have a patent granted and issued. If you are infringed, it would be wise to proceed immediately to get your patent issued so you can protect your rights and stop the infringer.

The Request for Examination

The new 21st Century Strategic Plan being enacted by the PTO includes a new fee structure and proposes certain new changes. One of the most obvious proposed changes is what's called a "Request for Examination."

With the proposed change, after the permanent patent is filed, you need to request that it be examined and send the appropriate fee. You may file the request with your patent application and include the fee or you may wait up to 18 months to submit it.

You may use this new law to your tactical advantage when (and if) it becomes a viable rule. If you're still qualifying the invention, you may wait until the later stages to pay the fee. By waiting you are stringing out the duration of your patent pending status, which may or may not be advantageous. If you want your application reviewed immediately, then you may submit the request with your filing fee upfront.

Strategy, Worldwide Patent Rights, and PCT's

Whatever your strategy may be, if you want to preserve your worldwide patent rights, file your provisional or regular, nonprovisional application before your invention is publicly disclosed. Once you have done this, you have one year to file a patent application in countries that are part of the Berne convention or take the more popular approach and file a Patent Cooperation Treaty (PCT). To preserve the foreign filing rights, the PCT must be filed no later than one year after the filing of either a provisional or regular, nonprovisional patent application in the United States, whichever occurs first.

The PCT allows you to designate foreign countries that are a part of the pact in which you may want to file patent applications at a later date. Most first- and second-world countries are part of the PCT today. By filing a PCT, you can stretch your overall time period to file foreign patent applications by up to 30 months. When patenting overseas, PCTs are commonly used to extend filing deadlines and preserve cash.

Don't let the few countries that do not recognize PCTs or patents at all stop you. These countries will not be able to export to countries that do honor the PCT anyway. More and more Third-World countries honor the patent systems of the developed countries of the world as a step to improving their international relations.

There is another alternative you can consider with PCT applications. First, you can file a provisional application in the United States and preserve your worldwide filing rights for a year. Then, prior to the end of that year, you may file a PCT patent application, instead of the permanent patent application in the United States. You would then file the U.S. application within the following 18 months. Doing it in this manner actually saves money.

You should consult with an attorney about the use of a PCT and how it affects your invention and your finances. If you and/or your company are absolutely certain that international patent protection is going to be sought, this approach makes a lot of sense.

WORLDWIDE STRATEGY

Independent inventors and small businesses ponder long and hard before filing applications for patent protection overseas. It's expensive and relatively difficult. If there ever is a time to call in the "suits," this would be it.
–Don Kelly, Intellectual Asset Management Associates, LLC

LICENSING LARGE ENTITIES

If you are going to license your invention to large companies, it's most likely that they have offices around the world. If you don't take care to preserve international patenting rights, your greatest licensee prospects may become your greatest disappointment. Take your time in developing your patent strategy BEFORE filing your first patent application.
–Andy Gibbs, CEO, PatentCafe.com

Considerations for the International Pursuit of Patents

Fewer than 10% of all patents filed in the United States are filed internationally. Most foreign filings are by large corporations on inventions that have worldwide appeal. The most important factor regarding your decision to file overseas is finances. The costs can be staggering! Overseas filings require translations, foreign attorneys' fees, and filing fees. The cost can be $100,000 or more.

One way to save a substantial amount of money is to file for a "European Patent," which allows you to file in nineteen countries simultaneously. There is no extra cost for translations or filings. A European Patent can cost as much as $30,000 for relatively simple inventions. However, this is inexpensive compared to individually filing in each European country.

If your intention to seek patent protection is on behalf of your company or to protect an existing or emerging product line, then it becomes a simple financial decision. However, if your intention is to license the technology, it's much more complex. The cost to seek out and license a foreign entity can be costly and time consuming. Who will travel there and search out licensees? Who will work with and train the licensee how to make the invention? Who will locate or train the sales/marketing team? Who will monitor the accounting? Who will follow up and make sure that there is no patent infringement by others?

The best way for an independent inventor or small company to license patents overseas is to have a U.S.-based partner with international exposure. Let it find the overseas licensing opportunities. Let them license the patents and pay the cost of patenting, etc. They will have a better understanding of the field of your invention and industry in the other countries than you will. They can save you a lot of time and money and turn a costly, time-consuming project into a winning situation for both of you. In exchange for their dedicated efforts, take a smaller royalty and/or up-front fee and let them run with the project. With a vested interest, they will have the incentive to make things happen.

Remember that the United States is the world's biggest marketplace. It still represents about one third of the world economy. Every company wants to sell its goods here. If you are successful in launching your invention here, you should have more than enough income as a result. Consider whatever you earn from foreign licensing as gravy.

How to Hire an Attorney

Many inexperienced businesses and inventors will consult a patent attorney upon the discovery of an invention and immediately pursue the patenting process. We know this is a mistake if the inventive matter is not understood or qualified. If you are advised that the first step is to file for patent protection, it's an error. There are other steps to take first. There's no need to budget your patenting expenses this early on. Spending a few thousand dollars before you know if the development and marketing of the idea will be feasible usually results in a waste of time and money.

There is little disagreement about the merits of patent protection and the need to hire an attorney, but hastily filed patents result in inadequate or wrong coverage. Your inventions simply must be sufficiently developed in order to have patents that cover the broadest and very best inventive subject matter.

ONE WORLD

We find that the markets for America's genius long ago outgrew her national borders. Inventors struggle along with corporations to compete in a world where the potential buyers of new products and processes live in nations around the world.
–Donald Grant Kelly,
Director, U.S. Patent Office

TRUST FIRST

Don't choose a patent attorney solely because he/she is the cheapest—you might get what you pay for. Choose a patent attorney you trust and feel comfortable with.
–Michael S. Neustel,
U.S. Patent Attorney

Early on it's wise to start the hiring process if you don't already have an attorney. It will not cost you a cent if you do it right. Just follow this simple procedure:

- First, when hiring a patent attorney, never pay for the first visit. You are interviewing him/her, not the other way around. When you call for the initial appointment, say, "I am interviewing patent attorneys." During the phone conversation, confirm that you do not expect to be billed for the interview. It is best to confirm in advance, before the attorney's daily log reaches the bookkeeper, who may automatically bill you. In this initial conversation, most attorneys will be up front and tell you they do not charge for the initial visit. There are more patent attorneys than there is business to go around and most are looking for new clients.

- When you meet, tell the attorney that you have a limited budget. Even if you are wealthy, don't give the impression that you have an endless stream of cash.

- If the attorney you select is not familiar with your field, tell him/her you do not expect to pay him/her for the time it takes to learn about your field. He/she will agree if he/she wants your business.

- Ask the attorney if you are going to be charged for all phone calls, including quick answers to simple questions. You do not want to see 5-minute phone calls turn into expensive monthly bills. Be careful about calling with a lot of questions. Instead, find the answers in this manual, other books, or on the Internet. When you do call with questions, keep them concise. Otherwise, each answer can cost you $100 or more! If you become friendly with your attorney, see if you can meet over lunch from time to time. You can have a lot of expensive questions answered for a $50 lunch tab.

- During the interview, ask for cost estimates to file and prosecute a patent application. They should include:
    - The cost to write the final application draft (provisional or permanent). Tell him how much of the original draft you will provide or that you will provide a copy of your provisional application, if you are writing it yourself.
    - The cost to prepare drawings to the PTO's standard
    - The filing costs
    - Prosecution cost estimate
    - Patent issuing costs
    - A total estimate of how much he/she thinks it will cost to get a patent filed, prosecuted, and granted

- End the meeting with a statement saying, "I will be writing and preparing the provisional application. I'd like to send it to you for review and you can tell me how much it will cost to edit it, fix it, and file it."

By making this last statement, you are, in essence saying there is no reason to require an upfront retainer. The fact of the matter is that if you follow the guidelines on writing a patent application provided in this book, your cost will be dramatically reduced.

If he/she still wants a retainer, ask him/her to start without one, since it will be awhile before any real costs are incurred. Let him/her know that you are not sure how long it will take you to finish writing the application as you have some qualifying to do,

NO TRICKS

Almost all of these techniques used to hire and qualify a patent attorney were contributions from patent attorneys.

HIRE EARLY ON

Following the Strategic Guide, you will normally want to hire an attorney fairly early on so that you have counsel as may be required. Keep in mind that he/she will not be spending much time on your project (nor will you be incurring any major bills) until sometime later. Usually he/she will not be needed for patent matters until the review of your final provisional patent application before filing or the review and writing of the claims for your final application.

When working with a patent attorney, make sure to give them all information relating to your invention regarding products, systems, processes, and so on. Failure to adequately disclose your invention in a patent may result in the complete loss of patent rights.
–*Michael S. Neustel, U.S. Patent Attorney*

such as completing the prototyping and testing, developing, etc. This could take weeks or months. Besides, if you have good credit, a retainer is just not necessary.

The interview, is a good time to talk about some of your marketing or licensing intentions. If there are areas in which you have questions, ask for his advice now, at no charge. It is easy to get a few pointers, if the attorney really wants your business.

In your search for a patent attorney, find one who understands your needs and with whom you communicate well. If you get along as human beings and business associates, you'll get along on a professional level as well. Beware of any attorney who tells you that the first thing you must do is hurry to protect yourself with a patent, especially if it is the more costly, regular patent application. In fact, an attorney who does not agree with the use of a provisional patent application should probably not be considered. If you have taken the precautions described in this manual, you will have more than adequate protection. Encouraging an inventor to pursue a costly regular patent application without qualifying the invention's commercial value borders on unethical.

After you make your choice of attorney, ask him/her to send a letter to confirm the rates. Then, send a letter confirming your intention to hire him/her and say that you look forward to working together.

You've just hired an attorney without any money up front or a retainer.

About Patent Infringement Protection

Another important consideration when selecting an attorney is his/her position on patent infringement protection. If you are a small company without a lot of cash or if you are expecting an infringement problem, the best time to ask about infringement protection is up front.

Find out if the attorney is able or willing to work on a contingency basis in the event of patent infringement. Many will, providing the scope of your coverage and the patent's integrity are sound. Remember that 70%–80% (depending on the field of the invention) of all court cases are ruled in favor of the inventor and as many as 90%, maybe more, are settled out of court in favor of the inventor. Having this commitment is like an insurance policy. It is one you can tout later if you think there may be an impending threat of infringement.

If you assume that infringement costs can run up to $250,000 to prosecute and into the millions to defend, a contingency plan is smart for most small companies. A workable split, which can be established early on, will probably be something like this for the attorney:

CONTINGENCY

For inventors, one consideration in hiring an attorney is to know that he/she can also defend them on a contingency basis in the event of infringement. Of course, the two key factors that warrant doing so are predicated upon receiving broad enough patent protection (or multiple infringement of more than one patent) and having an infringement that has enough profit to justify the investment.

- 30% if settled out of court
- 35% if court documents need to be filed
- 40% if it goes to court

Three other points you'll want to keep in mind are:

- In most infringement cases, you have to pay the actual costs for filing court documents. These costs are deducted from the settlement before the contingency split is made. These expenses may be high, but only if the case goes to court. Remember, most cases are settled before they go that far.

- If your financial condition is strong, you will probably not want a contingency arrangement with legal counsel. You can have all the award money, less costs, that should be far less than a contingency split.

- If you are a small company and have a contingency arrangement with strong counsel from an attorney who is successful in your field, you are in a sound position to develop your invention, mass produce it, and get it sold without the fear of infringement.

A contingency arrangement for patent infringement is almost always a smart decision for a small company or independent inventor. However, having an attoryney working for free is another issue altogether.

Having an Attorney Work for Free

Some attorneys may work for free or for a "piece of the action" to prepare and prosecute patent applications, but this is rare. They will at least want you to cover the basic costs, such as filing fees and copying and fax charges. The attorneys I have met who would consider these types of arrangements are generally not of high quality. You can find equity partners elsewhere, who will not be as controlling. Besides, it could cost you a small fortune just to have another attorney review your partnership draft.

Prosecuting Patent Applications

Your attorney has been hired and one of his/her most important functions is to prosecute your permanent patent application. The Patent Office wants to communicate with one person, which should be your attorney, not you. When filing a patent application, you will give your patent attorney "power of attorney" to file and prosecute the application.

Periodically, your attorney may send letters and faxes referencing your patent's serial number to request an update should it be necessary. He/she can also talk to the examiner by phone. For a charge, the patent office now offers an interview directly with the examiner via the Internet. What's important to know here is that your attorney is experienced in these matters and will know what to say and how to make the appropriate changes that may be required for acceptance.

First Office Action

After anywhere from nine to eighteen months later (it averages about one year), you will receive your first office action. If your application is rejected, don't panic. Most patent applications are rejected on this first office action. Even a patent application that contains highly unique inventive matter is rarely accepted with the first office action. Patent examiners almost always ask for qualification of the inventive aspects and will want to clarify the way the claims are written. When you fight for your claims, you are actually improving the quality and integrity of your patent.

When the first response is returned, do the following:

1. Carefully read your attorney's response to the patent office action. Your attorney may inform you that some or all of the claims are acceptable with only minor changes. If so, ask yourself, "With these changes, will the patent still have the scope that was anticipated?"

For instance, your patent may refer to inventive matter related to the manufacture of ice cream, but the examiner cites similar art related to freezing of bacteria for med-

YOUR PARTNER

Can you find an attorney to work for free? Probably, but would you really want an attorney as your partner? Many attorneys are deal killers when it comes to licensing, too. Would you want to take that chance? If you've established a relationship with an attorney and he'll consider it, then you've got the possibility.

Mary S. of St. Louis, Missouri (1851–1880) was a genius inventor who was poverty stricken. Lacking finances and confidence, she sold the rights to her inventions (most of them mechanical) to various male agents for as little as $5 each. These agents received 53 patents and a great deal of wealth! Mary S. died penniless at age 30. While the battle against sexism was not yet won, one principle is clearly declared. That was . . . women inventors were a force to be reckoned with in modern America.
–*Lemelson-MIT Invention Dimension*

ical purposes. In this case, it would be fairly easy to get clarification on your claims by adding words relevant to making ice cream for human consumption.

2. If the reason for rejection is not clear to you, carefully reread the examiner's rejections and the accompanying patent references cited. Many times they really aren't related subject matter. However, this is something that you may know much better than the examiner. This is particularly true if you are an expert in your field.

3. Jot down the reasons you believe the references cited do not apply to your invention. Be thorough and look at the rejection from many angles. For instance, is the examiner citing art that is far more costly to produce? Is it impractical to use the teachings of the prior art on your invention? Does the prior art not apply because you know from considerable experience that it is irrelevant to your invention's function?

4. Discuss the examiner's rejections with your patent attorney. He/she should be very experienced in this area and able to help tremendously. Patent Office rejections and objections will not intimidate your attorney, whereas it might intimidate you. Your attorney will most likely know precisely what the examiner is rejecting and why. Sometimes it's an easier fix than it appears.

5. Discuss the response strategy with your attorney. With your input on your invention's cost effectiveness and application and functional differences, a savvy patent attorney will know how to overcome the objections. Be ready to supply additional facts. To substantiate your point, you may have to prepare an affidavit on the state of the art in the industry.

Most patent applications are rejected because the examiner cites prior art from one or more sources that he/she believes makes your invention obvious. Another common basis for rejection is that your invention was "anticipated" by what is contained in a previous patent.

In both cases, you are in a better position to prove that the examiner's prior art references or anticipations are incorrect. You will frequently find that the references are irrelevant or that they pertain to archaic prior art and processes that could never be applied to your invention and its new technology.

Other common reasons for rejections and objections by patent examiners are:

1. **Application contains more than one invention.** As we discussed earlier in this chapter, your strategy may include filing a single application for more than one invention or various inventive matter. In this case, you will have to split out the application into two or more divisional applications. One invention is prosecuted immediately; the other(s) are prosecuted in the order you select after initial prosecution has been completed. Base your choice on your anticipated return on investment and the broader patent protection. You will probably want to prosecute the weaker applications later based upon their potential value as defensive patents. A defensive patent is one that covers alternative inventions that you will most likely not use. When patenting the alternatives, you are in essence defending the scope and position of your primary, more dominant patent.

You will probably never use defensive patents, but you don't want others to use them either. It is like having the two best patents in the field of your invention—developing the sales on the best one and blocking the sales of the next best one. Defensive patents can be an asset in your developing patent portfolio.

2. **Drawings that are incorrectly drawn or have misapplied numeric references.** This is a common occurrence and refers to drawings that the examiner cites are not definitive of the specification. This is usually easy to overcome by correcting the misapplied numbers or perhaps redoing the drawing so it's clear.

3. **Unclear references to the invention in the body of the text.** This is similar to the previous argument and is usually easy to overcome as well. Some rewriting to clarify the language in the patent body usually does it.

4. **Claims that do not adequately describe the invention.** This too is usually easy to overcome. They may have been written too broadly to pinpoint the inventive matter.

5. **New inventive matter is being cited.** If the examiner cites that your response to office action (your argument) brings up new subject matter, you may have to file a Continuation In Part (CIP). This allows you to rewrite the patent application and add clarifying drawings so that it conforms with the new subject matter.

Whatever the rejection is, you really need an experienced attorney to help you. It makes no sense coming this far to just let the prosecution cause your application to be narrowed down into a worthless piece of paper. It is unimaginable that any inventor wants his/her name to go down in history on a patent application that is nothing more than prattle.

A Second Rejection

After submitting corrections and arguments in response to your first office action, you may receive a second rejection. In such a case, repeat the previous process and discuss strategy with your attorney. Reportedly, in 97% of applications, the second response to action you take will be the final one, regardless of whether the patent is granted or not. However, even though the second action and your response may not result in the granting of a patent, it is not the end of the road.

A good tactic to try is to ask your attorney to arrange a telephone conversation with the examiner and frankly discuss the matter in more depth. Your attorney can request the examiner's advice about how to overcome the wording and fix the problem. If you have licensed the invention, that fact may be used as evidence to substantiate the merit of the patent and its claims.

A Final Rejection

If you receive a final rejection and all or part of your claims are being rejected or contested to the point that will ruin the patent's scope, you still have two good alternatives.

First, you can make a physical appearance at the PTO with the examiner and take your invention with you. Show the examiner how great and unique it is. Willingness to do this is an indication of your conviction and determination that you have an important invention. You can also arrange an interview via the Internet with the examiner, although it may not have the same impact as a face-to-face meeting.

If you do this, you'd be wise to have a patent attorney accompany you. Your attorney will be able to prep you on what to say and do. There's little reason to believe that

FACT

Ann Moore (b. 1940), worked for the Peace Corps in Togo, Africa in the early 1960s. She realized that native women carried their babies in slings, which provided a great deal of comfort and security for the baby, while keeping the parent's hands free. After returning to the U.S., Moore and her mother, Lucy Aukerman, designed and patented the "Snugli" (1969), a rugged, adjustable, pouchlike infant carrier. In-house production eventually led to a buyout from a national corporation.

between the three of you that you can't figure out how to make the patent application read to have a satisfactory scope. So often, it is just a readjustment on the language, a clarification of function.

Continuation-in-Part

A continuation-in-part application (CIP) is the label given to a specification of a second application that has a different specification from the first application, but is related in some way. Most frequently, the CIP has some extra material added to the earlier specification. CIPs are commonly filed after the initial office actions have been received and perhaps the inventive matter is rejected or consider unclear by the examiner.

This is a tool you may use when patenting to preserve a prior filing date. If the related subject matter has not been publicly disclosed, you may file a new patent application instead of a CIP. By doing so, you will have the latter priority date but will have extended the life of your patent protection since it would begin 20 years from the more recent date.

In other instances, a CIP may be directed to the same general subject matter as the earlier specification or a subset of the earlier specification. For instance, if you wanted to expand on a particular inventive matter that may pertain to a use outside of the invention cited in the patent application, a CIP would be the perfect instrument to accomplish that objective without losing your priority date.

Your Attorney Earning His Money

Responding to office actions is done using a certain protocol. We inventors are emotionally attached to our creations (no matter how objective we may say we are!). If we respond to an examiner, we might say something like, "You're crazy and don't have a !B#F!ST*LK clue about what you're talking about." Needless to say, that kind of outburst won't help your cause.

On the other hand, your patent attorney will respond tactfully and will negotiate the best possible outcome. They do this all the time. He/she knows how to compose a respectful response, following the courtesies that must be extended to the patent examiners.

This is why you hired your attorney in the first place. This is where your attorney really earns his/her money.

Patent Quality Management

With a lot of hard work, you will compile a patent portfolio and other intellectual property. The quality of this portfolio depends on the means and the system in which it is managed. Over time you will assemble a team of experts and together you will increase your influence in business.

Patent Quality Management (PQM) is a business management system that sets the foundation for a company's future. PQM provides a method to manage and promote a company's present intellectual property and to exploit new opportunities over the long term. There is substantial pressure for companies to install a PQM system; this stems from the fact that as much as 80% of a company's assets are its patents and IP. According to the *Wall Street Journal*, this percentage will most likely increase in the coming years.

When a company's value is based on virtually no patents or intellectual property and shareholders' returns on investment are based on dwindling profits of generic-based products, a precarious future is in store and management will be held accountable. In contrast, you are building a future that relies on your intellectual property, and if well-planned, your invention development activities will have a bright future.

The benefits of taking steps today to manage your IP assets or taking steps today to build a war chest of patents and intellectual property is obvious to those with vision. The fact is companies that don't heed the demands of the future will be unable to maintain a competitive advantage and will likely have a deteriorating market presence. Embracing PQM is a means to avoid a future disaster and take a giant step into the future.

The driving forces behind a PQM system are:

- To increase a company's worth

- To increase shareholder value

- To protect and exploit a company's present IP

- To create and develop future proprietary products protected by patents and IP

PQM is for all companies, large and small, that have proprietary product lines, with or without patents, regardless of whether the products are manufactured in-house or through subcontractors. It is also for start-up companies or R&D-based companies, such as software developers that want to protect their intellectual assets and plan and protect future developments.

The PQM approach is somewhat similar to Total Quality Management (TQM), but it has more future-thinking objectives. It is not a complex management system in the sense that TQM is; it is more of an adjunct to TQM with future-thinking objectives.

Once you're on your way, you or your company should look into the management of your future by installing a PQM system. By applying PQM tenets and techniques, innovative companies will excel. Instead of becoming the low-priced, low-profit generic supplier in a field, your company may maintain its position as a producer of higher quality, innovative proprietary products that support higher profit margins.

Read *Essentials of Patents* to learn more about Patent Quality Management and how it affects every department in a company, large or small. Visit http://www.frompatenttoprofit.com to learn more.

Patent Strategy, Licensing, and Venture Capital

There are other issues you'll want to consider as you develop your invention and patent strategy. For instance, if you plan to license your patents, pay particular attention to the details that may affect your patent's scope. A patent that is easy to design around will be difficult, if not impossible, to license. It's easy to overlook some aspect of the patent's scope that can allow a competitor to design around your invention. So be thorough.

You will also want to fight for your claims and get the broadest possible scope. Again, a patent application that has been fought for is usually a better one anyway and more difficult to contest later on in court. Fight for your rights and patent claims until the bitter end if you have to. See Chapter 5 for more discussion of this subject.

A solid patent position with broad scope can also improve your ability to secure venture capital, fund a company, or secure expansion financing. If you intend to use your patents as part of the assets of your business, be sure to fight for every last claim that will ensure its value.

THE MOST PATENTS

The company with the most patents continues to lead in innovation, registering the largest number of U.S. Patents ever issued in one year, 3,415 patents in 2003. It was the eleventh consecutive year that IBM was awarded the most patents. Its total portfolio is probably in excess of 30,000 U.S. patents. In 2003, Canon was second, with 1,992 patents, which marks a substantial increase in recent years as it changed to a patent-centric business philosophy.

The Most Important Strategy—Keep Inventing

An inventor's first patent is not usually a big moneymaking opportunity. The smartest thing an inventor can do is to continue to improve upon his/her invention and secure more patents. It is usually with one of these subsequent discoveries and the resultant patents that a breakthrough opportunity arises.

There are four substantial benefits that may result when you continue to improve your inventions.

First, sales will improve. A subsequent discovery—for instance, a means of making your invention easier to use—may turn a million-a-year product into a ten-million-dollar a year product! It happens all the time. In fact, you can almost expect it.

Second, if you've improved your invention and have additional patent protection, you will be in a superior position to defend against infringement of your invention. It will be more difficult and costly for an infringer to try to invalidate two or more of your patents, instead of just one. Think about the odds. If the odds of a patent holder losing a court case is 1 out of 5 with a single patent, then the odds drop all the way down to about 1 out of 20 with two patents, and further down to 1 out of 100 with three! Those attractive odds will also dramatically improve your ability to license and secure financing.

Third, with new improvements, you are extending the life of your product line or licenses. With a license agreement based on the "expiration of the last of the patent rights," you may add years to the term.

The fourth benefit to continuing the invention process is that you get better at it, and you will be clearly on the road to becoming an international expert in that field. At times, you'll find new opportunities with related product groups. With such a success, you may even be asked to invent something for a company or help design around and improve upon existing products. The more you invent, the more opportunities you will find.

WRITING PATENT APPLICATIONS

Writing patent applications helps you accomplish your primary
objectives: become an expert in your field, broaden the scope
of your patent protection, and save money.

Your patent writing strategy includes your ability to write the two types of utility patent applications that can be filed at the U.S. Patent Office: the provisional patent application and the regular, permanent patent application. The provisional patent application is the easier one to write since it is a greatly simplified version of the regular application. The regular patent application is then written based upon the provisional patent application's contents and is filed within one year of the provisional's filing date. You'll see how to write both of these in this chapter.

Strategy for Writing a Utility Patent Application

As we discussed in Chapter 6, in most cases you will file the provisional patent application first, followed by a regular patent application. Even if you skip the provisional and file only the regular application, there are several reasons you should write as much of the application as possible. These reasons include improving the integrity of your patent, discovering potential design-arounds, making new discoveries, and saving money. Your patent writing should dovetail with your invention development and thus enhance your ability to include material that is the most valuable and relevant to the invention's success and the patent's scope.

Since the regular patent application references one or more previously filed provisional applications, it is wise to have your regular patent application read on your provisional application as much as possible. In other words, if they both have the same specifications and drawings, the preservation of the priority date should remain intact. In addition, the laws regarding the writing of either patent application are essentially the same, except that the permanent one includes the claims and the abstract. So it's easy to convert the provisional into the permanent application.

An issue regarding provisional applications is how well written should they be? The answer is "as good as possible." Nevertheless, you could argue that a poorly written provisional is better than none at all.

The decision about whether to write and file the provisional yourself is yours. Before deciding, consider that the better the document is written and the better the subsequent regular patent application reads on the original provisional, the better your chances of getting a patent with the scope you want. It is recommended that you write as much of the provisional application as you can and have your attorney edit, rewrite, and file the final provisional draft. When you're good at it, you and your attorney may decide you can write and file your own.

The same is not true for filing the regular application. Your attorney is an essential part of that process. Most everyone in the patent business agrees that a highly qualified attorney should at least edit and write your final draft and include the legal claims.

We will reveal a methodology for writing patent applications that will put you on your way quickly, with the security of knowing you'll have a successful patent.

Model After Another Patent

Modeling after an existing patent is easy to do. You'll save yourself a lot of grief learning about your field and learning about what to include in your patent applications.

Patent documents are not copyrighted. Therefore, you can legally include the same (or very similar) language other patents use. You can even trace the drawings should you be able to use them. There's no reason to run out and purchase lengthy technical manuals telling you how to write a patent application. All you have to do is modify the language in existing patents.

Here's how you do it. Before writing your patent application, you will have conducted a patent search. As you review patents from your patent search, select one or two reasonably related patents as a model for your patent application. Copy the style and language. Try to model after a patent that has issued after about 1998, as the present format has changed somewhat and most of the recent patents should be in the new format. You'll see what this is later in the chapter. You'll also see how you can change and modify an existing patent to include your particular invention.

Draft Your Application on a Computer

You will undoubtedly edit your patent application draft several times. Don't be surprised if you write five or ten drafts before you have the one you want. This is particularly true if you don't have a lot of experience drafting patent applications. Without question drafting your provisional or regular application on a computer word processor saves hours, maybe even days.

A more important advantage to using a computer is that you can send your final draft to your attorney on a diskette or via email for the final editing. This saves your attorney a substantial amount of time, since it will not have to be retyped. It will be very easy for your attorney to open the file and edit it.

During the writing process, print copies and keep them in your file. These dated copies will be a part of your paper trail demonstrating your diligence. The preferred word processing program is Microsoft Word as it is easy to save in a format used by your attorney's word processing program.

Meet Provisional Patent Application Requirements

The U.S. Patent Office requirements for a provisional patent application and acceptance thereof include the following components:

1. Written specifications adequately describing the invention and the preferred embodiments. This is defined in U.S.C. 112, first paragraph. Simply put, this is what we will describe in this chapter in layperson terms.

2. Drawings per section 1.81 and 1.83. Use the drawing format as defined herein.

3. A provisional patent application cover sheet. You may download this from the U.S. Patent Office or use the one in the Resource Guide (see appendix).

4. A check for the filing fee (presently $80 for a small entity and $160 for large entities).

A declaration claiming small entity status is no longer required. When mailing your application, don't forget your check.

Follow Patent Application Layout

Writing a patent application (provisional or nonprovisional) is somewhat like writing a high school or college term paper, although more technical. Remember that it's particularly important your provisional patent application follows the Patent Office's rules with the correct format and content, or it'll raise a red flag and may be returned to you. Here are the guidelines you should use:

1. Draft the written portion according to the standard manuscript format on $8\frac{1}{2}$ inch x 11 inch white bond paper with 1-inch margins at the top, bottom, and right side; the left column should be set at 1.3 inches to accommodate line numbering.

2. Set line spacing at 1.5 spaces with line numbering along the left column, justified left.

3. Use an easy readable serif font such as Times New Roman 12, black color. The exception is the title of the patent application, which should be in a larger font such as 14 point.

4. Indent the paragraphs, but do not leave spaces in between them.

5. Center subtitle headings in bold type for easy identification. Do not leave spaces between them and the preceding or subsequent paragraphs.

6. Insert page numbers in the bottom footer (not shown in the example) and number all pages.

7. Use a basic drafting format for drawings. They should also be on $8\frac{1}{2}$ inch x 11 inch white paper with 1-inch margins at the top, bottom, and both sides. They should meet the Patent Office's standards for number and letter references (discussed later on). If the drawings don't satisfy the PTO standards, the application will not necessarily be rejected, but you certainly don't want to take a chance. Whatever you do, make sure your drawings clearly depict the inventive matter.

TEST TIME

Inventors should write their entire patent application—it's a test to see whether they really understand their invention. However, inventors should have a patent attorney review the entire application, especially the claims, before any application is ever filed with the patent office.
Andy Gibbs, CEO, PatentCafe.com

PROVISIONAL ADVANTAGE

One significant advantage to writing and filing provisional patent applications is that during development, if you make new discoveries, you can write and file another provisional application. When doing so, you can usually start out with the computer draft on a previous application and change it accordingly. This newly drafted patent application can then be either incorporated into the same final, permanent application or can be filed as a separate application later. Discuss this strategy with your patent counsel.

Since it is important that a regular, nonprovisional patent application reads on your provisional application, use the same format in both, with as much of the same wording as possible.

Use the Standard Provisional Application Format

There are several minor variations of the standard format. The one that follows represents a simple way for an inventor to write a provisional patent application. A patent attorney or patent agent editing the final draft may choose to alter the format slightly according to his/her experience, but the content is essentially the same regardless.

Your regular patent application also follows this same format except that you'll be including an abstract of the invention, which is printed on the first page of a patent once it is issued. The regular application may also include a reference to the previously filed provisional application. These elements you need not concern yourself with as your attorney filing the permanent application will know what to include.

In 1996, the USPTO amended Rule 77(a) concerning the order of elements to be included in a utility patent application. Based upon this rule, the elements (title and subtitles) for a provisional patent application should be as follows:

1. Title of Invention
2. Background of the Invention
   a. Field of the Invention
   b. Description of the Related Art
3. Summary of the Invention
4. Brief Description of the Drawings
5. Description of the Specific Embodiments
6. Drawings

As you write each section, you will be explaining the specific details surrounding your patentable subject matter so the reader may understand it. You will also learn more about your invention during the process. When you are finished, you should have an excellent understanding of how and why it works.

Use a Storytelling Format

One of the best ways to look at patent writing is to think of it as a story. The story automatically covers the categories in the standard format in the proper sequence. Your story will begin with the field of your invention, and in a time line sequence, cover the past, the present, and future. In other words, what was used (past), what is currently used (present), and what you have discovered (future).

The elements of the standard format tell a sequential short story. It's almost like "once upon a time." Try to think of your patent writing in this vein. It is as though you are storytelling by giving a brief history, talking about present usages, stating your invention's solutions, and describing exactly how your invention does what you say it does.

More specifically, the storytelling sequence you will discuss in your application should be:

1. **Background of the Invention.** The first paragraph under this subtitle defines the *field of the invention*. It is usually two or three sentences. The following paragraphs

---

EXPERIENCED INVENTORS AGREE

It is important for you, the inventor, to write the original patent application draft. Most experienced inventors agree that the patent-writing experience is invaluable to learning about your invention and patents in general. While writing, you will make new discoveries, find potential design-arounds, and develop alternative means of making or using your invention. You should cover these alternatives in one patent application with the broadest language possible.

MARKETABILITY FIRST

A good patent application should read like a story where your invention is the hero!
–Michael S. Neustel, U.S. Patent Attorney

should provide a *description of the related art*. Give a brief summary of prior art, those products (patented or otherwise) used in the past, and those that are presently used. Describe the problems associated with past and current art and why fixing these problems may be beneficial or desirable.

2. **Summary of the Invention.** Now that problems associated with prior art have been described, you tell about your invention and how it solves the problems.

3. **Brief Description of the Drawings.** List the illustrations you've provided in the application by giving them simple one-sentence descriptions.

4. **Description of the Specific Embodiments.** Last, describe each illustration in detail and explain exactly how the invention works. This is sometimes referred to as "Detailed Description of the Drawings."

This is the format for the written specifications as defined in U.S.C. 112 requirement for your provisional application.

If you want your attorney to write the entire patent application, he/she will ask you to tell the essentially the same story, so why not just put it in writing yourself?

Before you Begin . . . Write a Claim Statement

Before you begin, you should make a list of the inventive matter you want to discuss and illustrate in your application. One of the best ways to do this is to make a list of the claims you plan to include in your permanent patent application. You won't be including the claims in your provisional patent application, but the claims in your permanent application are the most important element. By listing them now, before you begin your provisional application, you'll have a much clearer idea of what you're going to include.

Read about the claims a little later in this chapter and then, write a claim statement you can use to guide your patent writing. Remember, marketability supports good claims. Here is how a statement might look:

1. The invention (or inventions)

2. The inventive elements

3. The method in which it is used

4. The special material in which it's made (You'll want to describe why and how it works best with this particular material.)

5. The variations on the theme

This statement may also include the process by which the invention is made and any individual elements that may be considered improvements. With the claim statement compiled, you're ready to organize and write your patent application. Without it, you are wandering around in a forest.

Writing a Patent Application

Once you have accumulated enough information and data to begin writing the application, set aside some quiet time, so you can think, analyze, and draw as you write.

The following patent application is an example of how your provisional patent application should look. It reflects all current patent writing rules and other pertinent laws as of the publication of this book. This patent application was eventually issued as U.S. Pat. 6,095,687.

You can use this example as a model for writing your patent application, which may be somewhat easier, since you'll be more familiar with your subject matter. Follow the instructions to convert this application to yours, with *examples shown in italics*. We'll use a cleaning device as the example.

Title of Invention

The title of your invention should be centered on the page, in bold type, all caps, and in 14-point font. It must be a descriptive title. It should not be a word such as your trademark or some gimmicky phrase. For example, we would not use the trademark M2K bag™ or Boxxer™ to describe the bag in our patent application. Instead, we use a descriptive title "Flat Bottom Plastic Bag." Another choice would have been "Square Bottom Bag," but we didn't want to intimate that it had to be "square" in structure.

Background of the Invention

This first heading should be centered, all caps, and single spaced beneath the title. The first paragraph satisfies the requirement to define the ***field of the invention***. Usually, this is only two or three sentences in a single paragraph.

This is where the story begins. Start by describing the field of your invention, such as remote controls, tortilla manufacturing processes or kitchen cleaning devices.

Begin your patent application like the model, change the words to your invention, and write the first two sentences accordingly, such as:

This invention relates to kitchen cleaning devices commonly used in households and commercial applications. More particularly, it relates to a cleaning device that substantially improves the ability to remove baked-on cooking residue and other hard-to-remove substances.

These two sentences will get you started in the right direction.

Next, you want to address the ***description of the related art***. Talk about the history of how your product, process, system, or so on was made in the past. The description of related art subsection makes up the balance of the ***background of the invention*** section.

This section is where your research and your patent search data are revealed and discussed in some detail. You'll use this research to illustrate the inefficiencies and problems associated with prior art. In essence, you're establishing the proof that your invention is novel, unique and useful.

In our application given on shaded pages that follow, we point out all the various methods of making square bottom bags that have been previously used and patented. The research we found actually revealed more than 30 patents, but all of them substantially increased the cost of manufacturing from 20% to 300% over paper bags. Obviously the viability of these technologies will severely limit the marketing potential. They will be confined to niche markets, or none at all, which is the case in almost all of these patents.

Our patent application points out four of the most relevant prior art patents and then discusses in detail how they negatively affect known manufacturing processes. We also point out how cold seals are generally considered "defective seals" when used to seal a bag bottom. This description clarifies our patentable matter and will be used in at least one claim.

In your application, you will want to discuss the relevant prior art as a means to prove your inventive matter. In your writing, you may proceed something like this:

Cleaning devices are well known. They include wire mesh prior art such as Brillo® pads, common household sponges, Scotch Brite® pads, and even wash rags. Cleaning devices such as Brillo tend to scratch metal surfaces . . . (and so on)

Make sure you include a list of the previous patents that apply. In particular, you want to include a list of those patents that may be somewhat similar to your subject matter, so that you may describe in detail why your inventive matter is superior. If there is more than one unique feature to your innovation, include them all. For instance, you may say:

All the aforementioned prior art cleaning devices tend to retain water after use. The retention of water tends to create an environment that may harbor bacteria. This is particularly true with spongelike devices and rags and even more prevalent when the device has been rinsed out after applying soap. As is well-known, this is a common occurrence.

If you plan to use a prior art drawing in your application, this is where you will discuss it. You'll want to do so only if it is necessary. Usually the drawings in other patent applications and their subject matter are enough to describe the inefficiencies of prior art.

The end of this section should lead into the next. The subject matter that you've discussed now sets up your ability to explain the unique inventive subject matter.

Summary of the Invention

In this section, you describe how your invention overcomes the prior art-related problems. You may begin this section by stating that your invention overcomes the problems associated with prior art. Then, you must explain how, but not in detail, which will be explained later on in the section Brief Description of the Drawings.

In our example of the flat-bottom plastic bag, we start by describing how the flat bottom bag of the present invention is made and its construction. In the example of the cleaning device, we may begin the summary like this:

The present invention overcomes two problems associated with prior art, more specifically, the prevention of scratching or marring the surface of the item being cleaned and helping to prevent the harboring of bacteria.

Due to the unique construction and the materials used in the present invention, the nylon mesh is bundled together to form a round cleaning pad; the mesh materials are soft enough to prevent scratching of surfaces, including coatings such as Teflon®.

Furthermore, the present invention helps prevent the harboring of bacteria because the nylon mesh material is sufficiently porous to allow it to dry out more completely after use.

The words "present invention" are commonly used to describe your invention. You must describe how your invention works, how the inventive subject matter accomplishes its objective. You can't just say "it does it" . . . without explaining how.

PATENTED SPACE TRAVEL

The first manned space capsule, Mercury, received U.S. Patent No. 3,093,346 in 1963 and was invented by Maxime A. Faget, Andre J. Meyer, Jr., Robert G. Chilton, William S. Blanchard, Jr., Alan B. Kehlet, Jerome B. Hammack and Caldwell C. Johnson, Jr.

FLAT-BOTTOMED PLASTIC BAG

Background of the Invention

This invention relates to plastic flat-bottom bags. More particularly, a plastic flat-bottom bag is illustrated where either a hinge, a cold seal, and/or a hybrid cold seal and hinge forms a linear folding axis for squaring out the bottom of the bag. This construction enables rapid bag manufacture of the resultant flat-bottom bag, making this design competitive for the first time with conventional plastic bags

Flat-bottom plastic bags are known.

In such flat-bottom plastic bags, it is necessary to form a folding axes so that upon bag opening, the flat bottom of the bag hinges along diagonal fold lines to "square out" and form an upstanding bag structure for the receipt of articles to be placed in the bag. In addition to paper bags in commerce, exemplary of the prior art is:

Platz et al US Patent 3,917,159 takes an already manufactured bag and folds the bottom so that it will "square out."

Le Fleur et al US Patent 3,915,077 utilizes individual heat seals to form the "squared out" portion of the bag.

Hanson et al US Patent 3,988,970 discloses a bag process where a square bottom bag is first produced. Thereafter, the bag is folded so that the plastic film emulates the folding present in ordinary paper bags. This folding process occurs after the square bottom bag is fully manufactured.

Ross et al relates to a plastic bag that utilizes among other constituents glue.

In the disclosure that follows, a plastic bag is manufactured which uses either "cold seals," hinges, and/or hybrid cold seals and hinges in plastic film for the rapid production and formation of a plastic bag. It is important that the reader understand both the "cold seal" and "hinge" terms and how they are limited in the disclosure that follows.

First, it is important to understand that in modern plastic bag manufacturing technology, speed of film processing is essential. In a typical bag manufacturing process, film passing through a machine for bag manufacture at rates up to 300 feet per minute is required for economic manufacture. Further, each manufacturing step must anticipate the subsequent opening and loading of the bag. As a consequence, steps taking an inordinate amount of time slow down the bag line speed and are generally unacceptable. Further, any bag post-production step—such as folding an already manufactured bag—is unacceptable.

Because of this speed requirement, so-called "hot seals" are generally not acceptable for any portion of the bag that does not require full strength sealing. In a hot seal, one film layer of a bag is permanently fused to another film layer of a bag. Hot seals take time to produce. Melting must be sufficient for complete fusing to take place between the bag parts. As a result, a hot seal is a significantly slower method of plastic bag manufacture than either utilizing a "cold seal" or forming a "hinge" within plastic film.

In a hot seal application transverse to the direction of film conveyance during plastic film bag manufacture, line speeds of the passing plastic film are usually limited to 150 feet per second. In the manufacture of the disclosed bag, line speeds in the range of 300 feet per second can be tolerated.

Hot seals produce permanent fusing of the plastic layers involved. Such permanent fusing can be detected by trying to separate the joined film layers. When an attempt is made to separate the two layers joined by a hot seal, either joined film layer may tear. The "hot seal"—composed of two layers fused one to another—does not tear and remains in tact.

In the plastic bag manufacturing arts, there has been known for many years a *defective* seal known as a "cold seal." Such a cold seal has heretofore been an imperfect hot seal. Usually, the imperfect hot seal would be located at the bottom of a plastic bag. When load was placed in the plastic bag, the seal would part—and the bag contents drop to the ground, usually causing content loss and damage.

Cold seals as used in the specification that follows are easy to distinguish. Where a cold seal is present, it imperfectly seals two plastic film layers together. This imperfect fusing can be easily recognized. Film separation at the cold seal allows the two imperfectly fused film layers to separate. Each layer separates from the other layer without loosing substantial structural integrity of the film. Unlike the conventional "hot seal," neither of the previously fused film layers tears when separation occurs.

Cold seals have been used with plastic bags for joining bags loosely in a bundle. Such joinder usually takes place at the top of the bag adjacent the opening. In the typical application, separation of a bag from a bundle of bags at a cold seal usually assists in the opening of the bag.

Until this disclosure, no one has made a "cold seal" an important structural element of a bag.

In addition to the "cold seal," this disclosure makes use of a preformed "hinge" with the plastic film of the bag wall. Such hinges are capable of rapid formation by impressing the rapidly passing and planar film along a linear boundary. This impressing at the linear boundary leaves the film predisposed to "hinge" or fold at the boundary.

The hinge that is here utilized is to be distinguished from a "fold." In folding, film is bent over and then creased so that it folds. Once this bending and creasing has occurred, the film is disposed to repeat the bending and folding along the crease.

The "hinge" here utilized is placed in the plastic film of the bag wall while the film is planar or in the "lay flat condition" and passes through the bag machine at high speed. The placed linear hinge is typically either at right angles to the direction of film motion or alternatively diagonal to film motion. Thus, the hinge here used is not to be confused with side gusset creases that result from creased folds made in the direction of film passage. When the film is hinged for the first time from the lay-flat state to the erect disposition of the bag, the hinge structure impressed in the film predis poses the film to bend along the hinge axis. Through the combination of the hinge-joining panels of plastic across the multiple flat surfaces of an erect bag, sufficient structural integrity is imparted to the bag to remain upright and open to receive and contain articles, such as fast food orders.

The reader should also understand that the difference between a "cold seal" and a "hinge" is not always precise. For example, where plastic film is passing a die at relatively high speed, and the dye comes down on two or more layers of plastic film, a hybrid "cold seal and hinge" can result. The cold seal will be recognizable by the imperfect fusing of the plastic layers one to another. The hinge will be recognizable by the reduction in thickness of the plastic film wall with the increased tendency of the film to fold at the hinge. It has been found that hybrid "cold seal" and "hinge" structures are produced in the bag production process herein set forth.

In the disclosure that follows, neither the "cold seal" nor the "hinge" incorporate or refer to bending and creasing film so that the film may bend again along the bend or the crease. In both the cases of the "cold seal" and "hinge," the film is at all times flat and planar when the "cold seal" or "hinge" is applied and introduced. This is known as the "lay flat" condition.

SUMMARY OF THE INVENTION

An expandable plastic film gusseted bag includes a front panel of plastic film having two substantially parallel side edges; and a rear panel of plastic film having two substantially parallel side edges. Two front panel gussets of plastic film are used. Each front panel gusset is connected to one of the two substantially parallel side edges of the front panel. Two rear panel gussets of plastic film are used. Each rear panel gusset connected to one of the two substantially parallel side edges of the rear panel. As in conventional bag manufacture, each front panel gusset connected to a rear panel gusset and usually each of the panel gussets are foldable between the front panel and the rear panel to provide a continuous bag periphery. A bottom joins the bottom of the front panel, rear panel, two rear panel gussets and two front panel gussets at substantial right angles to the side edges of the front panel, rear panel and gussets to close the bottom of the bag and leave only the top open. The improvement provides each plastic bag with eight first linear folding axes configured adjacent the bottom within each front panel and each front panel gusset and each rear panel and each rear panel gusset having one first linear folding axes. These folding axes extend from a joinder of a gusset at the bottom to the side edges of the front panel or the rear panel and the front panel gusset or the rear panel gusset at an oblique angle in the range of 45°. The folding axes are made by a process restricted to a group consisting of either a cold seal, a hinge, or a hybrid cold seal and hinge being placed in the plastic film while the film is planar or in the "lay flat" disposition. On opening of the expandable gusseted bag, the bottom seal of the bag disposed adjacent the gussets forms a generally triangular shaped overlay relative to the gussets to form a square bottom bag.

BRIEF DESCRIPTION OF THE DRAWINGS

Fig. 1 is a schematic representation of a "cold seal" apparatus illustrating the diagonal placement of cold seal in rapidly passing planar film for causing the "imperfect" joinder of two layers of bag material;

Fig. 2 illustrates the separation of two "imperfectly fused" joined layers illustrating that upon separation of the layers that each of the separated layers retains its structural integrity;

Fig. 3 illustrates a "hinge" as it is here defined being placed within a plastic bag wall while the film is disposed in a planar condition;

Fig. 4 is a expanded side elevation section of the plastic film at the hinge illustrating the linear impression in the plastic film which leave the film when folded for the first time with the predisposition to fold along the linear axes of the fold;

Fig. 5 is a side elevation view of the plastic bag article of this invention illustrating in particular the placement of the "folding axes" (either hinges, cold seals, or both) in the side wall of a flat bottom plastic bag predisposing the bag to open and "flat out" when used for the first time;

Fig. 6 is a perspective view of the bag of Fig. 5 being opened and "flattened out" for the placement of articles into the bag;

Fig. 7 is a so-called "side weld" bag shown having the hinge axes of this invention; and,

Fig. 8 is the side weld bag of Fig. 7 in the open disposition.

DESCRIPTION OF THE SPECIFIC EMBODIMENTS

Referring to Fig. 1, double bag unit D is shown passing along bag production line B. This double bag unit D is sealed at both ends along leading seal S1 and trailing seal S2. As will herein after be made clear, eventually, double bag unit D will be severed medial at medial cut C. This severance will allow an opening for each bag of the double bag unit D to occur in the center with flat bottoms at each end.

Double bag unit D is a gusseted bag construction. As such gussets G are attached between front panel P and rear panel R, they may be seen expanded in the view of Fig. 6 as panels 82, 84, 86 and 87 respectively. The reader will understand that the schematic of Fig. 1 is for the purpose of illustrating the placement of cold seals 20 and 21. All other features of the bag construction have been omitted.

Returning to Fig. 6, a typical flat bottom bag is illustrated. This includes the placement of cold seals or hinges 70 at the bottom portion of the illustrated and expanded bag. It will be developed that the hinging of the flat bottom bag illustrated in Fig. 6 can either be a hinge or a cold seal. For the purposes of Fig. 1 and 2, only the cold seals 21, 22 are considered.

Returning to Fig. 1, it is seen that die 24 having cold seal impressing ridges 25 and 26 overlies double bag unit D. For the purposes of Fig. 1, die 24 has been rotated up and away from double bag unit D so that ridges 25 and 26 are plainly visible. As is schematically illustrated, die 24 impresses on platten 22 with double bag unit D captured there between. Impressing of die 24 may be heated or done in the cold state. In any event there is an imperfect joinder of at least two layers of plastic material. To understand how this imperfect joinder operates, the reader's attention is invited to Fig. 2.

Referring to Fig. 2, two plastic layers 27, 28 are being separated from one another. Such separation easily occurs until cold seal 21 is reached. At cold seal 21, it will be found that the respective layers 27, 28 are imperfectly joined to one another. Easy separation will stop along the boundary of cold seal 21.

If one continues to try and separate respective layers 27, 28 beyond cold seal 21, such separation will eventually be successful. Separation will occur to seal 21. Both respective plastic layers 27, and 28 will maintain their structural integrity. In the usual case, the broken cold seal will not appreciably weaken the layers 27, 28.

A brief preview of how cold seal/hinge 70 functions to allow the opening of a flat bottom bag is in order with quick reference to Fig. 6. Referring briefly to Fig. 5, and then to Fig. 6, it will be seen that when the flat bottom bag, cold seal represents a point of increased resistance only to opening of the bag. This is sufficient to cause the bottom of the bag to "flat out" during the opening process.

I have also discovered that the placement of a hinge in location 70 of the illustrated flat bottom bag has the same effect. Specifically, and referring to Fig. 3, plastic film 28 is impressed with hinge H. Impression occurs between die 40 and platten 42. As before, die 40 has ridge 41 extending across the bottom of the die in the path of hinge H. Again, the path of hinge H is linear and constitutes a reduction in thickness of film 28 along the axis of hinge H.

Refer again to Figs. 5 and 6. Assume that instead of cold seals 20, 21, hinge H is placed at cold seal/hinge 70 at the bottom of flat bottom bag. Again squaring of the bag B will occur. In other words, either a cold seal such as cold seals 20, 21 or hinge H will predispose the bag article of Fig. 5 to open as the flat bottom bag illustrated in Fig. 6.

Taking the case of the open flat bottom bag, it will be remembered that the production here illustrated forms the bag completely in flat and planar state. When the bag moves to the disposition shown in Fig. 6, the bag is opened for the first time. This being the case, it will be realized that cold seal/hinge 70 need only be highly transitory. In short, they need to work only once.

Further, it will be understood that both the cold seal and the illustrated hinge do not have to have the full strength of the plastic film to operate satisfactorily. All that needs to happen is to have their respective formation add to the bag the predisposition to open and flat out at the bottom. Once this is achieved, the bag can be conventionally loaded and conventionally used.

It should be understood that the hinge or cold seal used with this invention can be hybrid. In this case, impressing of a cold seal may reduce the thickness of the plastic film giving the bag a tendency to "hinge" along the cold seal. At the same time, the vicinity of the hinge may result in an imperfect bond along the linear hinge axis with adjoining sheets of plastic film. This imperfect bond coupled with the "hinge effect" causes the film processed in the "lay flat" disposition to open and form preferentially a flat bottom contour.

Having said this much, the remaining structure of the illustrated bag can be described. Turning Fig. 5, the bag has open top 62, front panel 63, front panel gussets 64, 66 with central gusset fold 74, 80. The bag includes sealed bottom 68–72. In accordance with the preferred embodiment, hinges 65, 75 are impressed in the bag front panel 63 and rear panel 62. These hinges have been shown to react to a panel type hinging to impart additional strength to the bag.

Referring to Fig. 6, the open bag is shown with its flat bottom. It includes gussets 64, 66 extending between front panel 63 and rear panel 62. It will be seen that these respective gussets 64, 66 are in the full open position imparting to the bag a flat profile from the opening toward the bag bottom. At the bag bottom, characteristic triangular folds 90 will form. These characteristic folds dispose the bag with its flat bottom resting downward. In fact, after opening an empty bag from the disposition shown in Fig. 5, to that of Fig. 6, it has been demonstrated that the bag physically rests in the disposition of Fig. 6. In short, a clerk may load the open bag at the bottom without fear of bag collapse to the closed position.

In Fig. 7 side weld bag 200 has a top 202, a front panel 203, a rear panel (not shown), a die cut handle 205, a bottom gusset 204 and its center gusset crease 206, and side welded edges 208 and 210. At 45 degree angles in both outer regions of bottom gusset 204 are hinge axes 212 and 214 which respectively terminate approximately at the point where center crease 206 crosses side weld 208 and where center crease 206 crosses side weld 210. A vertical bi-directional hinge 216 begins at point 218 and terminates at point 220 and another bi-directional hinge 222 begins at point 224 and ends at point 226, both of which hinges have been impressed upon front wall 203, rear bag wall (not shown) and bottom gusset 204.

In Fig. 8, the side weld bag 200 of Fig. 7 has been opened and sets upright upon bottom gusset 204, Linear hinge axes 212 and 214 cause the lower outer regions 230 and 232 respectively to turn upright and help flat out the bottom gusset assisting in the formation of a flat base. In turn, bi-directional hinge 216 causes the upper outer region 234 to stand up and box out and side weld 208 to cooperate by turning inward. Hinge 222 causes its respective upper outer region 236 to stand up and box out and side weld 210 to cooperate by turning inward. Bag 200 is now ready to be loaded.

It is easy to see that cold sealing the bottom gusset in side weld bags has a similar outcome as cold sealing the side gussets in bottom seal bags described in Figs. 5 and 6.

Some attention should be given to the linear hinge axis which is at an oblique angle—most preferably 45°. It will be understood that variations in this angle can occur. For example, the angle can easily vary in the range of 40° to 50°. Further, both of the oblique axes do not have to be identical. They may each vary by differing amounts. All that is required is that the axes produce a tendency of the bag which is manufactured in the lay flat condition to hinge and open up with flattened out bottom.

More accurately, in the example, you would explain that the nylon mesh is a continuous piece of nylon material gathered to form a ball (later on, you'll explain how it was formed into a ball). You should explain or provide an example of the nylon mesh's characteristics such as:

. . . the nylon mesh of the present invention is much like that used on common bathroom puffs with the spaces defining the mesh being about $1/8$ inch square. This spacing may be as great as $1/4$ inch and still accomplish the objective. Too small of a space will create surface tension on the mesh material and therefore may prevent sufficient drying qualities.

While summarizing your invention, you are describing your product's unique qualities. Use these words if you can, just like we did in the cleaning device example.

Remember, you may borrow any terms and language you find in other patents as there are no copyrights on them. This is part of the wisdom of conducting a thorough patent search.

In the cleaning device example, we end this section by saying:

The gathering of nylon mesh material and fastening it in the center allows the material to be formed into a ball-like shape, suitable for use as a cleaning device.

Brief Description of the Drawings

In this section, you will list the drawings in your invention. Typically you will compile this list after you have written the next section, Description of the Specific Embodiments. However, you'd be wise to at least prepare a preliminary list of drawings and if changes are made after writing your patent application, make the necessary changes.

This section is usually easy to write, one sentence and one paragraph per drawing. It simply references the drawings included in the application. All you need to consider is putting them in their proper sequence. Try to start with a drawing showing how the invention is made or assembled and then follow with subsequent drawings showing the completed product. Last, show how it works if it's not obvious at this point.

It is usually best to describe the type of drawing you are using, such as we'll use in the example:

Fig. 1 is a perspective view of the present invention illustrating the gathering of the nylon mesh prior to forming the ball-shaped cleaning device.

The various types of the most commonly used drawings are explained later on in the chapter. Again, you may use language in other issued patents to find out what kind of drawings you may use in your patent application. There's no need to re-invent the wheel . . . just use—at times trace—commonly used drawings and language in other issued patents.

Description of the Specific Embodiments

This section will require some quiet time and much thought. Try not to be interrupted when writing this section and compiling your drawings. You must explain in specific detail how each facet of your invention is made or how it works.

You want to start this discussion at the beginning of the assembly or production process and continue in a logical sequence. For example, in our flat-bottom bag application, we sequenced as following:

1. We show how the cold seals are made (Figs. 1–4).

2. Then we show the final product (Fig. 5).

3. We illustrate how it works (Fig. 6).

4. Last we show two relevant variations to the flat bottom theme (Figs. 7 and 8)

Without question, this is the most difficult section to write. If you find that it a breeze, you are probably not explaining your invention in sufficient detail.

There are four challenges with writing this section:

1. You want to explain in detail, but not so much detail that it may limit the scope of your invention. You need not get into the physics of why something works the way it does, only explain why it does. For instance, with our cleaning device, you may point out that $\frac{1}{8}$-inch spaces that form the mesh material are sufficient to keep water tension low, but you don't need to explain the phenomena in scientific terms. Usually things such as size are not important to an invention; however, if it is, you want to make sure you include the ideal tolerances.

2. You want to write in a logical sequence to carefully explain exactly what is happening. In other words, explain from beginning to end how your invention works. You may jump around in your descriptions of the drawings for clarification purposes, but once you have, the next drawing you discuss should put you right back in sequence. In our cleaning device example, we elect to begin with a long piece of nylon mesh material and explain how it is gathered into a ball and show the final product last.

3. Accurately match your reference numbers (and letters) with the drawings. As you'll see in the subsequent section on compiling your drawings, this is best accomplished by using only even numbers so you can add odd numbers later on should you have omitted some important attribute.

4. You have to proofread the section to confirm its accuracy. This is not particularly easy since you'll find you'll most likely make several changes before you are satisfied with the outcome. Thus, the importance of using a computer is paramount in order to accomplish this in a timely manner.

This section alone may take several hours to complete and you may redo the drawings and the corresponding copy several times. However, one thing is for certain. Writing this section is one of the best ways to

FIG. 1

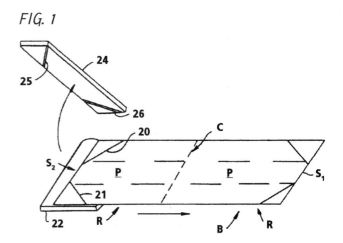

FIG. 2

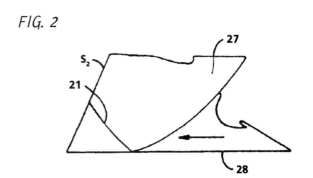

learn about your invention in depth. **It is usually during this process that you find flaws, problems, and alternative methods or design-arounds.** If you discover alternative design-arounds that may affect the scope of your patent, you'll certainly want to include them in your application.

When writing this section you will find that long sentences are common. That is to be expected, since legal writing often employs long, rambling sentences for clarity. A rule you can use when describing specific embodiments is to keep each related thought together in one continuous sentence, even though it may be a bit complex and run-on.

After this section and the drawings are completed, you are in the home stretch. Just make sure to proofread for accuracy.

Even if you have struggled with this section, the worst-case scenario is that your attorney may take a little extra time to edit and rewrite it for you.

What's important is what you learned during this writing experience and that you have hopefully covered the most important design-arounds.

FIG. 3

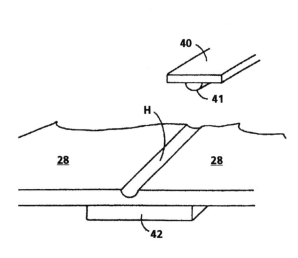

FIG. 5

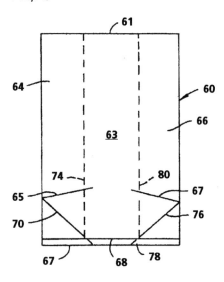

FIG. 4

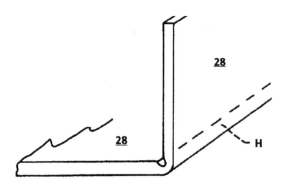

FIG. 6

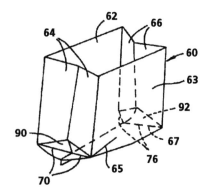

FIG. 7

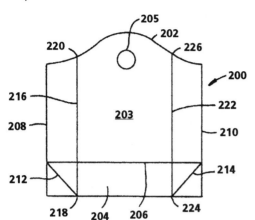

FIG. 8

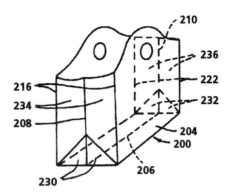

Compiling Drawings

Generally speaking, drawings for patent applications are not nearly as sophisticated or complicated as engineering or CAD drawings. In fact, they really are more like drawings that would be prepared in a high school drafting class, with an emphasis on simplicity. Perspective views are by far the most commonly used drawings, with plan and blow-up views probably second most common.

The crucial purpose of your drawings is to clearly illustrate the inventive matter. If they don't do this, you may be in for some expensive rework or outright rejection causing you to either re-file or file a CIP application.

In spite of this warning, there is no need to rush out and learn how to become an expert draftsperson. Nor will you need to purchase any expensive books on the technical aspects of drafting patent drawings because in most cases it's unnecessary. Most inventors tend to be fairly good at drawing and most inventors can prepare adequate drawings for provisional patent applications and generally speaking, the permanent ones too. However, if you have any reservations about your ability to draw or to effectively depict the inventive matter, you can have an experienced professional, maybe even a qualified college student, prepare your drawings for you. Let your patent attorney be the judge of whether or not your drawing ability is adequate. It's also interesting to note that many patent attorneys will personally be able to prepare or help you clarify the preliminary drawings.

Patent application drawings do not have to be the formal drawings required for issuance of a patent either. Upon review of your patent application, the PTO examiner will invariably send you a written objection citing your drawings are "not up to standard." This only means they have to be fixed. It does not mean your patent application or its content is rejected. You will have to have a professional perfect them and the best time to do this is when you have the formal drawings prepared just prior to the patent issuing.

The reality is you'd be wise waiting to have them done later on and spending the money then. The reasons are two-fold. First, during prosecution the examiner may cite the ineffectiveness of your drawings to depict the inventive matter. Thus, you will have a chance to rework or clarify them. Second, if some of the patentable material is being

SCIENTIFIC JOURNAL

The Scientific Journal is an inventor's journal that makes patent drawings much easier to compile. It has a special $1/4$-inch grid pattern with $1/32$-inch marks that allow you to easily construct your perspective patent drawings. The grid is designed so that when you copy your drawings on a dark setting the words "Do Not Reproduce" print through. Or, you can copy it on a light setting and no grid lines will appear at all. It's perfect for patent application drawings. See the Appendix for more information.

deleted or put into a divisional or CIP application, then some of the submitted drawings may be deleted anyway. So why pay upfront for formal drawings that might have to be re-done or are discarded later?

Organizing Your Drawings

It is best to compile your drawings while writing the section entitled Description of the Specific Embodiments. After you've done this, then you can go back and fill in the part entitled Brief Description of the Drawings. It's really easier this way because once you get into the storytelling mode it's better to finish the story with sequentially done drawings. The normal sequence in doing this is:

1. Show one view clearly illustrating your invention; this is usually a plan view or perspective. (At times, this may begin with the prior art figure, if there is a specific point you want to illustrate as a comparison.)

2. If necessary, show some side elevations (side views).

3. Show the specific embodiments that illustrate the inventive matter. This may be a blow-up view of one or more elements.

4. Show the invention in action. This should be one or more drawings that illustrate how the invention works.

5. Last, show any variations on the theme you believe are important.

At least one of the drawings should illustrate the "preferred embodiments." You are legally obligated to reveal the preferred version of your invention in the patent application.

Since the section on drawings and the corresponding written portion is such a fine way to learn about your invention, try to be as thorough as you can. Remember, you can model your sketches after existing patents—even copy or trace and resketch them.

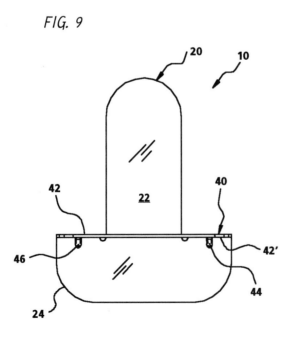

FIG. 9

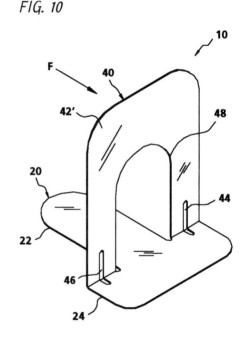

FIG. 10

Drawing Types

Learn the following drawing types and methods and you'll quickly be proficient in doing patent drawings. The most commonly used drawings in product development and patent applications are:

1. **Plan view.** The view typically is seen from directly above an invention, or in other words, is a "top view." It is one-dimensional and shows the basic layout of an invention. Fig. 9 is a plan view of a bookend.

2. **Perspective view.** These are by far the most commonly used drawings in patent applications. A standard three-dimensional drawing may show the invention or some of the important aspects that more closely resemble a real life perspective. See Fig.10.

3. **Blown-up view.** When a certain element in an invention is important but is small, it is best to illustrate it by using a blown-up view. See Fig. 11.

4. **Front, side and end views.** These may be helpful in illustrating the "head on" appearance of one or more of the one-dimensional views of the invention. They are not very commonly used in patent drawings, but they are certainly one type of drawing that you can use to help illustrate your invention. Fig. 12 is an end or side view of the bookend.

5. **Exploded view.** This view may be important for illustrating the connection between invention components, while also illustrating each of the components individually. An exploded view may be used with a side view, a top view, a bottom view, or perspective view. See Fig. 13.

6. **Flow chart.** These are becoming more important and typically illustrate how a methodology or system works. They may also be effectively used with computer- and software-related inventions. See Fig. 14.

FIG. 11

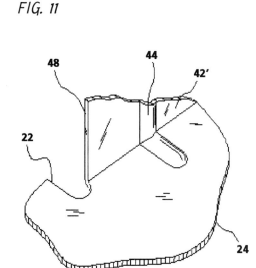

FIG. 12

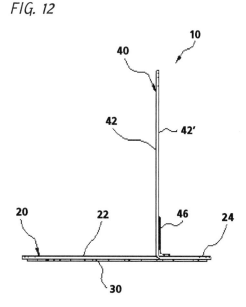

7. **Block diagram.** Block diagrams are useful for illustrating various electrical components without having to illustrate physical wiring connections. See Fig. 15.

Important: When compiling a series of drawings illustrating the utility of an invention, keep them in a natural, storytelling progression. It's a lot easier to understand an invention when the descriptive drawings are in a logical order.

Numbering, Lettering and Line Types

Your drawings and the components will be referenced using a numerical format. By adopting the following formats and suggestions, you'll quickly become adept at patent drawing. Use the figures for reference. The reference number guidelines to use are:

1. **Figure numbering.** Designate a figure number for a drawing if there is more than one on a page (or series of pages) that you are referencing with written details. Keep the figure numbers in their proper sequence.

2. **Lines with arrows.** Use a floating arrow-line (the number 10 in the figures) to reference the "invention in its entirety." Use an arrow-line touching the edge of a main component to reference "the entire component" (for instance, 20, 30, and 40 in the figures). These arrow-lines are illustrated in Figs. 9, 10, 12, and 13. In the flow chart in Fig. 14, lines with arrows represent the directional flow of the elements.

3. **Lead lines.** Use a lead line without an arrow to reference an individual element, part, or portion of a component

FIG. 14

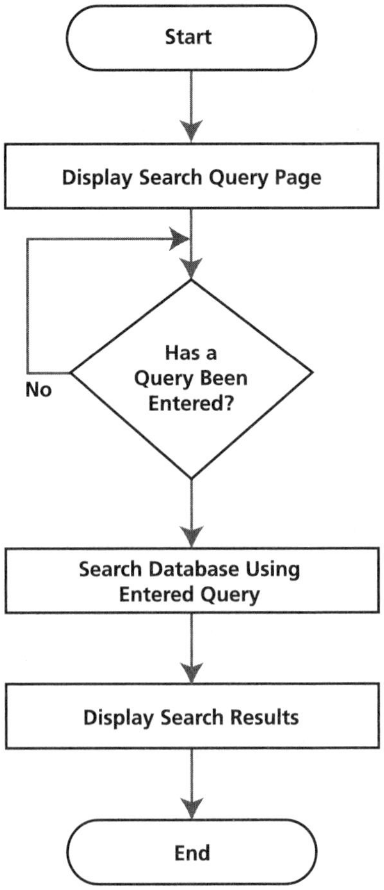

FIG. 13

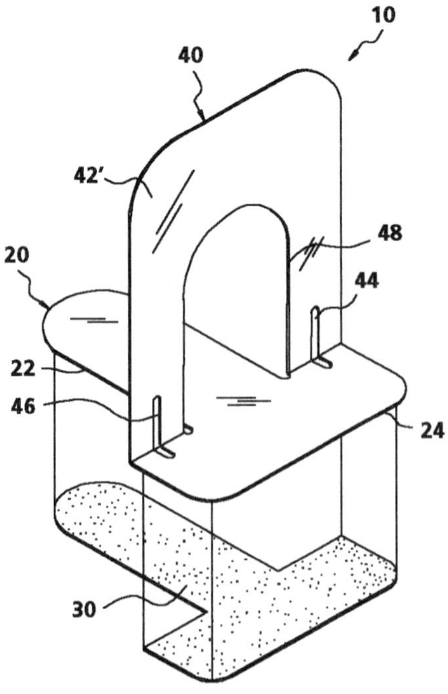

FIG. 15

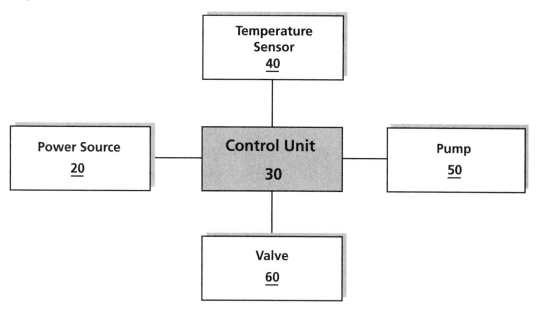

as illustrated in Figs. 9–13, and 16. In the block diagram in Fig. 15, lead lines show the connection between components.

4. **No lead line.** A lead line may be eliminated by placing a number directly on a designated component and underlining it. This is illustrated as 22 in Fig. 9.

5. **Numbering sequence.** When numbering components and elements, begin with the number 10 as in Fig. 9 to identify the invention in its "entirety" and then number each of the "main components" of your invention in a sequence of 10s (e.g. 20, 30, 40, 50, . . .). After all of the main components have been numbered, you then should number the "subcomponents" of each of the main components using only even numbers, based upon corresponding main component numbers. For example, if a main component was a base 20, the subcomponents may be numbered as a first portion 22 and a second portion 24, which together form base 20. See Fig. 10. It is often times useful to create an "index table" to keep track of your numbered elements. The reason for using only even numbers is that it is easy to overlook a particular element when describing a drawing.

Thus, as you write the description, and realize you've forgotten an element, you may go back and put in a number in the correct sequence. This, of course, would be an odd number. For instance, if you had forgotten to define an element that should have been described between the elements 38 and 40, you can add that new element as number 39. This is illustrated in Fig. 16. This way, the numbering is sequential as the drawing is being described. Otherwise, you would have to erase all the previous numbers to keep them in a proper sequence, or insert a much larger number (depending upon where you left off), putting the numbering out of order.

6. **Using a prime with numbers.** When your invention has a right and a left side, or a front and a back, use the same number for components on each side. On one side add an apostrophe (') after the number (referred to as a "prime"). For instance, the

FOR ENGINEERS

Use obvious symbols for things like electrical and electronic components, building materials, pneumatics, hydraulics, and so on. Your patent examiner is going to be an engineer, knowledgeable in the field and will readily understand the various components and symbols.

number on a front side of a component is 42, the number for the rear side of the component would be 42' as illustrated in Fig. 12.

7. **Similar and same elements.** You can also add letters to the numbers when referring to similar elements. For instance, the teeth in a gear could be illustrated as 38a, 38b, 38c, 38d and so on. See Fig. 16.

8. **Letter usage.** Letters may also be used to illustrate forces. They are helpful as a means to repeat an action. See Fig. 10. For instance:

"A" for "air flow"

"V" for "vacuum"

"F" for "force"

Letters are also commonly used to designate attributes on prior art drawings. For instance, in a prior art bookend drawing, "B" may refer to a base, "U" to an upright, and "R" to a reinforcing element.

Again, most patent application drawings are not highly technical CAD illustrations; they are simple drawings that sufficiently reflect the inventive subject matter. The ones we've included here should cover most simple patent applications. Should you wish to learn more about the various types of patent drawings, including those used for design patents, read *The Patent Writer*.

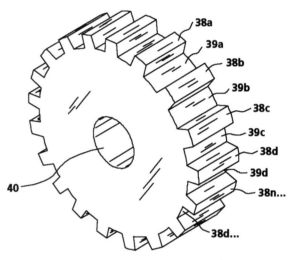

FIG. 16

USE YOUR TIME WISELY

You can waste hours trying to learn all the minute details about how to write a patent application and prepare perfect legal drawings. The result is almost always the same. Frustration sets in and you end up with an application that has an inadequate scope or is filed too soon.

Focus on making your application work with all the right attributes for superior patentability. Let your legal expert review and edit your provisional applications instead.

Completing and Reviewing a Provisional Application

At this point, you've completed your provisional patent application. Now you should review it carefully to ensure the content is thorough and that your application will not be returned. Here is a checklist:

1. The correct paper layout has been applied.

2. The legal format, including subtitles, has been used.

3. The inventive matter listed in your claim statement is clearly described throughout the written portion. Even though it is not required in a provisional application, a claim statement is by far the best guideline you can use to ensure you'll have broad scope protection.

4. You have included sufficient drawings to illustrate how the invention works.

5. The drawings use the correct numbering system, line style types, and so on.

Now, all you need to do is fill out the cover sheet and send it to your attorney for review and editing.

Claims

Claims are part of the permanent patent application. Although one or more may be included in a provisional application, it is uncommon. Your patent's claims are right at the heart of your patent protection. It is the single most important job your attorney has. Use an experienced one. Specifically, **the claims listed in a patent application define the scope of your invention.** Claims are listed in the last section of the patent application.

The claims you'll ultimately have in your final, permanent application essentially derive from the inventive matter you have put in your provisional patent application. Thus, you can see the importance of writing a claim statement early on. After compiling your statement and drafting the application illustrating the inventive matter that substantiates the claims, submit it to your patent attorney for editing and filing.

When your attorney files your permanent patent application within the next year, you let him/her write the application's legal claims based on your claim statement. Your challenge is to write them in simple English so your attorney can convert it into their legal form. You must spend some time and think very carefully about the claims because your patent's scope depends on it. The future of the invention and your company may very well depend on it.

You should take into consideration that some negotiation usually takes place with the patent office examiner. This is where your patent attorney can help a lot. Nevertheless, you want your claims to reflect the broadest possible scope you can literally imagine. Your attorney will have a claim-writing strategy in mind to maximize your protection when they get converted into their legal form.

To ensure maximum protection with your patent's claims, you want to include all the various forms of inventive matter as previously discussed in Chapter 6, whether that is a device, system, composition of matter, and so on. Then, you want to make sure they cover the greatest breadth possible.

Before you write your claim statement, read the following so you'll have a better understanding of just what a claim is.

Broad Claims

In summary, broad claims are what you almost always want to pursue. More specifically, broad claims:

1. Give greater protection.

2. Are powerful assets for totally new concepts such as Velcro®, ZipLoc® bags, and Kleenex® tissue.

The claim must be broadly, yet carefully, written to avoid design-around. For instance, words such as "materials" or "compositions" should be used in place of "wood" or "steel."

EARNING YOUR FEE

Drafting excellent claims is an art and takes experience. Learning the various ways to write and list claims is not easy. Having the claims you want—and need—will make or break your patent. Writing solid claims is where a patent attorney earns his/her fee.

CLAIM WRITING

Every experienced inventor I know uses the same method for claim writing. In simple English, he/she lists the claims as precisely as possible, to maximize the coverage. This is your claim statement. Then have your patent attorney write bulletproof claims in the application.

SPECIFICITY OF CLAIMS

Claims must have some specificity to them. For instance, your claims must read on the subject matter contained in a patent application and should not be interpreted as covering prior art subject matter.

3. Are vulnerable to subsequent improvements. Will the improvements be more important than the initial broad claims? For instance, the original ZipLoc® bags were a bit difficult to use, until they were made with the double and triple rails (a subsequent improvement patent over the original). They were also costly to make, until the manufacturer discovered how to extrude the rail profile right into the film, eliminating the entire step of bonding the two elements together. This, too, was a subsequent improvement patent.

4. Are usually in the form of independent claims.

Since broad claims are generally more powerful than narrow claims and subsequent improvements, they are frequently referred to as "dominant patents."

Narrow Claims

Narrow claims may not embrace the breadth of a broad claim, but when the subject matter is very specifically written and is considered the most advantageous variation in the field, a powerful patent can be the result. In summary, narrow claims

1. Provide less overall coverage, since they tend to relate to specific inventive details of existing products

2. Are commonly seen as improvements to existing patents and product groups

3. Are usually more vulnerable to design-around

4. Can, if they are the preferred technology, have a lot of power because of their specificity; hence, they can be very valuable

FILING PRO PER = A PATENT WITH NARROW SCOPE

The scope of a patent depends on the breadth of its claims. Inventors filing *pro per* (by themselves, without the assistance of a registered patent attorney or agent) invariably fail in this area. They settle for claims that are simply too narrow to give them any real protection. This is most unfortunate since at this late point in the inventing/patenting process, it could mean a complete waste of time and effort. Think about it. Do you want to spend 18 months to two years trying to sell a product that ultimately has very narrow patent protection and will be easy for larger competitors to design around? Of course not. Follow the Strategic Guide and you'll attain the scope you want with no added expense.

An example of the power of broad and narrow claims is:

• Patent 1 is the dominant patent that claims "a vehicle with one or more wheels."

• Patent 2 is an improvement patent that claims "a vehicle with four wheels which is steerable." (It has a steering system.)

The broad coverage of patent 1 covers all vehicles, including unicycles, regardless of whether they can be steered. In contrast, the narrower coverage of patent 2 covers only those vehicles with four wheels and steering systems. You could argue that patent 2 with the narrower claim is more powerful than patent 1 because without the ability to steer, patent 1 is almost worthless. Obviously very few vehicles would be in use today if they could not be steered!

Independent Claims

Claims that "stand alone" are called "independent claims." This means that the description in the claim does not rely on any other claim in the patent. They cover the broader inventive matter. There are typically one to four independent claims in a patent when it is granted. In simple English, here are three examples of independent claims that might be found in three different patents:

Claim 1.   A vehicle with four wheels

Claim 10.   A plastic bag with a reclosable bag mouth

Claim 20.   A nontacky adhesive tape

These independent claims obviously have broad coverage, but they are vulnerable to specific improvements. The improvements may be included in the same patent application as dependent claims.

Dependent Claims

This means that the description in the claim "is dependent upon" another claim in the patent, usually an independent claim. It may also be dependent on one or more other dependent claims, which is referred to as a multiple dependent claim. This is usually not advisable to do, however. With some planning, you can avoid this.

Dependent claims more narrowly define the inventive matter. When infringement occurs, it may be based on the independent claims and/or the dependent ones. Thus, dependent claims help the patentee broaden the scope of the patent protection.

Examples of dependent claims corresponding with the previous independent claims 1, 10, and 20:

Claim 2. The vehicle in claim 1 whereas any two or more wheels are steerable.

Claim 11. The plastic bag in claim 10 whereas said reclosable mouth is a plastic profile zipper.

Claim 21. The adhesive tape in claim 20 whereas the contact surface of said tape is a layer of ultra low-density polyethylene with a melt index less than 2, which layer bonds to other surfaces when exposed to temperatures in excess of 160 degrees.

Generally speaking, the best rule is to try to include the broadest possible coverage (independent claims) and to include as many improvements as you can imagine (independent claims). Using the preceding vehicle example, you can see how the following series of claims empowers the patent and the products that can be made from it.

Claim 1. A vehicle with one or more wheels

Claim 2. The vehicle of claim 1 whereas any two or more wheels are steerable

Claim 3. The vehicle of claim 1 whereas said wheels are made of rubber

Claim 4. The vehicle of claim 1 whereas any pair of wheels are at opposite ends of an axle

Claim 5. The vehicle of claim 1 whereas said vehicle is made of metal

The ability to write good claims comes with experience. What is more important is to have good patent counsel to write solid claims. You will be relying on these claims to protect your interests over the next 20 years. If there is one thing your patent attorney must do extremely well, it is to *make sure that you are protected with convincing claims*.

Sometimes a patent attorney will include more claims than you may expect. This might be because you may have overlooked some facets of the invention or perhaps the strategy is to accept some narrowing of the claims during prosecution. That is, your patent attorney's strategy may be to "blanket the world" and "settle for a continent" in an attempt to give you the coverage you want in the first place.

A Patent Office examiner's job is to thoroughly peruse, analyze, and evaluate the merits of your patent application and its claims. If every claim were accepted exactly as

---

CLAIM WRITING

Based on the review of an office action from a patent examiner, you may have an opportunity to actually expand upon the patent claim coverage. But don't assume this is automatic. Once an application is approved, you are in a position to file a CIP to apply the inventive matter to other related material and inventions. Your patent attorney or agent will know exactly how to do this and how much broader it may expand the scope. If you have been following the Strategic Guide model and have sales and profits generated, proceeding with a CIP may be an easy budgetary decision to make.

submitted, there would be no need to examine applications. It is the examiner's job to weed out claims that are obvious, anticipated or are simply far too broad and cover prior art as well. Conversely, it is your patent attorney's job to prosecute and negotiate on your behalf to get the broadest possible scope. In this light, the strategy of many patent attorneys is to begin with incredibly broad claims knowing that the patent examiner will be rejecting them anyway.

Regardless of the approach you and your patent counsel use, once the claims are written, you must review them and give the attorney your opinion and approval. It is your obligation to ensure that they accurately reflect your invention. Compare your simple English claim statement with the legal ones written by your attorney. Make sure you have not overlooked anything.

Supporting Documents

The permanent patent application requires two more additional elements: the Abstract and Oath of Inventorship. These elements are not required when filing a provisional patent application. You may also file a Small Entity Status declaration, even though it is no longer required. Many attorney still recommend it.

Abstract

The abstract is on the first page in your patent, once it is granted. It is a one-paragraph summary of the content. You will have a much better idea of what to include in an abstract after the entire application has been written. So wait and write it last.

An abstract has little to do with the power or scope of a patent. It simply offers the reader a snapshot of the inventive matter in the patent. The abstract sometimes does not accurately disclose the true content of a patent. This is particularly true if one patent has been split-out into two or more applications. In such a case, you may find that one abstract covers both patents, although one may refer to a product or system and the other to a process of making the product.

The abstract for the previous draft is as follows:

An expandable plastic film gusseted bag is utilized. As in conventional bag manufacture, each front gusset is connected to a rear gusset panel and each of the gusset panels is foldable between the front and rear panel to provide a continuous bag periphery. A bottom joins the bottom of the front panel, rear panel, two rear gusset panels and two front gusset panels at substantial right angles to the side edges of the front, rear, and gusset panels to close the bottom of the bag, leaving only the top open. The improvement provides each plastic bag with eight first linear folding axes, each configured adjacent to the bottom and within one of the front panel, front gusset panel, the rear panel, and rear gusset panel. These folding axes extend from a joinder of a gusset at the bottom to the side edges of the front panel or the rear panel and the front gusset panel or the rear gusset panel at an approximate 45° angle. The folding axes are made by a process from a group consisting of either a cold seal or a hinge in the plastic film while the film is planar. On the opening of the expandable gusseted bag, the bottom seal of the bag is disposed adjacent the gussets and forms a generally triangular-shaped overlay relative to the gussets to form a flat bottom bag.

A FINAL REVIEW

When you read the claims your attorney wrote in legal form, study them well to make sure you are covered based on your previous claims statement. This is your last chance to add drawings and descriptions to improve the scope of the invention. Even if it takes another week to rewrite and include, do it now!

Oath of Inventorship – Power of Attorney

When your regular patent application is filed, it will be accompanied by an additional document, your Oath of Inventorship, sometimes referred to as a Declaration of Inventorship. In simple language, the oath typically states that:

- You reside at the address listed on the document.

- You believe you are the original, first, and sole inventor.

- You acknowledge your duty to provide information material to the application.

- You claim foreign filing protection (if applicable).

- You are granting power of attorney to prosecute the application.

- You certify that all the above is true.

The power of attorney gives your patent attorney the right to file the application on your behalf and to subsequently prosecute it. Once the application is filed, the Patent Office will want to communicate only with the one person who is designated in the power of attorney.

Small Entity Status

A small entity is an independent inventor, a nonprofit organization, or a business with 500 employees or less (including all affiliates). Be careful, though. The law states that if you license a producer with more than 500 employees, your status will change to a large entity. If in doubt, consult with your attorney to determine your entity status. As you know, most fees, including filing, maintenance, and issuance fees for companies falling under the small entity status are one-half the cost for large entities. The proposed new examination fee is the same for both small and large entities, however.

It used to be a requirement to file a declaration stating that you or your company is a small entity. It is no longer required; however, it is still commonly done. To file as a small entity, all you have to do is pay the small entity fee, and it's assumed that you are one. If you wish to file a "Declaration Claiming Small Entity Status," your attorney can prepare one for you.

CHANGE OF ENTITY STATUS

What happens if you have filed your patent application as a small entity and after the patent is pending or has been granted, you license it to a large entity? The answer is: Once your entity status changes, then all subsequent payments will then be as a large entity.
–Bill Pavitt, Esq.

21st Century Strategic Plan

The new patent office initiative has some changes in the law that may affect how independent inventors and small entities write and prosecute patent applications. Some may even increase cost. What is important to note is that if you keep your inventive matter reasonably straightforward, there will most likely be no changes in the way you write your patent applications. Here is a list of the proposed changes to watch out for:

1. Multiple dependent claims having a substantial upcharge

2. Independent claims in excess of three substantially increasing costs

3. Claims in excess of 20 increasing cost

4. Applications with multiple inventions increasing cost

5. Long-term extensions for response to office action increasing cost

Generally speaking, with the new proposed laws, if you want to keep your costs in line, file patent applications that cover specific inventive matter, keep your independent claims to three or fewer, and don't write multiple dependent claims. Frankly, most patent applications fall under this realm anyway. The new laws are an attempt to stop companies from filing a single application with 50 to 100 claims or more and to extend the patent pending time period out several years.

PROTECTING YOUR RIGHTS

It is another problem-solving adventure . . .
at times disheartening, at times exciting!

After your patent has issued and the product has been successfully launched, you may be paid the biggest compliment of all—that is, patent infringement! In other words, competitors will want to copy your invention. Unfortunately, this may mean they may attempt to do so without paying you royalties.

Fortunately the law is on your side and there's plenty you can do about it.

Knock-Offs and Patent Infringement

You've piloted your invention to a successful birth, and now competitors may attempt to copy your invention and try to violate your patent rights in doing so. While copying your invention is a compliment of sorts, it can be a costly problem as well.

You may hear discrediting comments from others or even their attorneys but don't let them fool you. Stay cool, and remember who has the patent. Remember that most patent infringement cases are settled in favor of the inventor. Aside from software patents, which tend to be much more vulnerable, you will win 80% of the court cases. But most infringement cases are settled in favor of the patent holder before they go to court. As you will recall, the primary driving force behind an early settlement is the high cost of litigation for a company on the defensive end of an infringement suit. Far more than your cost as a plaintiff.

The only way for an infringer to get out of an infringement case is to invalidate a patent or license it from you. An infringer can easily run up a $1 million legal price tag trying to invalidate it. If that's not a big deterrent, what is? Add to that, the corporate responsibility issue become enormous. The exposure for shareholders can be substan-

VALIDATION

Infringement is the best validation that you have a great invention! If you've worked with your attorney in preparing a STRONG patent, you're in a great position to make money from an infringement.
–Andy Gibbs, CEO, PatentCafe.com

tial and major corporations may lose up to one-third of their stock market value upon notice of a patent infringement case. Read about how infringement can affect corporations in my book *Essentials of Patents*.

It can be painful to see another company blatantly copy your inventive ideas. Infringement on (theft of) your intellectual property feels the same as having your car stolen or your house robbed. Expect infringement to happen if you're successful, but keep in mind that you are well positioned to combat it.

So, if you planned your invention and patent development well, wrote a bulletproof patent application with a broad scope, and have a competent attorney on your team, you're in a position to profit. Exactly what do you do?

Handling Negative Comments

This is common. Keep in mind that many people who are going to try to destroy you and your patent are misinformed about patent laws. Some may still be thinking about archaic concepts such as, "I'll change it by 10% and get around it" . . . whatever that means. Others may attempt to intimidate you into thinking that you do not have a valid patent. Yet others are jealous of creativity and want to discredit the original creations of others. Last, some are motivated by greed and will try to profit from your genius without compensating you. Most of these people are generally unaware of the numerous court decisions in the last fifteen to twenty years that have been in favor of patent holders.

For some serious and yet light-hearted entertainment, here are some of the common comments you might hear and what they really mean:

"I've seen that before."

- They do not mean that they have seen *exactly* your product before, but perhaps a similar one. For instance, regarding your revolutionary high-speed tortilla machine, they may mean that they have seen a tortilla machine before.

- They have seen a similar process before, but it was five years ago and not with tortillas, but in a tile manufacturing plant. Keep in mind that most laypeople may not understand what makes one product patentable over another.

- If they truly claim to have seen what you created, get the details. If *exactly* what you have declared in your patent has been done before, your patent (or one or more of the claims) may indeed be invalid. Ask for proof. Where can you see one? Then determine if it is a similar, but non-infringing, product (95% chance), a recent infringement you already know about (3% chance), or previously unknown prior art product (2% chance).

"You can't patent that . . . it's too simple."

"You can't patent a hole in a plastic bag." Oh yes, you can! Frequently, people look only at the simplicity of a hole in a bag and not at the outcome it produces. If the hole does something unique, we know it can be patented.

For instance, the holes punched in the handles of a plastic grocery sack contribute to a bag design that self-opens and dramatically improves productivity. Furthermore, the holes in the plastic bag are of a special configuration that does not propagate tears in the handles of the thin plastic film, which of course, would be a disaster.

Negative comments about simple patents are really compliments! Why? Because you received a patent on something so simple, yet unobvious, and it does something so wonderful. In contrast, the positive person says the opposite: "Wow, it's so simple, why didn't they do it like this in the first place?"

"Your patent isn't valid."

This is a comment from someone trying to either put you down, has been listening to his attorney too much, or is trying to intimidate you. The person may actually be a little dense and unaware of patent law. Even if the individual who makes this statement knows something about patents, do not be concerned; it's just a ploy. And, to this person, it is a $1 million plus statement to prove his position. You can even let him know that. Besides, your patent application has already been scrutinized by experts and the U.S. Patent Office agrees with you.

"Well, until you get a patent, we'll go ahead and make the product."

This comment would be made during the patent pending process. It is usually made by a company much larger than yours. This is usually a bullying or a survival tactic and long term it won't work. Once your patent is granted, give them notice that they will be liable for willful infringement from that point on. Willful infringement is subject to treble damages. You can be in for a huge windfall profit!

"You just copied someone else's patent."

People who say this do not have a clue. If they are trying to insult you, consider laughing, because the insult applies to their own intelligence. Anyone with a minimum amount of patent experience knows that you can't copy another patent and get it issued.

Positioning Your Responses

Regardless of the negative comments you hear, having a patent (or patent pending) puts you in the stronger position. The burden of proof to invalidate your patent is on the infringer.

Your standard response should be to politely disagree. You can add that your attorneys disagree as well. You don't need to get into any details early on. You only want them to know that you want to effectively promote your invention in commerce and to create and protect jobs in America. Who can disagree with that philosophy?

When Infringement Occurs

Since the statute of limitations allows you six years to take action, you have plenty of time to respond. Don't panic. You may want to take several years to respond. Get the facts first. Here is what you do:

1. **Find out everything you can about the infringing product.** Get a sample of it, look it over carefully, and determine if it is indeed an infringement. You don't want to mistakenly press a claim for a product that does not infringe. This could result in a damage claim against you.

SINCERE FLATTERY?

Remember, infringement is the sincerest form of flattery for an inventor. It means someone likes your invention so much they are willing to risk being sued.
–Michael S. Neustel, U.S. Patent Attorney

2. **Send a sample of the infringing product to your attorney.** He/she can confirm your suspicions.

3. **Determine how big the infringement problem is.** After investigating the infringing companies, find out the magnitude of the infringement. How many units have been sold? What is the sell price?

4. **Find out everything you can about the infringing parties.** Since this includes all those manufacturing, using and selling products under the scope of your patents, make a list of all manufacturers, distributors, sales companies, individuals and end users. How many of these entities know they are infringing your patent? Are they doing it willingly? What are their intentions?

   Is an infringer the leader in the industry in which your product represents a serious threat to their sales? Is it an aggressive company trying to cash in on your success in a big way? Is it a small company trying to cash in on your success in a small way?

   Get all the inside information you can. Perhaps they are responding to their customers' requests for your products without regard to your patent(s). Take good notes on what you hear from others in the field. Try to get the infringer's annual report to determine its current economic strength.

5. **Try to determine how infringers will respond to your actions.** Try to learn if the infringing companies really need your invention or if they would switch if forced to do so. Will they try to protect their market share at all costs? Will they want to continue making, selling, or using the product in the future?

   After compiling this information, you'll be able to talk about an infringement strategy.

Patent Infringement Strategy

Keep in mind that patent infringement does not have to be a bad experience; in fact, it's usually not. The outcomes are usually positive, and in a few cases, they can even become a gold mine.

When infringement occurs, several issues will determine your best strategy. They are:

1. **Your invention development strategy.** Are you planning to develop, manufacture and market the invention yourself or do you want to license it?

2. **The size of the infringing companies and how many are infringing.** You will probably have an entirely different strategy if there are several small infringing companies as opposed to a few larger ones. Remember, infringement includes those manufacturing, using, and selling products under the scope of your patent. So even distributors, retailers, sales agencies, and end users of your innovation may be infringers.

3. **Your patent's integrity and the nature of the infringement.** Do you have a strong patent with broad claims that are being infringed? Is the infringement literal or does it fall under the Doctrine of Equivalence? If it is literal infringement, it's a lot tougher for an infringer to challenge.

GOOD NEWS

Ironically, patent infringement is usually good news. Sooner or later, the infringer is going to have to address the issue. In today's business climate, we know that patent infringement is a serious issue and record judgments have been granted in favor of the patent holder. Your job from the onset is to make certain your patent's coverage—its scope—is going to be interpreted as broadly as possible. Don't allow for any margin of error.

HOWE LOW WILL YOU GO?

The inventor of the sewing machine was not Singer, but Elias Howe. Howe had difficulty marketing his invention, but Singer—a brilliant marketer—went ahead and infringed Howe's patent and created an empire. Howe sued Singer and was granted a royalty of $25 per sewing machine! After all was said, Howe didn't do too badly for not having a means to market his invention.

---

4. **How much money is involved.** Is the infringement of minor or major consequences? Does it represent annual sales of $20,000 or $20 million? The larger infringement cases are the better moneymakers.

5. **The financial condition of the infringers.** Are the infringing companies able to defend themselves? Or, are they on the downside of their industry, trying to find a way to compete?

6. **Your financial condition.** Are you able to afford a patent infringement suit? If you cannot afford to hire an attorney, get one to represent you on a contingency basis. Providing that the scope and integrity of your patent are adequate, finding a contingency attorney should not be difficult.

Deciding on a Plan of Attack

Once you have accumulated all available facts, you will be able to determine the nature and timing of your action. In a strategy session with your legal counsel, you can make some sound decisions. Here are some points to consider:

- Use a strategy that meshes with your development strategy. For instance, if you are granting non-exclusive licenses, your tack will be a lot different than if you plan to sell the product in your company or go into business with partners. If you plan to sell the product exclusively in house, you'll be running up larger legal bills. The reason is that you'll most likely not be waiting long before taking action, and you'll most likely take action against all infringers as they begin offering the invention to industry.

  If you plan to sell the product yourself and license others within your industry (which incidentally may be a wise decision and help standardize the industry), you may want to wait a few years before taking action. In this way, the infringers can actually help "run-up" sales and standardize the technology. If you are trying to convert an old-fashioned industry with your modern patented products, it is a lot easier to do so with more than one company selling the change. Infringers can help you do this.

- Take a position that will improve your and your development team's position and may even improve the credibility of the patent. Weigh the short-term financial gains against your long-term objectives. If you have a limited budget, you can consider the friendly licensing of competitors right away, even though they may not be leaders in the field, to generate some income. In Part 4, Licensing, you'll see the time to do this is either before or during the patent pending process if possible.

- If you wish to grant nonexclusive licenses and there are several hostile infringers who you are certain want to continue to make, use, or sell products under the scope of your patents, you may want to start by picking a fight with one or two of the weaker ones who cannot afford litigation costs. By licensing them, you are setting a precedent and generating some income. If they are small enough companies, they are unlikely to want to spend their limited resources challenging your patent's validity. The income they provide can be used to tackle the larger infringers later.

- Can you cordially discuss the patent infringement issue with the CEO of the infringing company? Just because someone is infringing your patents does not mean he/she

WILLFUL INFRINGEMENT

If you can prove willful infringement, the damages will be substantially higher. Provable, malicious infringement has recently resulted in several gargantuan monetary awards in favor of patent holders. If you do not have the budget to sue the infringers, you can locate an attorney who is willing to take your case on a contingency basis. Proving willful infringement results in treble damages and usually results in paying attorneys' fees as well.

DRAWING THE SWORD

Usually you do not want to threaten litigation against a potential future licensee unless you absolutely have to. Ideally you'll want to resolve the infringement issue amiably before it gets to this point. A letter from your attorney may be just the means to accomplish this.
–*Michael S. Neustel,*
U.S. Patent Attorney

LEGAL ADVICE

Don't take infringement by another personally. Pursuing a potential infringer should be based on sound business decision making. You may try to license the infringing company or have it cease altogether. Be careful not to threaten litigation against a potential infringer unless you really are prepared to file a lawsuit since a threatened infringer may file a Declaratory Judgment against you.

–Michael S. Neustel,
U.S. Patent Attorney

A CREATIVE STRATEGY

In one particular case in which I was counsel for a patent holder, we sued a notorious small company that was infringing my client's patent. In order to protect itself, the small company hired a very large well-known law firm to defend itself. Not long afterward, the small company's cost to defend itself rose to over $1 million. Unable to pay, it was forced out of business. So, instead of stopping the infringement, the defendant forced itself right out of business altogether.

–Doug English

is your adversary. You may be able to amicably discuss the problem and turn it into a long-term, mutually advantageous license or joint development agreement.

- If you have to take on a megagiant corporation with significant financial resources and your resources are limited in comparison, you may want to take them on last. If you don't, you need solid proof of infringement and should have your counsel work on a contingency basis. You can also look for angles that may be easier to pursue. For instance, perhaps a distributor or end-user customer is not as large as the infringing manufacturer and you could start your infringement action there. The only problem with this is that usually the purchaser of an infringed product will have the supplier defend it. In either case, there will be legal counsel for the infringing distributor or customer, who will want to settle. It may be able to push some buttons to make it happen sooner.

- If you take on the megagiant corporation last, after defending your position with several smaller companies, and perhaps licensing them, you will be better able to beat the giant corporation if the case goes to court since you have previous licensees.

- If you have greater financial resources than the infringers, consider taking on the largest, most visual infringer(s) first. With that victory being made public, you will be in a powerful position to quickly stop the smaller ones and get them to enter into license agreements as well.

There may also be other factors to consider. Most important is to discuss the strategy with your legal counsel and determine how best you may profit from the experience in the long run.

Taking Action

You have a choice of the methods to give notice of infringement to others. Some are friendly approaches, others not so friendly. Here are four approaches you can consider in order of their friendliness:

1. **Write a friendly letter.** You personally send a letter to the infringing party offering a license and letting it know you believe it is infringing. You write the letter but have your attorney review it first. Offer to discuss the matter amiably with the CEO. If you are already friendly with the infringing party, there's probably no reason to do otherwise. If the infringing party was unaware of your patent, your letter then establishes a legal date of willful infringement from that point on. Thus, the clock is running for treble damages.

2. **Your attorney puts the infringer on notice.** Your attorney writes a letter to the infringer offering a license and letting it know the consequences of infringement, including willful infringement. This, too, can be effective, but remember this causes the infringing party to contact its attorney, who, in turn, will respond to yours.

3. **File a law suit.** Your attorney can file a complaint in court. With the complaint, you'll need proof of infringement, such as a sample you can show in court, or purchase orders, or whatever the case may be. Your attorney will ask for royalties and/or damages on sales. If you have previously notified the infringing company, for instance in a friendly letter offering your patent for license, your attorney will request treble damages and payment of attorney fees and court costs.

This approach is usually not as costly or as serious as you might think. First, the cost to file court documents is only a few hundred dollars. Most of the cost in taking this approach is your attorney's time spent on preparing the documents. But once they're filed and once the infringing company receives them, then it behooves it to seek a settlement in due course. From this point on, the infringing party either has to invalidate your patent (the $1.2 million bill), stop practicing the invention, or settle out of court. It's obvious what the best choice is.

4. **Cease and desist.** This literally means "let's fight." You can have your attorney file a cease and desist order, which, in essence, is asking them to stop production immediately. A court case is almost imminent in this type of action.

Stopping Importers

If a company is importing a product that infringes on your invention, you'll want to take legal action against it, which includes the stoppage and inspection of containers bringing in the illegal products. This action is done with the Department of Commerce. When the complaint is filed, it is thorough and is almost like presenting court discovery documents that qualify your position. You want to have a solid argument and proof establishing the infringing fact.

With such a case, the infringing product is held in drayage and subject to being destroyed. Until a decision is reached, it will be running up warehousing charges. This can become a very bad situation for the importer because not only is the present inventory confiscated but also all future shipments on the water may be as well. Thus, the biggest problem for an importer is the extreme cost to defend such an action and then usually losing the inventory as well . . . a double whammy. It may even have additional liability if it has a contract to honor with the manufacturing company overseas.

The Role of Your Attorney

However you approach a patent infringement issue, consult with your attorney and coordinate your activities through him/her. Evaluate your position and strategy before incurring legal bills. Ultimately, you will determine how to proceed to reach the most advantageous outcome. Your attorney's counsel will help you make your decision. Before you proceed, just make sure you have the facts right and you've uncovered all the information you can. You must know the extent of the infringement and who the infringers are.

What if Your Patent Infringes?

While not very common, it does happen. Your patent application may actually have slipped through the cracks and there may be an unknown dominant patent which it infringes. If someone sends you notice or points out a patent that yours may be infringing, don't panic. Here is what to do:

1. **Check the filing date.** Get a copy of the patent to find out the date it was filed, not the date it was granted. Does the filing date pre-date your patent? If not, theirs may be infringing yours. If your filing date precedes theirs, you are probably in a good position. However, the first true inventor is ultimately going to be determined based upon first-to-invent rights, not the filing date. This is where your journal records will be important.

BEING A LITIGANT IN A LAWSUIT

If you should be in the unfortunate situation of being a litigant in a lawsuit, try to picture a cow with the plaintiff pulling the cow's horns and defendant pulling the cow's tail, the judge standing aside watching with humor, and plaintiff's and defendant's counsel gleefully smiling whilst they take turns milking the cow. This will remind one that despite one's differences, it is in plaintiff's and defendant's best interest, though not their counsels', to quickly negotiate an agreement they can live with, albeit not with glee.
–Doug English, Patent Attorney

COUNSEL FIRST

Inventors should never contact a potential infringer themselves unless they have discussed it with their attorney. Doing so makes them look too cheap to hire an attorney and may set the inventor up for a lawsuit by the potential infringer.
–Michael S. Neustel, U.S. Patent Attorney

2. **Read the patent.** Start with the abstract, skim the summary, and read the claims. Does the patent really talk about the same subject matter or does it talk about a similar technology? Or, does the patent talk about subject matter that is different, but still covers your invention. In either case, do the claims read on your invention? As you read the claims, follow the language and decide if *every* element in any single claim covers your product. If it doesn't include all the elements, it doesn't infringe. Remember the Doctrine of Equivalence. For instance, if one of the elements refers to a water-cooled apparatus and yours is air-cooled, there probably is no patent infringement.

STILL GOOD NEWS

If after all your hard earned effort, you find that it is just not worth the effort to overcome the infringement of another patent, this should still be good news. What would be worse than continuing the pursuit of a project that will evolve into a serious problem? Instead, cut your losses and spend your time on developing the next generation or another product altogether.

3. **Check the maintenance fee payments.** If the infringed patent is more than four years old, call the PTO and verify that the maintenance fees are current (You can check this out by calling customer service at 703-308-5068 or 703-308-5069 or the voice response system at 703-308-5036 or 703-308-5037.) Maintenance fees are due at 3.5, 7.5, and 11.5 years. If the maintenance fees are not up-to-date, the patent is no longer valid. Confirm this with your attorney as there is a six-month grace period.

4. **Send a copy to your attorney.** If your invention appears to infringe, send a copy of the patent with a written explanation to your patent attorney for review. You need not request a written, legal opinion at this time. Just ask your attorney if he/she thinks there is a serious problem. Let your attorney decide whether or not the claims read on your patent.

5. **Order the file wrapper.** If your patent appears to infringe the other patent and your attorney agrees, your attorney should order a copy of the file wrapper (the entire file) from the U.S. Patent Office. After review, your attorney can clarify the scope of the claims and determine if there is some vulnerability to the patent and its claims.

6. **Qualify the owner.** Who is the owner? Is it an individual or a corporation? Most patent owners, whether they are large corporations or independent inventors, are not usually interested in legal battles. If their patent is not something they are actively pursuing and you want to actively pursue it, you may be able to reach an amicable solution.

With this done, you now have a basis to decide on your next step.

Infringement Solutions

If your product infringes, you have some options. They are:

1. **Modify or improve.** Is there a modification or improvement you've considered that would put you outside the infringing claims? When researching this, look for the operative words that must be included to infringe the claim. Which ones can you design around? Which ones are the crucial elements?

2. **License the patent.** Licensing the other patent may be of mutual benefit. Or, you may be able to cross-license one another. This means you license your patent and the other patent holder licenses yours. You attorney can play an important part in concluding this kind of deal.

3. **Abandon the project.** It's a last resort . . . but a possible reality. Sometimes it's best

just to cut your losses. It's not the end of the world to abandon the project. You were inventive to begin with, so don't stop. Invent the future or pursue another product line. After all, the company you're infringing probably won't be!

One last possibility is that a larger corporation that owns many other patents may own the infringed patent. It may have little interest in pursuing it because they subsequently developed other technology they consider a much larger market. It may think that the sales and profits are too small to develop. Hence, it may not even pursue patent infringement. Or, it may simply license the patent to you under favorable terms.

Problems Become Opportunities

In the invention development process, you overcame problems and turned them into opportunities. With patent infringement issues, you must show the same patience, creativity, and determination.

As an inventor, you have an obligation to yourself and your company or development partners to continue to improve and take your innovations to the next level. As long as you continue to strive to develop the next generation of products, you will automatically see your efforts surpass those of the competition. In your patenting and inventing efforts, remember:

The greatest rejection in patenting is to never have been infringed.
By experiencing the tribulations of infringement,
you'll learn the value of a patent.

In summary, infringement does not have to be bad news, and it usually isn't. Most of the time it represents substantial opportunity. As long you persist in your effort and strive to create the next generation and protect your rights on those you've already created, you'll profit.

OUTER SPACE PATENTS

According to U.S. Patent law (Title 35, Section 105), inventions made, used, or sold in outer space on a space object or component fall under the jurisdiction or control of the United States. However, as of now there is no provision for patenting inventions created by entities (human or otherwise) living in outer space.

MARKETING & MANUFACTURING

INNOVATION MARKETING

Marketing new ideas and inventions requires substantial
effort and takes a special kind of team.

The Right Invention Marketing Philosophy

Marketing innovations requires a philosophy much different from marketing established products or new ones that are spin-offs of the same old product line. Marketing new ideas and concepts requires having excellent manufacturing and sales contacts and involves a strong team effort. There are so many crucial details your team must consider at the onset of a new product launch. You are the pilot, the team captain, of the effort; however, eventually your manufacturing and marketing experts will take over the bulk of the responsibilities. They are the ones who'll really make things happen! Your objective is to help them get to that point of no return.

Before product launch, your innovation must be thoroughly tested in-house, packaged, and priced, with the appropriate brochures and specification sheets ready to go. This is a time-consuming task and one that your manufacturing and marketing team members should do, and can do, quickly.

Depending on your development strategy, much of this work may be done by your company's manufacturing department, marketing department, or outside sales agency. If you plan to license your innovations, then your manufacturing and marketing partners are your licensees. Or, the manufacturer may be a contract supplier to your licensee as discussed later. No matter what your strategy is, they'll proceed under your guidance as the expert in the product's niche.

There are some basic marketing tools you can use to get a better idea of the marketing positioning of your innovation. A new innovation is usually a new market oppor-

MARKETING PROVES IT

If your invention won't sell, it's worthless! Prove your markets, prove your profitability, and prove your distribution channel. If you have become a marketing expert in your field or have a marketing partner on your team, you have a high probability of invention success.
–Andy Gibbs, CEO,
PatentCafe.com

tunity. However, the product's positioning must be well researched or no market will be created. An innovation that has strong emotional and customer appeal must also have the right quality and the right price, and of course, it must be delivered on time.

Your manufacturing and marketing philosophy will dovetail with your customer-driven innovation philosophy, or in other words, always doing what's best for the customer. The attributes you have created in your CDI products will now be reflected in how your manufacturing expert produces the innovation and how your marketing expert takes it to market. Your marketing experts have to have the same commitment and a similar expertise in getting your innovation launched. They have to share the same philosophy.

CDI, Market Creation, and Establishing Your Niche

Market creation is the marketing concept most commonly used by entrepreneurial companies when launching new products. The term "market creation" is actually self-descriptive, but it may better be described as identifying *a niche that presently doesn't exist*. So, all new market creations start out as niches.

You create new markets by applying customer-driven innovation to your products. With a CDI-mindset, your product design emphasis will always be on creating niches that are difficult for competitors, large or small, to enter. One of your chief challenges early on is to identify the product's niche and then to design your innovation accordingly.

In his book *Thriving on Chaos*, Tom Peters writes about "niche or be niched." This was the trend in the 90s and into the twenty-first century and carried a clear-cut message for survival. Frankly, most companies that did not listen to his forewarning aren't in business today.

If you want to see an extreme example of carving out niches with an existing commodity, look at the flour industry. Think about it . . . just ten years ago you had your choice of two flour types: white or whole wheat. Now you have multiple options: 1) white; 2) pre-sifted white; 3) self-rising white; 4) white with 100% natural ingredients; 5) whole wheat; 6) natural whole wheat; 7) corn flour, and the list goes on!

Niches also tend to have higher profit margins. In comparison, high-volume commodity products tend to have low-profit margins. Next time you're in the supermarket, compare the prices on the above flour varieties, and you'll see the speciality flours all have a much higher retail price.

Over time, niches may grow into a commodity—just look at the personal computer! But that's your objective in the first place, to try to ramp up a new market so it becomes the standardized generic giant. The PC started out as a niche in the garage of Steven Jobs and Steve Wozniak.

By using a CDI design approach and tailoring your inventions to value-added niche markets, there's no reason you can't be successful for years to come. In fact, you'll only get better at it and continue to create more niches in the future.

Market Creation Tailored to Small, Flexible Entities

Market creation fits small, fast, flexible entrepreneurial companies better than it does the megagiant corporations. Unlike many larger corporations, inventors and small businesses are not bogged down with bureaucracy, large decision-making committees, and detached R&D departments. Small entities are inherently more flexible and naturally more entrepreneurial. They can invariably beat the large inflexible competitors to mar-

---

NICHE OR BE NICHED

Find your niche in a market and capitalize upon it before someone else does. Just like Tom Peters says, if a market is not yet niched, it will be. Why not niche it with your invention?
–*Michael S. Neustel,*
U.S. Patent Attorney

---

ket with new innovations. Small businesses can make decisions, gear up for production, and adjust and modify development much faster than their large generic competitors. While large corporations may take three years or more to launch a new product, small ones can do it in a fraction of the time and at a fraction of the cost.

Commodity product lines are usually dominated by generic giants. If you can secure a small niche (just 1%) in a billion-dollar commodity-oriented market, you will not be a threat to them. If the niche is based on short-run, added-value products, it will be much easier for the smaller producer to protect it. Focus on your niche, turn your inventions into customer-driven innovations, and then let sales rise via "customer pull-through."

The large, independent, and detached R&D departments of corporate giants can be easily outmaneuvered by small innovative entities with their hands-on R&D operations. Besides, the focus of large corporate R&D departments is usually *not* on customer-driven innovations, but more on manufacturing and engineering improvements. These improvements are frequently directed toward production processes and cost-cutting measures—not CDI.

In addition, large R&D departments frequently have a propensity to overengineer new products, sometimes to the product's ultimate death. By doing this, it is common-place for them to take years to release a new product, sometimes losing the all-powerful first-to-market position and exceeding their budget while doing so.

Another factor that contributes to the ability of small companies to out-innovate large ones is that many large companies hire uncreative people to maintain existing business, not creative people to create new business. Couple this defensive posture with their inherent bureaucracy and inflexibility and it is understandable that it is almost impossible for them to innovate and launch new market creations.

Small, flexible entrepreneurs are also better equipped to sell new innovative products into new markets, instead of trying to nibble on existing market share. Unlike their larger competitors, the focus of a small company's sales forces is on selling value-added innovations that have real customer benefits. These smaller (sometimes considered "guerilla warfare" type) sales departments, with their CDI philosophy, can almost always outmaneuver the megagiant corporations to seize new market creation opportunities.

However, one thing is certain: independent inventors and small companies cannot create new markets if their innovations are merely simple modifications of existing generic products. In the last several years in the United States, we have seen a continual onslaught of cheap generic products. Many of these products are sold by the giant domestic corporations, imported from all over the world. From 1990 to 2000, generic producers had been in a battle—literally a dogfight—struggling to maintain the last remnants of dwindling generic markets. Now that the house cleaning has been accomplished, they are turning more toward new product spin-offs that are based on further dissecting the existing market share. But the focus is not on innovation; it is on increasing market share.

Every inventor and small company entrepreneur must learn to create new products that are truly customer driven. They must learn to create innovations that the large, generic entities cannot easily duplicate. To do this, inventors need a narrow, yet flexible, development and design focus that differentiates them from the generic leaders in a given field. To be a successful small company in the world of invention and innovation today, CDI is by far the best approach.

AMERICA DOES IT BEST

The U.S. is flooded with success stories of start-up entrepreneurs—Pullman, Levi, Edison, Singer, Ford, Hoover, Jobs, Wozniak, Gates—because it is in our nature. The team of Franklin, Monroe, Madison, and Washington, et al, set the theme over 200 years ago by literally inventing a new country—America. Franklin was a supreme inventor (bifocals, taming electricity, and much more); Madison and Monroe were administrators, and Washington (who established the U.S. Patent Office in 1793) was the field marshal who pointed the way.

A GIANT CORPORATION THAT KNOWS!

Worthington Industries of Columbus, Ohio has about 6000 employees, yet no more than 100 at any single location. Mr. John McDonnel, Chairman of the Board of Worthington, says, "Get more than 100 employees under the same roof and the quality and service go to hell." As a result of being structured like a small, fast, and flexible company, Worthington Industries has become well known for its quality, service, and innovative ability to solve its customers' problems.

The Market Share Approach

The term market share refers to the percentage of the overall market a company owns. For instance, if the annual sales of a particular product group is $1 billion, and your company's sales were $230 million, then you'd own a 23% share.

In a mature market (or a maturing one), competition drives the major market leaders to try to outthink and outmaneuver their competitors and to capture additional market share. The central focus of these companies is to pick up additional market share at the rate of half-percentage points or hopefully more . . . a full percent to two. This is usually accomplished by major advertising and sales promotions of existing products or through simple modifications to existing generic products.

A small company may benefit from an industry leader's desire to increase market share by supplying an innovative new product or licensing its technology to the larger corporation. It's uncommon, but frankly, is becoming more acceptable all the time.

The key problems in this kind of approach is that the larger market share-driven companies usually don't look to innovation for new market share. They tend to be not invented here (NIH-type companies with big egos). However, this attitude may be changing somewhat as the companies realize that moneymaking opportunities can stem from anywhere in the world, and no one cares where the concepts first originated.

If you find yourself in a position in which your innovation relies on licensing or partnering only with the large megagiant corporations, you'll find yourself in a difficult situation. You'll find it very difficult to convince them that your innovation can boost market share. The better approach—market creation—is almost predominantly that of the smaller competitors and opportunity abounds for great ideas and inventions.

Shari's Berries

One of the most brilliant success stories of launching a niche product and then expanding and protecting that niche is Shari's Berries.

Shari was in the escrow business and began dipping chocolate-covered strawberries in her home to attract attention and gain clients. They became so popular that others started asking her to make some for their clients as well. It didn't take long before Shari had a full-fledged local business established and ideas for expansion.

Today, Shari's Berries dominates the chocolate-covered strawberry niche, has expanded its product line to include its patented Strawberry Rose® and many other chocolate delights, ships nationally, and has more than 27,000 Internet affiliates, a licensing program in place, and major expansion plans on the drawing board.

Check out Shari's Berries at www.berries.com.

The Right Size Marketing and Manufacturing Partners

Ted Levitt's product concept flower illustrated in Chapter 3 shows the product focus potential of both generic and innovative firms. To expand on this concept, the long-term strategy of traditional U.S. manufacturing companies focusing on the generic product tends to look like the flower in Figure 9.1. In contrast, the long-term activities of smaller inventive, innovative companies tend to look like the flower in Figure 9.2 focusing on the augmented and potential product. Companies focusing on picking up additional market share are those companies with a philosophy illustrated in Figure 9.1. Those companies with a market creation philosophy are represented by Figure 9.2. Ideally, this is where your focus should be.

**Figure 9.1
Companies Focusing
on Generic Products**

**Figure 9.2
Companies Focusing
on Market Creation**

Keep in mind that it is very difficult for large, cumbersome, generic-minded companies to change their production philosophy. It would take years and probably never happen! For a company with that kind of philosophy, it would be almost impossible for it to take on a new niche product. From the examples in Figure 9.1 and Figure 9.2, it is easy to see that there are excellent opportunities for inventors and small businesses with the smaller, flexible attitude. As a start-up business, Figure 9.2 should be your model to put your focus in exactly the right direction. When interviewing potential manufacturing and marketing companies for partnering or license, you also want to focus your efforts on those with the philosophy of Figure 9.2.

Figures 9.1 and 9.2 illustrates two different philosophies. A fairly easy way to identify these two profiles—generally speaking—is by looking at company size. Table 9.1 illustrates the size of companies, based on annual sales volume. As you can imagine, the larger the company, the more market share-driven they are. Following is a summary of the typical manufacturing and marketing philosophy in each size category. Bear in mind however, that category size may vary depending on the industry. For instance, the tiny "twist tie" industry may be dominated by a much smaller company, whereas the massive computer industry may be dominated by much larger companies. Nevertheless, this table and the size differentials of the various competitors almost always expose the market creation companies as the smaller ones.

Megagiants. The megagiant corporations tend to be the generic producers in any given field in industry. They dominate the industry and have in-house R&D departments and a defensive posture towards their existing products. They are rarely, if ever, interested in licensing in or partnering the development of market creation innovations and inventions from outside. Creating a new market niche means a potential loss of market share for them. Besides, they really don't know how to effectively conduct a market creation sales campaign anyway. If nothing else, it would be embarrassing to admit that an

NIH SHIFT?

Corporate America is becoming more aware of the potential of licensing inventions and technologies. In the book *Essentials of Patents*, we discuss in great depth the benefits of doing so and killing the "Not Invented Here (NIH) syndrome." It makes financial sense for a corporation to license in viable technologies, and it has a corporate responsibility to its shareholders to do so.

**EXCEPTION TO
THE RULE**

Microsoft is an exception to the rule, as other companies may also be. Although a megagiant corporation, Microsoft is a market creation company. It has been known to work with outside developers. In fact, it purchased its first DOS operating system from an independent inventor.

outsider did a better job at creating future innovations than their own, billion dollar in-house R&D team. You would be wise not to waste your time pursuing these companies as manufacturing or marketing partners, with one exception. That exception is an improvement on an existing product that is very clever and unique and has substantial patent protection. Even at that, it'll be hard convincing them it's worthy of pursuing.

Giant Corporations. Like the megagiants, giant corporations also tend to be high volume, generic-based producers. They, too, have in-house R&D departments and the same innovation- and invention-related problems as their megagiant competitors. Don't waste too much time with them unless it's a simple improvement they absolutely must have. The chances that they will license your innovations or partner in the development are not very good.

Large Corporations. These companies represent excellent licensing and partnering potential. Most large corporations are not the dominant leaders in their fields, and they tend to be more niche oriented, focusing on value-added products. They generally have small or no in-house R&D departments and are more willing to listen to outside inventors and innovators. Their top management is also more accessible, and it is to them that you will want to introduce your concepts because they are the ones who will have to approve a deal. This sized entity usually has well-established contacts and plenty of cash to invest into new product developments. You want to pursue them as they may be excellent market creation partners. (The exceptions might be those companies that are research-based in the drug or computer industry.)

Medium-sized Corporations. Just like large corporations, they too have excellent licensing and partnering potential. They are niche competitors in the market and usually eager to look for ways they can capture more sales via new innovative products. Rarely do they have R&D departments and if they do, they are usually small. They can make excellent licensing partners. Medium-sized companies frequently are younger and more aggressive companies that like to make things happen. The top management level is usually quite accessible and tends to listen to you. Likewise, they should have adequate cash to invest in new projects.

Table 9.1 Company Size and Annual Sales

Company Size	Annual Sales
Mega Giant Corporations	$5 billion +
Giant Corporations	$500 million – $5 billion
☐ Large Corporations	$50 – $500 million
☐ Medium-sized Corporations	$10 – $50 million
☐ Small Corporations	$3 – $10 million
Mom and Pops	Under $3 million

Small Corporations. These, too, can be excellent licensing and partnering candidates. Look for those that are newer, high-growth companies, busy building their niches. They may be eager to take on new projects like yours. Your ideas and inventions may be perfect for their growth plans. They almost never have R&D departments, and it will not be difficult to talk to the top people. The only problem you may find is that they may not have adequate financing to effectively launch a new innovation. But they usually make up for it with a willingness to try hard!

"Mom and Pops." Under most circumstances, you can eliminate them as potential candidates. They are usually too busy trying to survive or to grow. They may find it difficult to fund a new invention with their limited working capital. They also tend to have more local markets and may not have the ability to supply on a national basis. If your invention is simple and small volume, it may fit their product line, but then you must also ask yourself if the sales of your invention would be too small to consider pursuing.

Think Small

Smaller entities can develop and launch an innovation far faster than larger ones. But how large it too large? Worthington Industries is a large steel fabricating company that was in dire straights in the mid-1980s trying to compete against the Japanese imports. They discovered that to remain entrepreneurial in such a generic-based marketplace no facility should exceed 100 employees. This includes all work shifts, which invariably must work together toward a common goal. If your potential manufacturing and marketing partners or the division of a potential partner has more than 100 employees, you should be cautious.

Fast, flexible entrepreneurial companies should have the following attributes:

- A flat organizational chart
- Common CDI goals and missions
- Open and swift communication among *all* employees
- Hands-on approach to R&D, engineering, management, and supervision

If your manufacturing and marketing partners do not fit these criteria, you'll have to seriously question their ability to perform.

The Niche Marketing Philosophy

Niche marketers tend to be fast, flexible entrepreneurial companies. The more they focus on customer-driven innovation, the more niches they'll create. Over time their market position may grow and become the dominant share. At this point, niche marketers will attack themselves and break that dominant share out into more niches. If they don't, a competitor will. Thus, these kinds of companies have a keen understanding of niche marketing.

Figure 9.3 is a chart identifying marketing philosophies. Make a list of the suppliers in your field and try to determine their position on the chart. Those companies in the shaded region of the chart tend to make the best partners. They will be more receptive to hearing about and discussing your innovations.

INVENTORS ARE UNDERDOGS

If you find that you prefer to cheer on the underdog in sports activities, you probably have the true spirit of an inventor. Starting out as a niche is like being the underdog in the marketplace. But don't worry, just like in sports, there are many excellent examples of being the underdog and being the champion in the marketplace.

Figure 9.3 Niche Marketing Philosophy

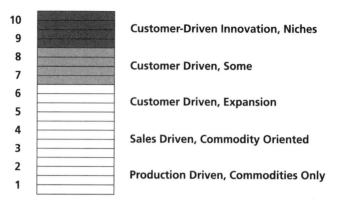

In an industry that is dominated by giant commodity-based manufacturers committed to large runs, look for the niches you can capture by partnering with companies that have smaller, more flexible manufacturing processes. It may take a while to find the right marketing and manufacturing partners to fit your product's needs, so be patient. Having the right partner is so very important, especially since the future of your innovation depends on it!

The smaller, more flexible manufacturing entities can also quickly run down the learning curve, thus improving manufacturing cost efficiencies and realizing profit sooner. In fact, it is not uncommon for niche manufacturers to have lower scrap rates and faster changeover times than their large competitors.

Your Marketing Expert Becomes Your R&D Liaison

The sales/marketing team that heads up your product's sales, whether that is your internal sales department, an outside manufacturer's rep group, some other form of marketing partner, or a licensee, is important to your effort. It is up to your sales team to gather R&D information so you may improve your innovations and, at times, develop new ones. A smart sales entity is effective at gathering information and serves as your innovation liaison. However, they have to gather the right information. They simply can't come back to you and say, "The customer doesn't want it." They have to tell you why and what the customer would like, so you'll know what to do to fulfill the customer's needs. This is one of the most important ways your sales/marketing team member can help you.

In the early stages of new CDI developments, sales associates should work closely with the customers. This hands-on activity serves as a valuable training vehicle, too. With industrial and commercial applications, their initial activities may include gathering information from the customer's operations, manufacturing, and even accounting departments. With retail-oriented products, this will mean close contacts with the managers and supervisors who sell the products and the end users themselves. The objective is to identify problematic areas, which we know are really CDI opportunities. The more sales gets involved with the customer, the more you will learn and be able to assist with new CDI developments in the field.

Your sales team can arrange key meetings between customers and retail personnel. They can also arrange the initial testing of new innovations and monitor the results in industrial and commercial applications. After its first success, the team will be in a position to use this process, and much of what they learned, with other customers. After a while, your sales force will become the premier experts in the field. They will be called on to counsel customers, new and old, on all important future product developments.

If your marketing expert is the marketing department in your company, the best way to manage these responsibilities is through a Patent Quality Management (PQM) system. PQM is a systematic means that shows your marketing department how to be aware of customer's true needs and problems as well as future trends. The marketing department then reports them back to you so you may turn their reports into new product development opportunities. This also gives you the opportunity to protect those future developments with valuable patents.

Identify Target Markets and Which One to Capture First

First, you'll want your sales team to identify all potential target markets in which your innovation may be useful. Second, you'll want your marketers to narrowly define the initial target market they'll be pursuing.

You may have identified numerous opportunities in various related fields. However, you should identify just one for the first product launch. Your marketers should even pick the first customer in the initial target market, because they'll know which one will be most eager, most vulnerable, the easiest, or the most profitable. Then, they'll coordinate the first market with you so you can cross the t's and dot the i's. Last, they'll follow up with additional customers and a major sales marketing offensive.

It is very important that you have clear-cut specifications for each target audience. Your invention's specifications—size, shape, etc.—may change slightly with each target market, depending on needs. Your sales experts will help identify these needs.

Categorize Your Niche

You'll want to be able to state your unique market position by clearly defining your niche. Products have a tendency to get sorted out into niche categories and if your invention doesn't have a category, frankly a retailer or distributor won't know how to sell it. You know the niche and can describe it in lengthy sentences. But now you must clearly define it in just a few words. The fewer, the better.

Successful marketing campaigns focus on a particular category and then exploit it. If you cannot describe you niche's category in a few words, the customer is not likely to know if your product will fit his/her requirements.

Remember that customers identify with categories quite easily, even though they may not even be aware of it. They will identify with your product and feel good about buying it knowing they made the right choice. Whether they feel safer or more beautiful or they know they can trust the performance, they will identify with a brand name that owns a certain niche. Here is how some companies have identified their niche markets:

- Nike: The basketball shoe
- Rockport: The walking shoe
- Volvo: The safety-minded automobile
- Hyundai: The low-priced automobile
- Harley Davidson: The road bike
- KTM Sport Motorcycles: The off-road bike

IDENTIFY YOUR NICHE

Inventor Shari Fitzpatrick didn't identify her chocolate-covered strawberries as an alternative to chocolate candy as many may believe. Instead she looked at them as an alternative to flowers. This makes tremendous sense from two viewpoints. First, her delicious strawberries have a higher retail price than candy. A dozen chocolate roses cost a lot more than a box of chocolates. Second, her competitors are flower growers. It is this kind of creative thinking that can identify an innovation's unique position in a less competitive market.

EXPAND THE NICHE

Before WWII only 25% of homes had refrigerators. After WWII it soared to 95%! With market saturation in the early 1950s, the refrigerator manufacturers looked for other niche markets and began developing air conditioners for automobiles. This new product expanded into window box air conditioners for homes and businesses, and then the central air conditioners we know today.

KTM SPORT MOTORCYCLES

This is a great example of a niche that's not commonly known, unlike its counterpart Harley Davidson. However, if you were looking for a real off-road motorcycle and started researching it, you'd find out that it's not Honda and Suzuki and some others you might think . . . but KTM, an Austrian company.

- Prego: The thick pasta sauce
- Barilla: The *al dente* pasta
- Arm and Hammer: The baking soda toothpaste
- Rembrandt: The whitening toothpaste
- Swatch: The hip watch company
- Rolex: The best watch, period
- Lo-jack: The only auto alarm police use

The niche category will define how your marketers will sell, promote, and advertise. They may even try to convince everyone that they should want nothing other than your niche. Of course they won't, but your product line will be solidly positioned as the leader in the field, and you will personally become the leading expert in the niche. Your vision of its future will be decidedly superior to your competitors, virtually ensuring your continued success.

Marketing Opportunities

Traditional marketing approaches for retail products are usually through retailers and, at times, distributors. Industrial/commercial products are usually sold through trade supply distributors or, at times, direct to the end user. There are also several other alternatives that may be considered; however, keep in mind that high-volume, long-term profits and stability are almost always via the traditional retail or industrial/commercial approach.

Retail Sales

Most retail products are shipped through some form of distribution and then to the individual retail outlets. This is the model that is reflected in the pricing structure model shown in Table 9.2 later in this chapter and more or less discussed throughout as the standard marketing approach. In many respects, it really is not important whether a distributor is involved or not. The decision to purchase a new innovation is almost always made by the retailer, not the distributor. Distributors rarely "sell" what they carry and for the most part only take orders. So, your principal focus is to sell your product to the retailer and then let the retailer tell you through what means of distribution it should go. An experienced marketer in the field of your invention will know how to arrange this. As you know, most major retailers, such as Target Stores, Walmart, Home Depot, Staples, Walgreens, and so on, use their own internal distribution system and rarely purchase from outside entities.

There are some exceptions on this distribution model. Some industries tend to sell their products direct to the retailer, such as industries involved in clothing, furniture, appliances, and other large, obvious product lines such as boats and automobiles. Your marketing expert will know how your product will be sold. If you're a start-up business and plan to do this yourself, you'll need to find out how it's done.

Industrial/Commercial Sales

Many commercial/industrial products are also sold through distribution of some form; however, many are not. This is represented by the pricing structure in Table 9.3 later in this chapter. Generally speaking, most raw materials, machinery, and high volume components (These parts are usually referred to as Original Equipment Manufacturer [OEM].) the manufacturer uses in the manufacture or assembly of its products are sold direct. But common items such as wiring, fasteners, packaging, and other internal supplies such as cleansers and tools are sold through distributors. More and more office supplies are purchased through a combination distributor/retailer such as Staples and Office Depot. Larger volume purchasers may receive discounts on certain items.

The decision to take on a new product is usually through the end user, not the distributor. If the supply of your item goes through a distributor, you'll still need to have someone market your invention to the end-user customer. Most distributor sales people have tens of thousands of products in their line and won't be able to remember yours. Whatever the method through which your innovation is sold—through distribution or direct to the industrial/commercial end user—your marketing expert must know how to handle it.

Private Label

Instead of supplying your product through traditional distribution sources, you may elect to supply an existing company whose product line is already being sold to established retailers and through established distribution. When doing this, you'll be marking the products with the established company's trademark instead of yours.

Private label is also a common marketing scheme used by large retailers who carry their own distinctive brand, such as Target Stores, Staples, Walgreens, and most major supermarkets.

The most important consideration is ensuring that the profit margin is adequate to justify the extra middleman. However, if you're doing private label to ship directly to a major retailer, it probably won't be an issue.

Regardless of how private label merchandise is sold, your marketer will know the approach and will be able to advise on price levels.

Private label can be a very attractive approach for any company that wants to produce an innovative new product, but doesn't want to market it. It is similar to licensing, but since the inventor's company is the supplier of the finished product to the marketer, no royalty from sales is typically paid. Instead, profit is realized through sales.

Infomercials

Increasingly, more and more products are being sold on television using infomercials. Infomercials may range anywhere from innovative new household products to money-generating schemes to oldies but goodies song collections to a new, slick fishing lure.

Infomercial experts say that in order to be successful, you are going to need a real slick gimmicky product that has some very attractive profit margins. Ken Robbins, a well-respected product development consultant, says that billions of dollars worth of products are sold annually through direct response TV, a.k.a. infomercials. Accordingly, here are the requirements that Ken says every TV marketer looks for, in order of importance:

1. The product is a new-to-the world consumer product.

2. The product is highly demonstrable or has raving testimonials.

3. The product appeals to the profit motive, vanity motive, or ease-of-use motive.

4. The product is patented, trademarked, or copyrighted.

EASE OF MARKET IMAGE

The primary reason industrial/commercial markets are easier to pursue than retail markets is that the product and the market is usually easier to develop. Retail products typically require a large investment in packaging and advertising and have two- and three-tiered marketing approaches (retailers, distributors, and the end user). In comparison, the industrial/commercial market is usually sold in bulk packaging such as cardboard cartons, does not require mass market advertising, and is sold direct or through a single distributor.

HEAVY ADVERTISING REQUIRED?

When pricing for retail applications, a simple rule of thumb that can be applied is that the retail price is about four times that of the net cost. The higher the price of the item, the lower the ratio. But with products that require heavy advertising, a much higher ratio . . . as much as a 6 to 1 or 8 to 1 . . . will be required. TV and radio advertising is expensive and is actually a costlier marketing avenue than normal distribution.

HOME SHOPPING SHOWS

Generally speaking, these shows can generate a lot of sales. But the problem for a new innovation is that the amount of inventory required to be on one of these shows is huge. It may be as much as $250,000 or more and will need to be shipped to their warehouse up to two months before the show airs. For inventors the best way to begin is to develop the innovation for other markets first. Then, after you have a cash flow from those sales, you may approach the HSN-type companies.

5. The product is available nowhere else.

6. It is priced between $20 and $300.

7. It does not require professional installation.

If you believe you have a new innovation that meets these criteria, the infomercial market may be worth pursuing. However, keep in mind that this approach to marketing can be very expensive and generally requires a much higher profit margin in order to justify the high cost of television. With an infomercial, you'll have to put out the cost of preparing the script, the studio rental, and any professional actors or actresses you may need. It can be costly, certainly upwards of $100,000 to create.

Home Shopping Channels

This too has become a popular means of marketing new products. It is similar to an infomercial, except that products are promoted and sold as part of many other products in any one filming segment.

This approach to marketing is different from infomercials in that there is no requirement to film in a studio. The hosts of the home shopping networks will promote and sell the product for you. Bear in mind that 90% of all products sold via the home shopping channels are targeted toward the female audience. So, if you're thinking about promoting your power tools on TV, think again.

There are two big drawbacks to using this approach, however. First, the network will want a substantial discount off the sell price . . . easily as much as 60% to 70%. Second, you will be required to ship a substantial inventory to their warehouse and whatever doesn't sell is going to be returned to you.

Nevertheless, if you're in a position to accommodate their requirements, home shopping channels are fairly easy to deal with. You can contact these companies on their websites and contact a local representative, or you can submit your products directly to them. You can also check the patentcafe.com website for access to the various entities seeking out new innovations.

Catalog and Mail Order Companies

Catalog companies provide another popular method of merchandising and marketing new innovations. One of the best ways to research catalog companies in the field of your invention is to search online at google.com.

Various catalog companies in various industries have different approaches to marketing their products. Larger companies like Land's End and Eddie Bower focus on developing their own product line. Some companies like Sharper Image and Hammacher Schlemmer focus on innovative new products and actually pursue them. Your innovation must be well-developed in order to be considered. Other catalog companies, such as Graingers, just sells what it would otherwise carry in its local warehouses. Yet others may specialize in a certain field and may charge you to include your product in their publication.

Keep in mind that most of the large high-exposure catalog companies like Land's End prefer to sell their own brand name merchandise, or products that have been made for them under private label. If you are just launching an innovation and look to the catalog trade as a potential marketplace, you'll most likely be looking at those smaller catalog companies dedicated to a certain trade, and remember, they will probably charge you to be in their catalog.

Catalog sales companies promoting their products are far too numerous to even consider listing. Use google.com to find those companies in your field and you'll be able to quickly determine if they are interested in carrying your innovation and what costs there will be.

Internet Sales

When checking out the Internet for catalog companies, you'll find that many of them are also on the Internet. Thus, you may try to tap this market as well. Whether a company like Hammacher Schlemmer wants to include your innovation in its catalog and/or on their website really doesn't matter. They'll make all the arrangements if they like your innovation enough. In addition, there are a multitude of Internet mall websites that can be contacted.

One way to establish your product on the Internet is to have your own website that promotes your invention. On the website you'll typically want to include helpful information that is relevant to your product and the problems it may solve. For example, the *From Patent to Profit* website at www.frompatenttoprofit.com provides a great deal of information for inventors and entrepreneurs and provides the ability to purchase inventor supplies such as journals, books, and patent writing and patent searching software. A website such as Shari's Berries addresses a gift problem by offering an exciting alternative to flowers.

If you do not already have a website for your product line, you can set one up fairly easily. Here's how:

1. Find and purchase a URL directly with Verisign (www.verisign.com) and its affiliate Network Solutions. It is best not to use a service to do this for you. An unscrupulous domain registration company may quickly register the name you seek, forcing you to purchase it from them or to select another inferior one.

2. Locate a service to post your website. There are several, including ones that have the ability to process orders, verify credit cards, and even archive certain data, such as addresses. The cost can vary from about $250 a year for this type of service up to $2,000 per year for a website that has its own automated credit card processing system with direct deposits to your bank. You can locate any number of services such as these through Verisign, MSN, or google.

3. Build a website. You may either purchase your own software or hire a company to do it for you. In either case, bear in mind you'll have to do much of the work anyway since someone has to write the copy that'll be included on the pages. Software such as Microsoft Front Page is fairly simple and includes templates you can use. Cost of the software is about $100. Cost to hire a professional website designer can easily range from $2,000 upwards, depending on content and sophistication.

4. If you do not presently have the capacity to take credit card orders, contact the bank you currently use for your business and have them set you up. Most credit card processing companies are similiar. The advantage of using your own bank is that it's local and easy to contact if you have questions or there are problems.

INTERNET PROWESS

The acceptance of the Internet by consumers eclipses all other technologies that preceded it. Radio took 38 years before 50 million people tuned in. Television took 13 years to capture the same audience. It took 16 years after the first PC kit was sold to reach the 50 million people mark. But when the Internet was available to the public, it reached the 50 million mark in only 4 years!

5. Promote your website through user news groups and submit them to all the major search engines. There are services you can consider that charge on a per-hit basis, such as Looksmart and Overture. You can also look to establish links with other sites that are either noncompetitive or would have a natural cross link to your website. One of the best ways to promote a specialty website, however, is through user news groups and establishing relationships and swapping links with other websites. You can also improve exposure by writing brief articles that address the problem your invention solves and submitting them for Internet publications. You can submit the articles to newsletters related to products in the field of your invention.

Keep your website very informative and expand it as needed. Be patient as it takes time to get noticed. The most important thing to remember about Internet business and about business in general is that the best advertising is by word of mouth. One excellent way to get noticed is to join the various related chat forums, where your expertise in the field of your invention may be shared.

Creating a Price Structure

After you have qualified a sell price with your sales/marketing expert, you will want to determine your pricing structure. If your innovation is a commercial or industrial application, your cost model is much simpler than for a retail item. In either case, you'll need to know this fairly early on so you can make sure the innovation is going to be marketable. After all, the number one criteria a potential customer considers when purchasing a new product—when making a conscious decision to switch from some current product they are purchasing—is price.

Regardless of whether your innovation is a retail item or an industrial/commercial item, royalties and commissions are usually paid on the manufacturer's sell price either to the distributor or direct to an end-user customer. It is very difficult to try to collect royalties in any other way. How would it be done? Would you invoice every end user separately? Would the distributor collect them and then pay you? Keep it simple and collect a fair royalty from the manufacturer's sell price. At that point, it is an easy transaction based solely on large-volume sales.

When pricing a new product, you always start with the retail or the end-user customer's price. This will be based upon what the market will bear. The best way to do this is to discuss it with your sales/marketing expert. Then, make sure your manufacturing partner can do what is necessary to get the product made at the right cost.

Most new inventions should be priced close to a competitive product or substitute. You can't assume that a new product will be so great that consumers will spend double the price just to enjoy its nominal benefits. This may be true with products that are substantially better, like a Mercedes over a Ford. However, this must be prequalified by your marketers up front. Don't waste your time pursuing an innovation that will cost vastly more than existing products unless your marketers and test market verify it.

The worst way to determine the sell price of an new product is the "cost up" approach—this is, to calculate how much it costs to manufacture and then mark it up accordingly. The proper way to determine the sell price of a new product is usually best done backwards. In other words, competition will determine the entry price level. Your marketing experts can most likely do this for you very quickly.

KISS

Keep your pricing scenarios and structures simple. Let your manufacturers pay all the associated costs, since they must do the billing of the finished product. They are used to doing it. Just let them know what to include in the cost and they can quickly figure it out for you. If you are a novice at business, don't get into an administrative nightmare and try to buy, sell, warehouse, pay commissions, and ship.

Remember, most competitive suppliers in an industry are priced about the same. It does you no good to simply make something and offer it for sale at just any price you want, believing it'll sell. If it is priced too high, you may have no sales, regardless of how advanced the technology may be.

Make sure that if there is one company manufacturing your innovation and another company marketing it, as may be the case with licensing, you include any royalties and sales commissions as part of the cost. Specific strategies regarding licensing are covered in detail in Chapters 13–16.

Retail Pricing Scenario

This differs from industrial and commercial applications, since it will usually have a larger mark-up to cover the higher costs of getting a product to market, frequently through a distributor, and to the retail outlet so the end user can purchase it off the shelf. In the pricing model in Table 9.1, we begin with a retail sell price that has the right psychological effect. While $10 may be appropriate in an industrial application, the psychology of $9.95 works a lot better in retail. It just sounds better and we've all become accustomed to it.

In the retail scenario, the ratio of the retail sell price to the manufacturer's cost is usually about 4 to 1 or greater. Notice that the dealer discount is 40% of the list price (the retail price) and that the manufacturer's sell price to the distributor is about 55% off the retail price. This is a fairly standard margin in most industries. It is referred to as a "50 and 10 discount." In other words, you take the retail price and subtract 50%, then subtract another 10% and you'll have the wholesale cost. By subtracting 50% and then another 10%" off that cost, it is the same as subtracting 55% from the retail price. The dealer price is commonly referred to as the wholesale price, as opposed to the actual wholesale cost.

If a large discount store like Target or Walmart is purchasing in volume, they will want to eliminate the distributor (considered a middleman) and buy directly from the manufacturer. These large retailers have their own internal distribution systems and have no need for the middleman. The fact is, they rarely purchase from a middleman. By purchasing in large volumes, they also tend to discount the retail price. It is easy to see that a large retailer can buy in volume, discount retail prices by 10%–20% and still have a reasonable profit margin. For instance, purchasing a product at a cost of $4.48 and selling it at $7.96 (20% off its retail price of $9.95) gives a profit margin of 43.5%, which is a higher margin than a small retailer would make on the same item.

In the model in Table 9.2 the two left columns illustrate a typical pricing scenario, one that would be applicable to the drug store or automotive and hardware trade. Use the right side column to determine the discount structure and pricing of your innovation. Start with the retail price that your marketing expert has qualified. If you do not yet have a marketing partner, then you'll want to use your best guess and have your marketer verify it later on. In either event, you must find out what current discount structures are in the field of your innovation so you'll know what to expect. Assuming that dealers are going to purchase directly from you at a 40% discount is a big mistake. They'd be much more interested in purchasing from companies in which they are already doing business.

DISCOUNT OR MARK-UP?

Most retailers and distributors supplying retailers talk in terms of "discount"—not "mark-up." You should too. You must know the language and discount structures used in your trade, or you'll really be considered an amateur.

Table 9.2 Retail Pricing Structure

Retail Pricing Structure					
Example		Use the column on the right to determine your sales price structure. If you are unaware of what your industry standards are, find out soon!	Your innovation		
Amount	Discount		Amount	Discount	
$9.95		Retail sell price			
$5.97	40%	Dealer cost (typically less 40% to 50%)			
$4.48	25%	Distributor cost (typically less 20% to 30%)*			
$4.39	2%	Less terms (usually 1% to 2%)			
$3.94	10%	Less freight (usually 5% to 10%)			
$3.80	3%	Less royalties (usually 2% to 7%)			
$3.58	5%	Less sales commissions (usually 5% to 10%)			
$3.36	5%	Hidden costs (usually about 5%)			
$2.24	25%	Manufacturer's cost with profit (15% to 25%)**			
		Compare your present cost			

* all of the subsequent discounts are taken off the distributor cost, which is the same as the manufacturer's sell price.

** start-up profit margins must be larger in order to offset start-up costs, such as dies, molds, cost of financing, and so on.

If you plan to be the manufacturer, this gives you a clear-cut idea of what the sell price (the distributor cost) will be. In such a case, there may not be a royalty payment, unless you've established a separate entity that holds the intellectual property. Thus, your company—the manufacturer—is likewise paying the other company (in which you may the sole owner) a royalty for use of the IP. This approach can come in handy if you plan to spin off the manufacturing company at a later date. Obviously, the royalty stream would then continue.

Also, if you plan to import the product, or have it made by a subcontractor and then, in turn, plan to supply the distributor at the distributor cost, you'll still have to make sure that all your costs to do so are covered. Later on, you'll see how a typical import cost will usually require a 20% to 25% profit margin to cover handling, brokering, duty, re-packaging, and so on. This extra cost should be in addition to the below listed start-up costs. Thus, the import price in the model shown in Table 9.2 should be about $1.68, which is $2.24 less 25%. This computes to about a 6 to 1 ratio, retail price-to-import cost.

Once you've completed this important step, you'll be able to talk more effectively about your innovation.

Industrial/Commercial Price Scenarios

If your innovation falls in these categories, then your pricing model is easier to prepare. Typically, you'll need to find out current price levels of competitive products purchased from industrial distributors or from competitors who supply direct to industrial accounts. In such a model, you start with the dealer cost and simply apply a distributor discount. Some examples of industrial/commercial products would be industrial chemicals and cleaners, utility gloves and apparel, tools, and other supply items such as fasteners and plastic bags.

In the industrial/commercial model, the manufacturer bills the distributor, who, in turn, reships smaller quantities to its customers. All you know is that the sell price to the distributor must include all royalty, commissions, and related costs.

If an industrial/commercial product is purchased through a distributor, you may have a set price to that distributor based on a certain volume. Many times, the distributor will set its own resell price structure. For instance, if your wholesale cost is $4.48, the distributor may have a one-case price of $9.96 (a 50% profit margin), a five-case price of $7.46 (a 40% profit margin), a twenty-case price of $6.40 (a 30% margin) and may include on its price page a "request for quotation on quantities over 100 cases." It's really not your concern what the distributor sells the product for . . . as long as you're paid for your product. See Table 9.3.

Some industrial/commercial products are shipped direct from the manufacturer to the customer. In such a case, you would typically have a discount schedule based on volume. The higher the volume, the lower the sell price. There might be three or four price breaks based on obvious packaging configuration and case or pallet counts. This is a natural way to offset the higher cost of shipping and handling smaller orders, which can easily be 20% to 30% more. Your marketing expert can once again help determine this pricing structure.

Make sure that any pricing a distributor gives you or your marketing experts includes the terms used in the industry, plus freight, rebates, and so on. There tend to be many hidden costs with distributors.

Other Pricing Scenarios

If your product is sold direct to the end user or only through private label or as an OEM product, your pricing model is usually much simpler. Then, it usually becomes a question of how much does it cost (if you are out-sourcing it) and how much are you going to sell it for. This may be the model with innovations such as those sold only on the Internet or sold via some form of direct sales through your company. Your marketing expert still should provide feedback on competitive price levels, so you're competitive with others in the trade.

When creating a different pricing scenario, be careful to first assess if your company will be doing the billing or if it will be billed direct from the manufacturing entity. In such a case, you will be collecting your profit from the manufacturer in the form of royalties, commissions, or both.

WHAT HAPPENED TO PRICE ELASTICITY?

Petroleum products tend to be affected least by price elasticity. The key reason is that the demand is so great that consumers are at times forced to pay the high prices if they want to use their cars for work or pleasure. The more a product becomes a necessity and the fewer the number of suppliers the lesser the effect of price elasticity.

Table 9.3 Industrial/Commercial Pricing Structure

Industrial/Commercial Pricing Structure				
Example		Use the column on the right to determine your sales price structure. If you are unaware of what your industry standards are . . . find out soon!	Your innovation	
Amount	Discount		Amount	Discount
$5.97	40%	Customer Price		
$4.48	25%	Distributor cost (typically less 20% to 30%)		
$4.39	2%	Less terms (usually 1% to 2%)		
$3.94	10%	Less freight (usually 5% to 10%)		
$3.80	3%	Less royalties (usually 2% to 7%)		
$3.58	5%	Less sales commissions (usually 5% to 10%)		
$3.36	5%	Hidden costs (usually about 5%)		
$2.24	25%	Manufacturer's cost with profit (15% to 25%)		
		Compare your present cost		

Building Barriers to Entry

BUILD BARRIERS TO MARKET ENTRY

Patents establish barriers to discourage an inventor's competitors. In other words, competitors can't enter the marketplace riding the inventors' technology. But it's important to remember that viable strategies other than expensive patent procurement exist for establishing market entry barriers. "First to market" and "brand leveraging" are two examples that can work well to build barriers to market entry.
–Don Kelly, Intellectual Asset Management Associates, LLC

One of your objectives when establishing a niche is to protect it by building barriers, preventing others from entry. Intellectual property protecting your innovations is usually your first defense, with patents usually being the most important, and trademarks and copyrights following.

After establishing your patent potential, you'll want to build additional barriers to entry by establishing trademarks and by having distinctive packaging. You want consumers to think of your trademark and identify with your packaging when thinking about your niche. This is one primary objective early on. One example of this is the natural tea market. When you think about this niche, you immediately think of the name Celestial Seasonings (protected by trademark) and its packaging (protected by copyrights and trade dress). If you think of the high-volume soda market, the first name that comes to mine is Coca-Cola and its red can.

Your trademark should reflect the added value your innovation carries. This is a task that'll take some consideration. Rarely does a first name or a first packaging concept become the one you'll use, so be patient. Could you imagine if Celestial Seasoning Teas had started out with a name such as "Mo's Natural Teas" and packaged the teas in plastic bags?

Selecting the right mark that fits the quality and appropriate performance of your product can be challenging. In a start-up business or the launch of a new product line, you should consider a professional packaging designer and a commercial artist who specializes in trademarks. If you are licensing your innovation, then this may not matter as your invention may end out with the licensee's trademark and may be consistent with the licensee's other packaging.

Even if you plan to license your innovation, try to establish some trademarks you can include in the license agreement. When you do, you'll be extending the life of the license agreement since trademarks never expire as long as they are in use.

Establishing a Trademark

Trademarks can become more valuable than patents! Some excellent examples are KLEENEX™, ZipLoc™ and Styrofoam™. When your invention becomes a success and even more so after years of improvements, it will be well known by its trademark. Create a trademark early on and make it part of your marketing plan.

Trademarks are generally easy to establish. Read more about the registration process in Chapter 6. Simply put, here is what to do to get your trademark established: Whenever you use your trademark in literature, put a "TM" superscript at the end of the word(s)like this: MYMARK™. Trademarks are similar to copyrights; the minute you start using the unique mark, you are claiming ownership. Putting a "TM" at the end in a local offering may establish your position in a local statewide market; putting it in interstate commerce establishes it in a national market.

Eventually you will want to register your trademarks. You can file your marks online with the USPTO (see appendix) or an attorney can do it for under $1,000.

Creating a Trademark

Innovative people are usually not short on ideas for trademarks. You can develop your own, but it may take experience to create really good ones. Here are a few suggestions you can use:

- Initially use the project's nickname—in other words, the same code word that has been identifying the project all along. For instance, "FLAGGER™" for a new marking system, "STICK-EMS™" for a new label maker, or "BIGMOUTH™" for a new bag with a large opening.

- Listen carefully for words people use when referring to your invention. You may be able to use some combination or offshoot of them. Can they be joined together in a catchy way? Common words themselves cannot be trademarked, but a unique spelling or combination can be. For example, BOXXERS™, DIZPENSERS™ or SOFT-EEE™.

- If you use common words together in interstate commerce without being contested or used by others for a period of five years, you can register them as trademarks. Examples include "DUAL-TAB™", defining a bag with two tabs, and "BIG-RED™" for a type of fire truck.

- Borrow ideas from other trademarks, but make sure that you change them sufficiently so they do not appear to be in conflict or cause confusion.

- Brainstorm with others on your team to possibly produce some good choices.

- Ask your children if your invention is toy related or geared for a younger crowd. They can surprise you with their creative ideas.

ARE TRADEMARKS MORE VALUABLE THAN PATENTS?

After several years, the value of a patent starts to decline and the value of a trademark will increase. If you can create a trademark and make it part of your license agreement, your license may endure even after the expiration of the last of the patents. You can use the patents as a means to establish the trademark.

PATIENCE

Trademarks tend to fall into place sometime well after the product development has begun. It is frequently some word or combination that those working on the project adopt. Don't be too concerned if you don't find a brand name you like right away. It may take a year or more to discover this magic mark.

- Hire an advertising agency.

When developing trademarks, you can also consider service marks that illustrate how a product is offered to market.

Becoming the Low-Cost Producer

Once a niche product has been established, there is another very important way to protect that niche long term. Become the low-cost producer. Once a specialized product reaches a certain high-volume mark, it will be difficult for competitors to enter the market.

By the time they may have an interest in entering, you will be busy expanding the product line to include similar offerings, sizes, colors or what have you. You'll also be in a position to lower prices should you elect, and thus make market entry less attractive for a competitor to consider.

Long term, having a trademark and being the low-cost producer in the niche will ensure profitability.

Selling to Competitors Through Private Label

PRIVATE LABEL KING

The most prolific private label use is in the supermarket trade. It is followed by the drug store industry and the discount store trade. In virtually every industry—automotive supplies, electronics, health foods, nutrition, even sporting goods, computers, and cell phones—private label is becoming more prevalent than ever.

One excellent way to protect your niche and expand sales, is a variation in what Tom Peters refers to when he says, "either niche or be niched." Essentially, if you don't carve out your market into small niches and capture those as well, then others may try to enter the market and deteriorate your market share.

If you are concerned about competitors entering your established market, why not try contacting them and suggest being a supplier to them? You can produce your products under private label, which is another way of saying that you'll package the products with their name and trademark.

You will be in a highly advantageous position to do this if you are the low-cost producer. Usually, you'll probably want to use some variation on your established theme. The benefit to the competitor is that it will instantly have quality product, shipped at the right price, and shipped on time. The benefit to you is that you'll be able to profit from your competition, while expanding overall market appeal for your niche.

Packaging

Great packaging can make a mediocre product sell well and poor packaging can make a far superior product fail dismally. Just like a trademark, your packaging must fit the innovation. It must be the right quality and the right image for the product.

With any retail item, you have only three looks to attract a consumer and make the sale . . . and it must be done in a matter of seconds. They are:

1. **Catch the consumer's attention.** This is usually accomplished by the package's color, design, and other unique attributes such as a different shape or a piggybacked special offer. Catching a consumer's attention is done in literally a fraction of a second or the opportunity is lost.

2. **Draw the consumer in.** Once attention has been caught, the package design must draw him/her in so the product may be identified and further interest generated. Typically, this means the customer will look at the brand name to see what it is and perhaps what it means. This is where great artwork can also make a difference because

with attractive artwork, the eye moves around to take it all in. If the customer does not like what he/she sees when drawn in, then he'll look for other products.

3. **Qualify the purchase.** When the customer has been drawn in, he/she will then read more and decide whether or not to purchase the product. This is done by reading the back or side panels of a box or bag, reviewing pictures to see how easy a product may be to use or install, and checking out a guarantee; it may also include seeing how much the product costs.

These are the three chances you get to make the sale. If there are any important issues, such as the ease of use or a guarantee, make sure you convey that information on the container. If you don't, you'll lose the sale to a competitive product or through indecision.

Packaging your innovations begins at the point of purchase container and follows right down into case counts, master cartons, and palletizing. Every aspect must be taken into consideration, so that the product can ship without damage, palletize and stack (usually three high), and occupy the right amount of shelf space on a retailer's shelf.

Retail products should be packaged in acceptable norms for similar products in a group. For instance, you would not want to package vitamins in a plastic bag when the industry uses hermetically sealed bottles, nor would you want to package nails in a box since they should be sold in bulk.

You should have some idea of the kind of packaging that fits your innovation. If not, you'll need to do some research to see what fits best. You can gather this information fairly easily by talking to sales reps in the packaging field. If you are a start-up company, contact a local packaging distributor (look them up in the yellow pages) and let the rep know what you're considering. This is one person who can help your team effort tremendously and can steer you and your marketing and manufacturing experts in the right direction.

If you are licensing or partnering your innovation with an outside marketing company, it will determine the type of packaging and it will be responsible for artwork, design, and related costs.

The three most common packaging types (excluding those for foods) are plastic bags, chipboard boxes, and blister packs.

Packaging in Plastic Bags. Plastic bags (which may hang on a pegboard hanger) are used for inexpensive, lightweight items. They don't typically convey a high-quality image, but may if they are printed with great graphics. They can convey an even better image if the graphics are printed on a chipboard insert.

Most plastic bags are made from low-density polyethylene plastic. The bags you use should be from high-clarity film. Cellophane or polypropylene can convey a better image, but cost more. Be careful with cellophane, as it tears easily.

There are two types of headers used on bags for hanging on gondola pegs. One is made from chipboard; the other is a plastic reinforced header. Chipboard is much classier and easy to assemble in small quantities. Plastic-reinforced header bags cost less, but the minimum purchase quantities are larger, usually a minimum of 40,000 to 50,000 bags. Since they are not as stiff as the chipboard headers, they tend to droop at the sides and over time can look used and worn on a pegboard display. Chipboard headers are superior in almost all aspects of plastic bag packaging although they cost considerably more.

ANCIENT INVENTION

The loom has been used since about 7000 BC. The most commonly used weaving materials were wool, linen, and hemp. Early woolen cloths found in Scandinavia are quite rough and irregular, but cloths dating from the Roman Ages are much finer and closely woven with colors and intricate designs such as stripes, diamonds, and animals.

Packaging in Boxes. Boxes usually convey a better image than plastic bags. The cost is usually much greater as well. A product packaged in a $.03 plastic bag might cost $.15 in a box. But the reality is that U.S. consumers like to purchase their retail products in boxes. An example is retail trash bags, which are sold throughout most of the world in bags, but are sold in boxes in the United States.

There are many varieties of boxes used in retail packaging. A packaging distributor sales rep can help you determine the type most suitable for your product and the related costs. Keep in mind that it can take several months to get the packaging and artwork designed, manufactured, and shipped. Prepare for these delays. Here are some basic considerations when using boxes for packaging:

- Most high-quality printed boxes you see in supermarkets and retail stores are made of inexpensive chipboard. Lower-quality boxes use gray liner board on the inside, since this is the color of the recycled material that's used. Higher-quality chipboard usually has an inside white liner. There are various thicknesses available . . . the thicker the board, the higher the quality and the higher the price. Board thickness is important in conveying the right quality image and in stacking the boxes on a retail shelf.

- Chipboard boxes may be attractively printed and then treated with a special coating on the outside. Usually ultraviolet (UV) coating is very shiny and the best choice for a high-quality image. It also helps protect against scuffing and discoloration from neon lights, sunlight, and so on. Varnish is also used as the outer layer, but it is not as glossy. At times, a matte finish, which is also a form of varnish, may convey the right image.

- Cardboard can also be used for heavier items, but it usually cannot be as attractively printed. White cartons cost more than standard brown (referred to as Kraft). There are two ways of printing cardboard carton for retail applications. First, you can print on the carton itself, which is usually going to be a lower-quality image since the outer liner board on cartons is fairly porous and not conducive to high-quality printing. Another reason is that carton manufacturers have traditionally printed only shipping cartons, which require a lower-quality print copy, and they simply don't have the ability to print higher-quality images. However, this may be overcome by printing a high-quality sheet and then gluing (or laminating) it to a carton's side. Carton thickness also determines quality and price. There are standard thicknesses, such as C-flute and D-flute. If you're looking to convey a higher image than chipboard, the flashy, narrow gauge referred to as E-flute may be attractively printed on an outer laminated sheet; it conveys an excellent image. It'll cost more than chipboard, but it's a sturdy, good looking box. If you're unsure of the right carton type, thickness, and image for your innovation, check out similar products on a retailer's shelf and you'll get some ideas.

- There is growing use of plastic boxes in retail applications. They are usually made from a stiff, high-clarity polypropylene material or styrene plastic. The cost of this packaging is still much greater than paperboard boxes. Printing is usually done on a printed insert placed inside the container.

Blister Packaging. A popular alternative to using a box is blister packaging. One of the chief benefits is that the consumer can see the product. Another advantage to retailers is that blister packaging helps protect against theft. Small items that would be easy for a shoplifter to slip into a pocket or purse are much more difficult to steal when packaged in a large, bulkier blister pack.

Obviously the cost of a blister pack is higher than bulk packaging, but if retailers want blister packaging to protect against theft, you'll have to accommodate them.

There are thousands of companies throughout the U.S. that do blister packaging. Some are dedicated to short runs, some longer runs. Many of these companies will also be able help you make inexpensive, prototype packaging if you have the artwork already completed. You can locate them in the yellow pages.

Case, Master Packing, and Palletizing

Retail items must be in logical *case packs* for the retailer to handle. You'll have to find out some industry standards in your field if you don't already know this. This may be 12 per carton, 24 per carton, or with smaller, lightweight items as many as 100 per carton.

Frequently, these case packs will be packaged in *master cartons* for distribution purposes. A distributor will purchase larger volumes in master carton lots. The master cartons are subsequently broken down and reshipped to dealers in case lots. This is more commonly done with smaller items such as paint brushes for the art supply industry, fingernail polishes for the health and beauty aids (HBA) industry, or packages of brads for the hardware trade.

Larger case packaging, such as that used with larger boxed items like cereal (as many as 24 to 48 per case because they are so lightweight) or gallon bottles of anti-freeze will not be packaged in master cartons. They'll simply be palletized instead.

In industrial and commercial applications, packaging your invention is relatively easy. Eye-catching packaging is usually unnecessary. The primary consideration with industrial and commercial packaging is to ensure that the product arrives without damage and can be easily handled once it arrives. Your manufacturing and marketing experts can help determine what's best.

When determining case packs, master packaging and/or industrial/commercial products, here are a few considerations:

- Most cases or master cartons should not weigh more than 30 pounds. In some industrial/commercial applications, 50 pounds may be acceptable. The exception to this rule would be items like bags of cement, 55 gallon drums, or bulk raw materials. Whenever possible make your case count one that is convenient to handle.

- Don't scrimp on carton quality to save a few pennies. Carton crushing is always a potential problem and can result in returned merchandise.

- Bulk packaging is for OEM or strictly industrial applications. It is not popular or desired in most commercial or retail applications.

- Use carton sizes that can be effectively stacked on a pallet.

Palletizing of cases or master cartons should be done so that pallets may be stacked three high (and sometimes more). This means cases and master cartons are typically full with no leftover air space inside. A typical palletized product weighs no more than about 1200 pounds. If the anticipated volume of your innovation is relatively large, establish a pallet count and stack (referred to as "tie and high") with cartons that fill out the pal-

---

EYELINER INVENTION

The earliest archaeological evidence of the invention of eyeliner can be traced to the urban civilizations of Egypt in 4000 BC. For centuries they used cosmetics to beautify a woman's eyes for aesthetic or religious purposes. In southern Iraq and Egypt, men and women painted kohl around their eyes. Kohl is made of ground lead sulfide or antimony sulfide and is still used today. The eyeliner made their eyes look larger as a means to protect them from the evil eye.

THE TELESCOPE

Several different people claim to have invented the telescope. Many historians believe it was invented in 1608 by Hans Lippershey, a glassmaker from Holland. He discovered that placing two lenses inside a tube made distant objects seem much closer. The great Italian scientist Galileo (1564–1642) is also credited with inventing the telescope, though he may have gotten the idea from Lippershey. Today, telescopes have much more powerful lenses, and many other versions have been developed such as radio, X-ray, and gamma-ray telescopes, capable of sensing and measuring things like invisible forms of radiation.

let as much as possible and so each layer is level. You can't have variations in the layers if pallets are going to be stacked. At times, corner braces are used to help protect the palletized-product and improve stacking ability. Your manufacturing department's shipping department can usually help determine the best count and configuration for the tie and high, but make sure your marketers confirm it meets acceptable industry standards.

Most pallets are 40 inches x 48 inches in size. Some retailers will require a certain type or quality of pallet, so make sure yours meets the specifications or a large shipment may be rejected. Also keep in mind that using used or poor quality pallets reflects poorly on a product's quality and the company shipping it.

Package Designers

If you are planning the launch of an innovation with a start-up company or introducing a completely new product line, a talented packaging designer can help significantly. Experienced package designers can help match the right colors to the right demographic market. They can create the right image, the right feel that'll attract customers.

When seeking package designers, interview several and review their work as their backgrounds and abilities may be quite diverse. Costs can vary too, but try not to let the higher cost of an excellent designer deter you. The extra cost may mean the difference between millions in sales and may be the critical factor in capturing your niche. Find one that you feel confident can properly position your innovation for the intended target market. See the appendix or check out our website for updates on qualified designers.

Marketing-Related Buying Influences

Many buying influences based on product design and customer-driven innovation principles are addressed in Chapters 2 through 4. However, there are several other marketing-related buying influences that may need to be met in order to make an innovation acceptable to the trade. Here are just some you may need to consider:

- **Shipping.** Are there some special shipping considerations required by a distributor or retailer? These considerations could include using #1 pallets, a pallet return program, calling for appointments, a certain size pallet, pallet count restrictions, bill of lading, proof of delivery documents, and so on. One slip up here can result in a refused shipment, or at times, can result in the warehouse people asking the purchasing department to stop carrying your products.

- **Accounting.** Many accounting systems require order confirmations to be sent upon receipt of a purchase order (PO). They may require special terms and at times billing in duplicate or triplicate. They may pay bills though electronic payment systems only, or have an upcharge to write a check. Some industries have rebate systems that will have to be considered. It may be payable on a monthly or quarterly basis. In any case, your marketers will want to know all this up front as it will affect your pricing structure.

- **Operations.** Does your product require some internal training of retailers in order to sell it effectively? With an industrial product, you may have to provide training as well. Retail operators may decline handling a new innovation if training is required.

If training is minor, how are you going to do it? How will you train a chain store's employees throughout the U.S.?

- **Buyers.** Obviously the first consideration is an adequate profit margin by a distributor or retailer. However, simply making a sale with a buyer is not usually enough. There'll be several other vender qualifications that must be fulfilled, such as proper case markings and UPC codes, verification of insurance, and completing a new product worksheet to get your product entered into the company's computer (many times in order to generate a purchase order). Sometimes you may have to consider slotting allowances, which can be very costly. For test market application, these can frequently be waived, but you may need to consider them for a future ramp-up.

These are just some of the buying influences you may encounter. If you are not familiar with those buying influences that may affect your product, you'd better meet with your marketer and make sure they are satisfied.

THE AMERICAN DREAM

Utility patents generally give an inventor the most protection and hence have the most value. Utility patent protection is the chief focus of this manual and is what every inventor should seek.

Chapter

10

MARKETING AND MANUFACTURING STRATEGY

Make it a team effort and that alone will make it easy.

Y ou may have one of the greatest, most unique innovations of the last twenty years, but it will be a complete failure without the right manufacturing and marketing strategy. This strategy involves forming a team that includes four primary partners: an inventor/innovator (you), a manufacturing expert, a marketing expert, and a legal expert. You may use other experts, such as product and package designers, engineers, and prototypers, who'll play secondary roles assisting in development. Forming the primary team partnership is essential to the innovation's success regardless of how you plan to take it to market.

Who is on your team will be determined by your marketing strategy. Thus, without a strategy you really can't move forward to build your team. It would be akin to building a sports team and not knowing what kind of team you will be assembling. You don't want to assemble a team consisting of gymnasts, swimmers, and golfers only to find out later you really wanted to build a basketball team.

Treat Manufacturing and Marketing Separately

The main reason you have to treat manufacturing and marketing individually is that they are two entirely separate functions. You simply can't rely on manufacturers to market your innovation, nor can you rely on marketers to make it cost effectively. If they had their way, manufacturers would only make large quantities of their own generic products. And most marketers are not sensitive to the important issues affecting high-quality manufacturing.

The sooner you separate these two entities in your invention-related activities, the sooner you will be successful. You cannot invest all your time with manufacturing

THE CHALLENGE

The challenge of getting a product from idea to market can be daunting. It's a process fraught with difficult decisions demanding reliable information and advice at every stop.

–Don Kelly, Intellectual Asset Management Associates, LLC

MONOPOLY

Charles Darrow, an unemployed heating contractor, invented the board game, Monopoly, in 1933. The many toy companies he contacted repeatedly rejected him. They said the game was too complicated to learn, took too long to complete and "who ever heard of a board game that kept going round and round. There is supposed to be a beginning and an end." He struck out on his own and began marketing the game. Two years later, Parker Brothers saw the light, bought out his rights, and today, it is the most popular board game in the world.

experts speculating about whether your invention will sell. In addition, you cannot invest all your time with marketing people who don't know how to get your innovation made cost effectively.

These two experts must work closely with you to determine what will be the invention's best, most cost-effective attributes and how it will be sold. While finding a manufacturer to make a product is easy, finding the right one is not so easy.

But where most innovations fail is a lack of marketing. It is frequently neglected until after a lot of money has been spent on prototyping, or worse yet, production has begun. When the prototypes receive a poor response in sales presentations, the inventor wonders why. Marketability is the most important aspect of inventing and, as illustrated in the Strategic Guide, you really should have your marketing expert play a part in product development early on . . . not after you've gone into production.

Two Strategies and Seven Approaches

Before building a team, formulate a reasonably clear idea of what you believe is the best way to take your new invention to market. Stay flexible and open minded as you may make changes along the way or you may encounter some interesting offers.

Generally speaking, there are two basic strategies to consider. They are:

1. Sell the innovation through your company

2. License it

Selling Through Your Company

This means you are either going to sell it through your existing company (hopefully already well established in the trade), or you're going to initiate a start-up company. In either event, this means you'll need sufficient cash to develop the innovation, build up inventories, and perhaps hire additional employees as may be required.

Here are four variations on this strategy:

1. **Your company manufactures, and your company sells.** You are an established company or a start-up, you plan to produce or assemble the innovation in-house, in your facility, and either use your existing marketing team to sell it or build a new one. Your manufacturing expert and your marketing expert are company employees. This approach requires the largest cash outlay, especially if you are creating an entirely new company, but it does give you the best control over the ultimate outcome. Keep in mind that this approach also gives you more sales and marketing versatility later on. You may be able to expand sales by incorporating some of the following approaches as well.

2. **Your company manufactures, and you hire manufacturer's reps.** If you are confident about your company's ability to produce the innovation with the required quality and its ability to become the lost-cost producer, but you lack the ability or finances to build an internal sales team, you can employ outside specialists. Manufacturer's representatives work on a commission basis, and many are generally very good at introducing entrepreneurial new products. This approach requires only your cash outlay for manufacturing and building inventory and can save a lot time since you do not need to build an internal sales team. This approach will usually require

your financial assistance for trade shows and other promotions, such as advertising. However, the cost of hiring and employing a national sales manager, salespersons, and staff are eliminated.

3. **Your company manufactures and you supply private label marketers.** This is similar to the previous approach, in that you'll produce the product, but you'll play a secondary role in marketing the innovation. So, again, instead of hiring an internal sales team, you strike an agreement to supply one or more existing companies that are already suppliers to the trade. The suppliers purchase the item from you, mark it up accordingly, and resell the item as though it were its own product. You will either be supplying these companies under their trademark or under private label and marking all products accordingly. You may also include the supply as an OEM item, and they'll include the item on lists of existing products they sell. In either case, you serve as the manufacturer and maintain a certain degree of control based on your patent protection.

4. **You outsource (or subcontract) manufacturing, and your company sells.** If you are strong in marketing (hopefully in the field of your innovation!) but do not care to go into manufacturing, this is your best approach. In this approach, you are partnering with an existing manufacturer to produce your innovation to your specifications and with your labeling and trademarking and to ship the product to your company or perhaps drop ship the product to your customers as required. You'll be writing purchase orders that typically the manufacturer ships against. This is an enviable position since you are most likely going to bring in orders and commitments at an early stage. Existing marketers in a given field who know their customer base and needs tend to have a great deal of success. Knowing market potential and having the ability to get order commitments from existing contacts will also improve your ability to have a subcontractor help in the prototype process and to invest its capital for start-up. Why? The reason is simple: instead of paying for dies and molds and so on upfront, you can give the manufacturer something much better. You can give it a large purchase order that you can draw from over time. This is a win/win situation. This approach preserves a substantial amount of cash outlay and keeps you in close contact with the all-important customer.

In addition to these four approaches for selling the product yourself, you can also license it. If this is your long-term intention, then you will not want to contract with a sole marketer.

Licensing Your Innovations

This alternative alleviates most of your start-up costs. In a licensing scenario, the majority of start-up costs may be born by your manufacturing and marketing partners. The key to licensing is to focus on marketing since that is where the money is. Typically, you'll license a company that is active in the field of your invention and has the ability to sell the innovation in the marketplace, usually as a product that represents a product line extension. In other words, it will be one additional product it will be offering its customers, distributors, and retailers. Licensing your innovations and patents is discussed in detail in Part IV, including how to license them before and during the patent pending phase.

OPPORTUNITY KNOCKS

A critically important thing to keep in mind is that an invention is a business opportunity, no more and no less. It's not a lump of gold with inherent value in itself. The quicker inventors grasp this fact, the quicker they will take a business approach to invention development and protection.

–Don Kelly, Intellectual Asset Management Associates, LLC

STAY FLEXIBLE

While marketing one of my first inventions, I came across an opportunity to join forces with a competitor. They wanted my creative mind and patented products and had money too. We worked together, embarked on an exciting journey, ramping up sales to $10 million a year in just a couple of years. In your invention activities, keep an open mind and always look for options and possibilities. You never know what offers you may receive.

–Bob DeMatteis

COMMON GROUND

Question: What do Mr. Potato Head®, Viewmaster®, Frisbee®, Lionel Trains®, Operation®, Colorforms® and Play Doh® have in common? Answer: They were all created by independent inventors that either licensed their creations or established some form of strategic business partnership.

1. **You license a manufacturer and it markets the innovation.** This is the best way to eliminate almost all or most of your start-up costs. You license a single entity and the licensee arranges for the manufacture of the innovation—either in one of its plants or through one of its existing subcontractors—and then markets it to industry. You may license to one or a few exclusive licensees or to several nonexclusive licensees. With this approach, your job becomes one of coordinating support behind the scenes with the manufacturing and marketing departments of the licensees. Unfortunately, this scenario is not as easy to find as you might expect.

2. **You license a marketer and authorize a manufacturer to supply it.** This approach also alleviates the majority of your start-up costs. The marketing partner in this scenario may be: 1) an existing supplier to the trade; 2) a company that wants to get into the trade, or 3) a well-positioned manufacturer's representative group. Licensing a committed sales expert in the field of your invention frequently makes it much easier for you to find and authorize a manufacturer. This is accomplished by providing a written authorization to the manufacturing entity, making it aware of the patented subject matter or by granting a limited "manufacturing only" license. This type of license or manufacturing authorization gives the manufacturer the rights to make products falling under the scope of your patents but only for sales through authorized sources—which is the marketer you have licensed.

3. **You license a company and it sublicenses the rights.** This approach is uncommon, but there are applications and good reasons to do this. First of all, a certain entity may be well-connected in a particular trade and may have some alternative objective. For instance, the licensee of your invention may profit from the sales of raw materials it supplies to a licensee. Or, it may profit through the sales of machinery used to manufacture your innovation. By giving the supplier the right to sublicense, you will most likely have a broad, committed effort to find licensees and get the innovation launched. In such as case, the licensee will be nonexclusively sublicensing the technology or licensing various customers based on the field of the invention or in a certain region. This can be a particularly effective approach when a company wants to promote your invention internationally and you do not have the contacts or the financial ability to pursue it.

OTHER CHOICES

There are a few other choices to consider regarding your approach, but they are not ones you would normally pursue. As we discussed in Chapter 1, you can develop your invention, patent it, and then sell it. However, your time will not be well spent, since there are so few buyers who will have the same vision as you do. Another approach is to just wait and see if anyone starts to infringe your patent and then try to license him/her.

Whatever approach fits your innovation best, you'll want to zero in on one or two that fit your position and then be open to other offers as they may present themselves.

Financing Your Inventions

Depending on how you are going to develop your inventions, you're going to have to consider financing. One thing is certain: if you've done a good job of creating a dazzling, high-impact invention, others will be willing to invest their money and resources. Early on you may use your or your company's own capital to finance the effort, but sooner or later the effort will require a substantial infusion of cash to convert your concept into a viable product or to launch a start-up business.

There are several excellent ways to secure financing or offset most of the development costs. In fact, you can use your innovation's ability to secure financing as a bellweather of its potential marketability.

If you are building a start-up company or major expansion within your existing company, you must budget the product's development. If you're using development partners, subcontractors, marketers, and licensees, you'll want to line them up as early on as possible so they may invest their time and money on inventory, packaging, marketing, and so on. Keep in mind there is a huge difference between budgeting for a new product and budgeting for a new company.

This section refers only to budgeting the invention and not a start-up company, albeit the two may be related. The chief difference is that with a new company, you'll require a complete business plan, which is not necessarily the case with a new product introduction. Some companies will require an internal business plan to justify the pursuit of a new product, but generally speaking, your job as an inventor/product developer will not require that, unless you are also the marketer and the head of the company.

Any marketing partner you have may do its own business plan that fits its company's needs. At best, you may prepare a prospectus, such as the one described later in this chapter.

Early-Stage Financing

In order to develop an invention and prove the concept, almost invariably some capital is required. However, if handled wisely, these costs can be minimal and your development partners may bear most of it. Whether your partners are subcontractors, licensees, or your company's internal manufacturing and marketing departments, they'll have a budget they can apply to your project.

Seed money is usually funded internally—that is, by the inventor or his company. If an invention of a small company or individual developer is fairly sophisticated, then early-stage financing will be required shortly afterward and development partners may play a much greater part. Seed money and typical early stage-financing are as follows:

1. **Self-financed.** The seed money comes from your or your company's bank account and through the existing cash flow.

2. **Personal loans.** You may borrow money from friends and relatives. This can be risky however, because if they want a major piece of the action, you may lose substantial control and you may lose substantial ownership interest by the time secondary financing is required.

3. **Bank loans.** Seed money and, more importantly, early-stage financing may be in the form of bank loans collateralized by other assets you or your company has.

4. **Credit cards.** Many projects have been financed through the use of an individual's credit line.

5. **Accounts payable.** Financing through extending the terms of your accounts payable is another way to finance a product development. For instance, if you can extend the terms on some present payables totaling $10,000 per month, you have effectively borrowed this amount of money for product development.

6. **Investment angels.** Great ideas can attract investment angels. Typically, investment angels may invest from $100,000 to $1 million in a promising opportunity. They'll want a percent of the company and will most likely want certain key management personnel to be their appointees. They will be looking for ownership in the company

CREDIT CARDS

Credit cards originated in the United States during the 1920s when oil companies and hotel chains began giving them to repeat customers. The first international credit card was introduced by Diner's Club in the 1950s for use at twenty-seven New York restaurants. The customers paid Diners Club on a monthly basis and the restaurant owners paid the company for the use of the service. American Express launched its version shortly after, but the new plastic money did not become popular until the development of the magnetic strip in 1970. Today, so many credit cards are in circulation that it is impossible to list all the names.

and patents of about 15% to 25%. They do not expect a short-term buyout; instead they expect the company to soon be funded by venture capitalists (VC) or some form of initial public offering (IPO).

7. **Early-stage venture capitalists.** These are like angels, but generally speaking are interested more in development loans to prepare a product for a market so sales may be generated. They will want a substantial percentage of your company, usually 50% or more, and will expect a short-term buyout, usually two to three years. The buyout may be either from profits generated, a new set of VC funding, or an IPO.

Bear in mind that early-stage financing can substantially dilute your ownership. However, that may not be a negative if the value of the company and its patents is significantly increased. Obviously 25% of a $30 million company is worth a whole lot more than 100% of a $1 million company—or nothing at all.

The greatest problem innovators have when trying to secure early-stage development funds is not how much they'll get or whether or not their ownership is going to be diluted, but rather finding an angel or VC that is willing to provide funds. Rarely do venture capitalists or angels lend on patents alone. It is almost always the marketing potential that they'll consider. If you are unable to show them the marketing potential, your innovation is just not well-developed, or you're not an expert in your field with a solid team behind you, your chances are slim.

Partnering with Manufacturers and Marketers

One of the best ways to secure early-stage financing—or the equivalent thereof—is to have your manufacturing and marketing partners invest their capital. Regardless of whether it is your company's internal manufacturing and marketing departments, outside subcontractors, suppliers, or licensees, they'll most likely have a budget to devote towards product development.

Here are some of the contributions they can make:

1. **Manufacturing partners.** They can contribute not only their knowledge and knowhow but also may make a commitment to spend their funds, providing there is a long-term advantage to doing so. If this is your internal manufacturing department, the incentive is obvious. It may be able to fill idle production lines, fill up unused capacity, or expand its current production facility. Outside manufacturing sources may also be interested for the same reasons and will expect some form of long-term manufacturing contract. If your current outsource manufacturers are not interested, it may be time to reconsider your relationship and search for new manufacturing partners. The primary ways your manufacturing partner can assist you is to help with prototyping and production of sample product, defer start-up production costs such as dies and molds, and even help build inventory. Frankly, it would be suicide for any existing producer to refuse to help you launch a new innovation if you have an established working relationship. If you do not have a subcontractor and are pursuing the outsourcing and subcontracting of your innovations, then their willingness to contribute should be a major consideration in your selection.

2. **Marketing partners.** The same holds true for marketing partners. They too may profit from the launch of a new innovation. Sales means income, increased commis-

sions, sales overrides, and profits. Your internal marketing department, an outsourced marketing company, or manufacturer's reps can provide valuable developmental input to help with a new product launch, arrange packaging, and provide all your sales/marketing documents at a fraction of the cost you'd incur and in a fraction of the time. They may also be willing to arrange for the purchase of sample product in order to test market the innovation. Just like with manufacturing, your marketing partner will expect some form of long-term sales arrangement. When interviewing these entities as potential team members, make sure you find out what they can and can't do for you. This should play a big part in who you hire to be your marketing partner.

3. **Licensees.** For the most part, having a licensee should defer all your start-up costs. The licensee should accept responsibility for all start-up costs, including those previously listed for your manufacturing and marketing partners. It may even be advantageous for a licensee to pay patent-related expenses. This approach is not as easy to do with several nonexclusive licenses, but it may be if you start with one dedicated licensee that has a short-term start-up exclusive. The advantage is that in exchange for its initial developmental investment, you can grant a short-term exclusive for one to three years, depending upon the nature of the invention. Read more about this in Part IV.

No mater how you approach it, your two most important team members, your manufacturing and marketing partners, should be contributing to the development effort with assistance in prototyping, start-up molds, dies, machinery, packaging, and even inventory in a timely manner. If your present partners or those you are currently interviewing are not willing to do this, you're talking to the wrong people.

Other Investment Partners

In addition to the traditional manufacturing and marketing partners, there are a few other entities that may be able to help finance a new innovation. They may be considered Joint Development Agreement (JDA) partners and are harder to find. However, if you are already an expert in your field, you may know some of these companies.

1. **Raw material suppliers.** If your innovation represents a means to significantly increase the sales of raw materials—for instance, plastic resin or steel sheeting, then the raw material suppliers may have an interest in providing development funds. At the very least, it can provide your manufacturing partner with raw material it may use to make the initial test product.

2. **Machinery suppliers.** An innovation that provides the means to sell more machinery or modifications to existing machinery may induce a company to assist in development financing. Again, at the very least, it should be willing to provide start-up machinery or modifications in order to get the innovation tested and launched.

3. **Parts suppliers.** With innovations that require assembly, you can ask various suppliers to participate in return for long-term supply contracts. This may include everything from hardware and various components such as wiring, fasteners and so on, to software used to operate the innovation. It may even include others such as packaging suppliers and even artwork.

Think creatively when you think about those JDA partners who'll be involved in the

MILLIONS OF PATENTS

Patent number 1 million was issued in 1911. In 1930, plants were also made patentable; on August 18, 1931, plant patent number 1 was issued to Henry Rosenberg for a "climbing rose." In 1935, patent number 2 million was issued. In 1936, design patent number 100,000 was issued. In 1961, patent number 3 million was issued. Then, in 1975, patent number 4 million was issued; in 1991, patent number 5 million was issued; and in 1999, patent number 6,000,000 issued. At the present rate, U.S. patent 10,000,000 should issue around year 2022.

project and what they can contribute. It's in their best interest to help you be a success, so they'll profit once the product is launched. To explore what suppliers may be able to help all you have to do is ask.

Second-Stage Financing

After early-stage financing has resulted in a product launch, major financing is frequently required in order to take the company to the next level. In this phase, you are looking for funding to build inventory, advertise, and push sales way up . . . such as may be done in a national marketing push. This is typically where venture capitalists may fit in or some form of bank or government financing.

At this stage you want a major influx of money, but not in the form of an IPO, which may take several months to arrange. Besides, an IPO is usually a better bet once some decent sales have been generated and there is the promise of a whole lot more!

Frequently this second-stage funding is also used to cash out an early stage venture capitalists or perhaps other start-up loans.

Bank Loans

Second-phase financing (and seed money or early-stage funding for that matter) can be financed through banks. However, keep in mind that any loan will be secured by traditional means of either collateral financing or equity financing. Banks are not in the business of "lending on patent potential;" instead, they are in the money business. They simply want to lend money at a profitable rate and they want to know how they are going to be paid back.

If this is your plan, you'll need to provide all the standard loan information, such as a balance sheet, income tax records, cash flow statement, and so on. The bank may also require a business plan showing that your sales are on track. The business plan should also show where you are going to spend the loan money and the debt service to pay it back. In your company, your Chief Financial Officer can deal with this. If you are working alone, you'll need to prepare or copy the required forms and prepare your own business plan should you not have one.

If you are an individual planning a start-up venture or planning to propel your existing start-up to the next level and want to secure bank financing, the requirements are the same except that the bank will be looking at you more closely on a personal basis. Your personal credit history—past and present—is important. If you are seeking development money and are not in good standing with your personal finances, you're sending the wrong message to the bank.

Government and SBA Loans

The Small Business Administration (SBA) is active in business finance. Loans may range from $50,000 up to several million dollars. Interest rates are significantly lower than those with traditional bank financing.

Most SBA loans are secured through existing banks. While an existing bank's loan policy may not match up with your company for a business loan, it may match up for an SBA loan. One chief difference between a standard bank loan and an SBA loan is that the government guarantees payment of a certain percentage of the SBA loan, up to 75% to 85%. Thus, the loan-to-value ratio may be more advantageous for the borrower.

The SBA will loan for several reasons including export working capital, international trade loans, and pollution control.

In such a case, the associate bank processing your SBA loan will still be reviewing all pertinent documents and be scrutinizing your assets and collateral (both company and personal). There tends to be much more additional paperwork required, as can be imagined.

SBA requirements tend to change, but, for the most part, they are eager to lend money to qualified persons. In recent years, the rates have been incredibly low. So if you can qualify, you should pursue the opportunity. For more information about SBA loans and prequalifying, visit their website at: http://www.sba.gov/financing/.

Another form of government loan is on a state-to state-basis. Every state in the United States has loan programs available to promote its businesses. They may also have funding available for needed projects, such as energy- or environmental-related concerns.

Last, there is also a large amount of loan money available on a more regional basis from local cities and counties. If you are planning to enlarge your business or are looking for the right place to relocate, you may find substantial assistance available if your business will be providing local jobs and industry. This is particularly true with some of the Economic Development Corporations (EDCs).

EDCs are prevalent throughout the U.S. and may provide financing, tax incentives, operating capital, and even training. Some EDCs focus on environmental innovations that may affect the local region. This is just one more reason that any innovation you are introducing should include the CDI benefit of being good for the environment. It may not only help sell your innovation, but may also help you finance its development.

The best way to find out what agencies are in your state, county, and city is to contact your local SBDC and let them tell you. If they are uncertain, they can research for you and find out what is available. You can also contact your county offices to see what loan programs may fit your application.

Government Grants

There is a limited amount of government grant money available for specialized projects. One of the more familiar programs is known as the SBIR (Small Business Innovation Research). The grants given out by the SBIR have strings attached however. First, the government determines what projects may be worthy, not you. Second, the government will have the rights to any patents you've developed once the project has been completed. The nature of the projects is determined by several departments, but the projects are generally defense, energy, aerospace, environment and science related. SBIR grants can be up to $750,000.

Another program is the STTR (Small Business Technology Transfer Program). This program is similar to the SBIR and focuses on generally the same subject matter, albeit somewhat more narrowly defined. The terms and conditions of these grants are similar to those of the SBIR. Grants may be as great as $500,000, depending on the nature of the technology.

Both the SBIR and the STTR are administered by the SBA. You'll have to submit a detailed proposal should you be interested in pursuing one of these grants. You can find out more about these programs and which projects they are seeking at http://www.sba.gov/sbir/. It may be a long shot, but it is worth investigating.

PENICILLIN

In 1928 Dr. Alexander Fleming, a bacteriologist from England, was studying *Staphylococcus*, a bacteria that causes pus to form. He went on holiday and accidentally left a pile of plates containing the bacteria on his lab bench. When he returned he found that mold had grown on many of the plates. Fleming dumped the moldy plates into a sink of disinfectant, but luckily, he forgot one of the plates. This small oversight led to one of the most important discoveries of the twentieth century. Fleming was intrigued with the leftover dirty plate when he noticed that where the mold grew, his *Staphylococcus* bacteria did not. Somehow the mold stopped the bacteria from growing. Fleming reasoned that perhaps it could stop bacteria from growing inside humans too. In that age, this was a huge discovery because infections like meningitis, tuberculosis, and diphtheria caused many deaths. Further experiments resulted in an extract that Fleming called "mould juice." In 1929, the mold was finally identified as "Penicillium" so Fleming changed the name to Penicillin.

The computer mouse we know today was first invented by Douglas Englehart of Stanford University in 1964. With the rise in the use of computers in the early 1970s, an office computer called the Xerox Star came with its version of the mouse. In 1983, Apple included a mouse with its Lisa computer. The name comes from the original mouse's two buttons that resembled eyes and the long cord that resembled a tail. Englehart also came up with a larger, foot-operated control, called a rat, that was never commercialized.

Another national grant program is the National Collegiate Inventors and Innovator's Alliance (NCIIA). This initiative is a private philanthropic entity, founded by famous inventor Jerome Lemelsen, that offers grants through colleges and universities. It will grant up to $2 million annually for worthy causes. Grants are generally from $1,000 to $20,000. Find out more at http://www.nciia.org/.

In addition to the federal grants, most states also have grant money available, much of which is available though universities. You'll have to do some research to see what's available in your state. The best one-shop stop to begin your research is the SBDC in your region. If it is not aware of state grant money that may be available, it has the resources to find out.

One last note . . . before thinking about these grants, you'll have to be prepared to submit a complete proposal outlining the project's viability. In other words, you'll not only have to have done at least some preliminary development on your own to prove the concept is viable, but you'll also have to employ some administrative skills compiling the information required in a proposal. At times, these proposals can be a six-inch high stack of paperwork, or in other words, hundreds of pages. Much of that may not be written proposals, but completed forms and support documents. It would be much like a highly detailed prospectus, such as the one illustrated at the end of this chapter.

Venture Capitalists and Angels

Venture capitalists (VCs) usually fall under two categories, early-stage VCs and second phase (more commonly referred to just as "VCs"). Apart from the early-stage VCs that provide seed money and want a rather quick buyout, VCs for the second phase invest for major product launches and ramp-up. They are investing in companies with product lines that show substantial market potential, not patents. However, patent protection is almost always key to the endeavor.

Early-stage VCs will want your management team to be approved. The dollar amounts may be up to $2 million or so. A buyout in one to two years is standard. Like all VCs they invest in market potential, not patents. Besides, at this early phase, patents probably have not yet been issued. Nevertheless, you'll want to have the patentability potential well substantiated (qualified) with a complete report from legal counsel. You'll also need a qualified business plan to submit for their consideration.

VC investments for the second phase are typically several million dollars. They invest only in established companies that are ready for a major market advance. They want to be paid back and want the lion's share of the businesses ownership. VCs rely heavily on your providing them with a complete, detailed business plan. Once again, the plan must show how the VC is to be paid back. It must include an exit strategy that shows an IPO or some form of major financing.

VCs also tend to lend in areas that they know. Some are high tech only, some biotechnology, and others may be in general industry sectors. When searching out VCs, you'll want to find the right match because you'll simply not be considered otherwise.

Typically, before you consider approaching a VC, you'll have to have a qualified innovation, one that's already on the market producing at least some income with the promise of much more. You'll need a qualified team in place, too. A top marketing expert and manufacturing expert (or a subcontractor if that's the case) as well as legal counsel must be in place.

Angels are much like early-stage VCs and are also specific to an industry. The same business plan you'd submit to early-stage VC would be submitted to an angel. Angels do not request their investment be returned; however, they tend to play an active part in your start-up endeavor, usually through having one of their key management contacts run the company. This individual will be an experienced executive who'll be the decision maker for the newly formed company. Just like early-stage VCs, they will expect some form of subsequent financing to take place in one to two years so that they can profit from their ownership share.

Whether you are considering VCs or angels, you want to be prepared. One of the best places to find VCs and angels is through your local SBDC. They'll be familiar with those entities and individuals in your area. If they aren't, they'll be able to research fairly quickly to identify them. Once you begin the process and have talked to a few candidates, you'll be able to compile a list of others by asking them for additional contacts that may fit your invention. VC investments account for less than 2% of all business loan-related opportunities.

Financing Partners You Should Not Consider

There are also some financing partners you should stay clear of. You want to avoid any potential of misgivings or misunderstanding. In summary, there are two types of financial partners you'll want to avoid:

1. **Friends and relatives.** If you are operating on a shoestring budget, try not to use borrowed money from your aunts, uncles, and so on. They will most likely expect to be paid back with some form of interest and in a reasonable time frame. Your family relationships can be damaged should you not be able to pay in a reasonable time frame. More difficult than paying back a family loan may be dealing with the friend/relative who expects some ownership in the venture. This can become very sticky when you're making a proposal to an angel, a bank, or VC, and there's an indifferent, inexperienced relative behind the scenes who is trying to call the shots.

2. **Attorneys.** Patent attorneys are usually a pleasure to work with, but they rarely become financial partners in a new innovation. Even if they were interested, they'd probably want a disproportionate percentage ownership. Besides, when you think about it . . . do you really want an attorney to be your partner anyway?

The reality is if your invention can't attract the interest of existing means of financing an innovation within your industry, you have most likely not done your homework well enough or the marketability of the innovation is not sufficiently qualified to attract these entities. Companies and JDA partners that are already in the field of your invention can contribute substantially more value and more money than any friend, relative, or attorney.

Strategy Is a Team Effort

First, you physically cannot do everything yourself. Even if you could, it is doubtful that you would be skilled in every facet of bringing a new innovation to market. You may be a clever inventor, but are you an expert in manufacturing, marketing, and patent law as well? Even if you are, do you really want to work 25 hours a day? Of course, this is impossible, but it would probably take 25 hours a day to manage the production, sales

INVENTION ASSISTANCE COMPANIES ARE NEVER TEAM MEMBERS

They might say they are . . . but it's really a one-way relationship in their favor. Why? They get all their money up front and chances are you get nothing in return. Nor can they act as an intermediary to locate your manufacturing and marketing partners. Their dismal success rate is proof. Your team members—manufacturing and marketing experts— will be spending their money to develop your ideas and inventions . . . not the other way around!

and marketing, and legal aspects of your new invention—*in addition* to your full time job of overseeing the product's development! Sooner or later, in the course of developing your invention, you will want to involve other team members so that you can really make things happen.

As you know, the more important reason you'll want partners is that they can help enormously in defraying costs. If you did it all yourself, the start-up costs for a new product innovation would probably be at least $100,000. If you are an existing small business, the cost savings is huge since you'll most likely have some equipment and a marketer in place. Thus, it's easier for a small business to spend a little extra time and money to help develop the concept.

If you are a start-up or individual product developer, think about the money you would spend buying manufacturing equipment, renting a facility, hiring people, flying around the country to make sales presentations, and meeting with suppliers. By using some business savvy and finding the right partners and building the right marketing and manufacturing team at the onset, you can defray all, or almost all, of these start-up costs. Frankly, if you are in such a position, there is no reason to spend much (if any) of your money to get your concept to market. Good ideas that are attractive to others will generate plenty of interest from innovative marketing and manufacturing entities. You just want to find the right match.

The primary reason they will be interested is simple: your team members will have a lot to profit by assisting you. The right team members will have the interest and the means at their fingertips to help propel your project forward, which would otherwise be very costly to you. Finding partners does not have to be painful; everyone wants to get in on ground-floor opportunities. You just have to find the right partners, those willing to invest their time and money on some of the start-up expenses.

You may already have an idea of some partners you want on your team or your future team members may be people or companies you have never met. But one thing is for certain: *the team members you select must be experts in the field of your invention* and they must have the right attributes to get the job done. You will be establishing long-term relationships with these individuals and you will be relying on them for your future. With you as the team captain and parent of the invention, these experts will help you induce the birth of a "baby."

The Right Team Members

First, think of the team you are developing as a new franchise football team. You are the team captain and you want to have the finest team players you can. You do not want "has beens," rookies, or team hoppers.

You want an all-star offense (innovators are inherently offensive) consisting mostly of seasoned professionals. Your sales team should consist of running backs and wide receivers who will score points (get orders). You want the Joe Montanas, the Emmitt Smiths, and the Jerry Rices on your team. The offensive line (manufacturing) will support the backs and wide receivers by manufacturing top-quality products. You want 2000 pounds of human muscle. The front office (your legal team) will make sure that the contracts you negotiate with these top-rate players are based on winning performances.

To assemble a winning football team, you look for quality football players who fit the style of game you intend to play and who have proven track records. You do not look for golfers and basketball players, nor do you look for minor leaguers playing in semi-pro leagues.

Just like you would assemble a professional football team, you want to assemble a professional team for your innovation with experts in the field. They must show the enthusiasm and spirit to make things happen.

As the team captain, the three primary team players you'll want on your team are:

1. **A legal expert.** This is typically a patent attorney who will file and prosecute your patent applications. He/she may also oversee and advise on your licensing activities, if it is going to be pursued. The same law firm will also protect you in the event of infringement.

2. **A manufacturing expert.** You'll deal with the manufacturing manager, who may be a general manager or VP of Manufacturing. As you know, the manufacturer is either a department within your company, an existing subcontractor, or a licensee (or an authorized subcontractor for a licensee). While it might take a start-up company months, even years, to gear up for production, the right existing manufacturer can do it in a fraction of the time. The right manufacturer can help make your prototypes and first working models quickly . . . so you can get to market in a matter of months.

3. **A marketing and sales expert.** This team member may be the sales/marketing head of your company such as a National Sales Manager or a VP of Sales. This team member has to be an individual who has direct contact with his/her sales team and potential customers of the new innovation. It may be a present supplier in industry that you plan to sell. It may also be the individual heading up a manufacturer's representative organization you've employed, or plan to employ. Last, it may be a licensee. It is not, however, so-called marketing companies that create an image or artwork. It is the responsibility of your marketing expert to employ this kind of person, if needed. Nor is it a distributor. Again it is the responsibility for your marketing expert to arrange distribution. And, it is never an invention promotion company that promises to help you find manufacturers and sales agents.

You will devote most of your time and attention to your manufacturing and marketing partners. The three of you will work together to get the innovation manufactured with the right attributes and then sold and ready for ramp up.

When Do You Build The Team?

By now, you understand the importance of involving your sales and manufacturing partners as soon as possible. Once you feel confident that your idea is a good one, dedicate your time to assembling your three chief partners. The sooner you do this, the sooner your invention will be born and the sooner it will have all the right customer-driven features and benefits for the all-important end user. To get a better understanding of the timing on building a team review Chapter 2 and the Strategic Guide in the appendix of the book.

Early on, manufacturing partners can avoid potential manufacturing problems. Sales and marketing experts can provide valuable input about what customers really want. With the right manufacturing and marketing team in place, you will have a much greater chance of quickly and successfully launching your new innovation.

LEMELSON AND TEAM BUILDING

This famous inventor knew the value of teamwork. His foundation sponsors the NCIIA in dozens of colleges and universities in the U.S. with grants to help college students develop their inventions into licensable and marketable innovations. The establishment of "E-Teams" ("E" stands for excellence and entrepreneurship) is the primary mechanism in this effort. Lemelson's inspiration for the E-Team model was the collaborative interdisciplinary development teams used during WWII to rapidly develop innovative technical solutions to urgent wartime needs. E-Teams were pioneered at Hampshire College in Massachusetts. They consisted of students formed into groups to develop solutions to real world problems with commercial application. Since Lemelson knew the importance of commerce, E-Teams invariably become linked to the business community through relationships with nonacademic mentors and advisors, thus fostering connections between the institution's most entrepreneurial graduates and local business and technical communities. It's no wonder Lemelson was as successful as he was.

PROBLEMS AND OPPORTUNITIES

One of the biggest problems today still remains a huge opportunity for the future—that is, spam. Whether it is in the form of email, snail mail, faxes, or unwanted phone solicitations, millions still seek an acceptable means of getting rid of it.

Most inventors look for manufacturing partners before they look for marketing partners. However, it would be wise to have your marketing partner in place first, since how will you know what the best sales attributes and design will be without the expert's input? Having a marketing partner in place will also give you the opportunity to have him/her assess marketability though the use of prototypes. With pre-sales commitments, you'll be in a much better position to strike a working arrangement with a manufacturer. Providing a manufacturer with real time orders is a strong inducement for it to pay for certain start-up costs such as dies and molds. The value of doing this is covered in detail in Chapter 2.

Manufacturing, Marketing, and the Role of Your Attorney

Regardless of your development strategy, your attorney generally doesn't play a big part other than qualifying the inventive matter in your patents. If you are going to be seeking manufacturing and marketing partners, it's best to have your attorney on the team first. The meetings you're going to have with prospective manufacturing and marketing partners will most likely include discussions about patentability and scope of the patents. With your legal counsel in place, you'll be able to answer these inquiries effectively. Having your legal counsel in place prior to meeting with prospective development partners also demonstrates your professionalism, and serious intent.

Your attorney will also help you with your licensing activities and any manufacturing agreements that may be required. In Part IV, you'll see how to use your attorney to give his/her blessings to your license agreement without it becoming a costly, legal negotiation.

Make sure you learn all you can about licensing and patenting before engaging legal services. At $250–$500 per hour, attorneys are expensive teachers. Remember, it will still be up to you to make the decisions that are best for you. If you employ legal counsel to negotiate a license, it can easily run into the tens of thousands of dollars. And unfortunately, most of the questions have nothing to do with the law, but have to do with making smart business decisions. The best strategy is to learn all you can beforehand. Part IV will give you a broad overview of licensing so you can get started.

The Partnering Process

The search for partners usually goes through two stages. First, there is the initial stage of discussing your innovation and talking about it solely from the viewpoint of its being a moneymaking opportunity . . . nothing more, nothing less. When doing this, you'll not be disclosing any inventive matter. You will only be talking about the features and benefits and the potential marketability. No confidential relationship is required.

The second phase begins when a prospective partner shows interest and wants further discussions. This begins after confidentiality agreements have been signed. In this phase, the prospective partners provide input as to the manufacture and marketing of the concept . . . whether or not they believe it can be cost-effectively manufactured and whether or not they believe a market exists or can be created. These discussions determine whether they are interested in being your development partner or licensee, as the case may be, and what they can bring to the table. Remember, if they are only willing to work with you on a pay-as-you-go basis, they are probably not the best candidates for you.

Apply the following guidelines and you'll quickly be on your way to finding the right partners.

Learn About Presentation Skills and Presentations

When contacting and talking to prospective partners, you want to convey the right image and you'll need some good presentation skills. First, if you're not sure about your personal appearance, how you should look, and what to wear, make some changes now. If you don't have good presentation skills, you can develop them.

Your personal appearance is important, particularly with first impressions. This doesn't mean that you have to go out to purchase expensive business suits. Generally speaking you'll be dealing with other professionals in the trade. How do they look and how do they dress? You're going to feel out of place if they're dressed to the hilt and you're in jeans, T-shirt and tennis shoes. Try to conform to dress codes typical in the industry of your invention.

Bear in mind that your good appearance also conveys confidence. If you feel like you look good, you'll be able to make an excellent presentation. To start with make sure that whatever you wear is clean, neat and pressed with shoes that match the look. A sloppy personal appearance sends the message that your work habits are the same. For men, a sport coat and matching slacks accompanied by a conservative shirt and stylish tie will mean a lot. It will automatically put others at ease. For women, a conservative business suit will convey the same first impression.

If you are uncertain about your dress, consult a personal shopper at one of the major department stores, such as Nordstrom or Macy's. These individuals usually have a lot of experience in helping you create the right image. They'll be able to help you select a variety of matching shirts and slacks that will make you feel comfortable and professional. If you're a woman, they can help you find the right kind of business suit to convey the right image. Most can also give you pointers and suggestions on other aspects of your personal appearance, such as hair length and style, make-up and even finger nail polish color.

For the most part, presentation skills are based on what you say and how you say it. You'll typically want to start the presentation with a basic background and history of yourself and the invention opportunity. Then, the presentation will go into a brief history of how the innovation came about and end with your invention solution. Then you'll want to present your innovation to the prospective partner; this may include handouts, drawings, and a demonstration of how the innovation works. Of course, any disclosure of confidential subject matter at a meeting would be disclosed only after a confidentiality agreement has been signed.

The more you can visually illustrate your invention and the opportunity it presents, the better impression it will make. In an initial meeting, all that may be required is a new product summary (you'll learn about this important document shortly). A second-phase meeting should include any number of appropriate instruments, such as technical drawings, market analyses, computer-generated designs, Power Point presentations, DVDs, prototypes, and so on. The better prepared you are with these presentation aids, the better your reception will be.

LEARN ABOUT THE INDUSTRY

Learn all you can about the industry your product falls into, and how the industry's products are categorized, sold and marketed. Research the industry's manufacturers and their channels of distribution. Don't forget to learn about the industry's retailers. Learn what it takes to successfully sell a new product to a retail buyer, by learning about their needs and problems.
–Paul Lapidus, CEO, NewFuntiers

Know What Partners Look For

Partnering is a two-way street. While you're looking for partners to help develop and offset start-up costs, the partners want something in return as well. So, when you're interviewing and evaluating them, they, too, will be evaluating you. There are three primary considerations and several secondary one. Here's what they'll be looking for:

1. **Marketability.** Of course moneymaking potential is at the top of the list. You'll have to show this to any potential partners before they'll consider anything else.

2. **Patentability.** If they are going to be your partners and invest their time and money, they'll want protection. You better make sure you've qualified your patentability ahead of time and that your attorney will be able to back it up.

3. **Your commitment.** No one wants to invest the time and effort if you're only sending out feelers. Potential partners want to know that you are in this for the long term and that you show the determination and persistence to do what it takes to make it happen.

In addition to these three considerations, there are several other factors that potential partners will consider. You'll hear about them during the discussions and you may need to qualify them. They should include everything you've been considering up to date, such as raw materials, environmental impact, cost to gear up for production, ease of use, sales projections, competitive analysis, and so on.

Position Your Invention

Before you start looking for prospective partners, you'll want to think about how you are going to approach them and what you're going to say. First, how do you plan to position and present yourself and your invention? Do you know what you're going to say? Do you know how to say it so you'll get the right response? In a way, you can say that this is a time to use some basic sales techniques.

It's important you approach your discussions in a professional manner. You'll want to be positive and display substantial knowledge about your invention, its market, and its patentability in order to impress a prospective partner. This is the time when your research will really payoff and there may not be any more important reason than this.

Also remember that you will be dealing with a lot of sharks and barracudas, who may want to try to design around your inventions. In these initial conversations, keep your wording sufficiently masked.

Follow these six guidelines and you'll be in a position of strength:

1. **Establish yourself in a position of strength.** If you have experience in inventing or product development, say so. If not, focus on CDI benefits and the substantial amount of time and research you've invested. Your superior knowledge about the innovation, the field, and the target market will greatly improve your credibility.

2. **Talk about money.** This means you'll want to thread the moneymaking potential of your innovation throughout your discussions. This is the chief consideration in gaining the interest of potential partners and can be done without revealing any confidential inventive subject matter.

3. **Use key words qualifying your patent protection.** Using the right words and phrases will show others you know what you're talking about. Here are some examples:

   • "The new product I've been developing increases loading capacity by 20%–30%." Continue with, "This provides a **distinct market advantage** over existing products. In the **patent application** we reveal two systems that accomplish this, one I'd like to talk to you about." (This type of conversation asks for confirmation that the extra loading capacity has marketing appeal.)

- "Interviews with twelve end users indicated that they would pay an additional 10%. Of four retailers we interviewed, three said **they would test market the product** as well."

- "Our **patent disclosures** cover four unique state-of-the-art improvements; one has benefits for your company."

- "The innovation can be produced for **about the same price as competitive products**; however, there would be the start-up expense of making new dies and platens."

- "When discussing **patent strategy with my attorney**, we believe it is best to . . ."

- "We found that all the testers who used the innovation for a week had **built up enough trust** that they wanted to purchase it for all the computers in their office."

- "The **patent search and legal opinion reveal** that we will be able to secure broad patent protection."

- "My **attorney** will not allow me to discuss the new innovation without a signed Confidentiality Agreement."

4. **Take the offense.** You have the carrot and you can dangle it . . . just don't be misleading. Show your confidence in the product and your determination to see it successfully launched and you'll attract the best partners. Ask your prospective partners about their willingness to invest in the project and to devote their time and money.

5. **Remember, you are building a partnership.** Don't view team building as a negotiation. You are seeking long-term relationships. In a way, it's like getting married. Would you negotiate your marriage with a potential spouse?

6. **Keep the customer in focus.** Every decision you make should reflect what is in the best interest of the customer. Make sure that all the customer-driven attributes you've applied can be delivered by the manufacturer and are verified by your marketers. After all, it will ultimately be customer pull through that keeps the sales going.

Talk About Your Innovation in Masked Terms

Your first contact with a prospective partner—either in person, in writing, or on the phone—will start out discussing nothing more than the moneymaking opportunity. You will begin by talking about how you realized there was a problem with existing products or that no products were available to address a particular problem. Explain how you discovered a unique solution and why you believe your invention is a good profitable opportunity. All this can be said without having a signed confidentiality agreement in place.

Talk about the size of the market it could impact, but don't be overzealous. Use comments such as, "It is useful in a $150 million market and can save 20% over existing paper and plastic products. I think we could capture a niche as great as 2%–5% in a few years. I believe with a more aggressive sales campaign and the right investment, it could

DON'T SAY YOU'RE PATENT PENDING IF YOU'RE NOT

This only shows inexperience and a lack of integrity. If you say you have a patent pending and you really don't, what are you going to say when it is time to file? That you have been lying? You might destroy your credibility. Besides, it is a fraud, which is punishable by fine. You can always say you are in the process of writing the patent application, which gives more power than just having a disclosure.

IF YOU ARE PATENT PENDING

When you do have a patent pending, it shows others you are moving ahead. Refer to your product as having a "patent pending" or "patents pending" if there is more than one. If you already have a patent, say so. If they want a copy, give them one. If you are sending out letters of inquiry, send a copy of the patent abstract along with it.

be even greater." This will raise their curiosity . . . and, as of yet, you've not revealed the inventive subject matter.

Don't talk about your innovation being a "must have" invention. These are few and far between and sound too much like pie in the sky. If they want to call it that . . . let them. The last two "must have" inventions were probably the wheel and the discovery of fire and it's doubtful your invention is anything close. To say that you are going to revolutionize the industry and take over all $150 million in sales in three years shows inexperience, a lack of business acumen, and a huge lack of credibility. It's not going to happen. Established businesses are not going to roll over and let you put them out of business.

It would be equally incorrect to exaggerate the size of the market. If you do not have exact figures, do some research and find out. Pie-in-the-sky estimates will cost you time and money and raise false hopes. Here are a few ways you can learn the size of your market:

- Ask your marketing expert.

- Visit the research section at the local library or ask the librarian.

- Get sales figures from a single company, find out what percent of the market they represent, and then calculate the total market.

- Ask a manufacturer.

- If it is a retail item, ask a store manager or a district manager.

- Do an Internet search.

The best way to prepare for these initial introductory meetings is by preparing a new product summary, which you will soon learn all about.

Talk About Customer Benefits and Test Results

In Chapters 3 and 4, you learned about customer-driven innovation—CDI benefits—and how they apply to making outstanding inventions. Now you can use some of this same language in your conversations, letters, and new product summary. It is easy to do, without revealing confidential design attributes or the inventive subject matter. The use of facts conveys serious intent and a professional, knowledgeable attitude. Following is a list of the types of phrases you can use in your meetings. Using these examples, prepare a list that describes your innovation:

- Environmentally sound, reduces waste, made of recycled material

- Saves time, space, and money

- Safe to use, produces no toxic emissions

- More convenient than ever . . . takes two minutes to assemble instead of fifteen!

- More comfortable than the most expensive alternative

- Dramatically improves accuracy

- The most entertaining machine on the market

- Easy to learn

- Appeals to the senses

- Improves looks

TALKING ABOUT YOUR CREATION

Many inventors believe there is nothing like their invention to compare it to. That it is not possible to talk about it in masked terms without revealing the exact subject matter. This is rarely the case. For instance, if you just discovered the automobile, would you not talk about it in terms of being a means of transportation that does not require the use of horses or oxen? If you just discovered the television set, would you not talk about it being like the movies, only inside your home? Use your creativity and find a way to describe your inventions in masked terms. Until you do, you'll never be able to find qualified marketing partners.

- Improves productivity
- A three-step system
- People friendly

If you have tested some prototypes (no matter how crude), you can talk about test results without revealing specifics. When you refer to testing, say that it was done *"in the lab"* regardless of whether the lab is in your garage or in an actual testing situation. Or, if you conducted a survey, cite the example as such. With test result data or qualified surveys, you'll want to provide concrete information. Here are some examples:

- Fourteen Elmendorf tear tests showed the 33% recycled content did affect rip resistance by more than 2% to 3%.

- 82% of all testers observed filled the 20%–30% larger capacity.

- Thirty-four tests showed it takes an average of 2.17 minutes to assemble.

- All but one of seventeen subjects surveyed agreed it was far more comfortable than all the others they have used.

- Accuracy of 280 test cuts was improved by 7.23% over traditional means.

- All fifteen subjects surveyed laughed out loud. (This may not be concrete in the classic sense, but it can be a powerful statement with the right innovation.)

- Twenty-two tests revealed that instructions were not necessary.

- Ten out of fourteen subjects surveyed felt it improved their looks.

- Productivity in the test facility improved by 11%

- All 82 employees surveyed showed they could figure out how it worked within two minutes without any training or additional help.

If you don't have any test results or survey reports, you may try to compile some. Also look for support documentation in the industry to substantiate your claims. Once you've compiled this list, you're ready to prepare a new product summary, one of the most valuable development aids you'll have.

Prepare a New Product Summary

If you can't discuss the moneymaking potential of your innovation without revealing the inventive matter, you'll never go anywhere. It is essential to be able to do this in the early stages when attracting manufacturing and marketing partners. One of the best ways to prepare for initial discussions and to know what to say and even how to say it is by writing a new product summary. This summary is a one-page report discussing the background and the marketability of your innovation, without revealing any inventive matter.

Preparing a new product summary (sometimes called an invention summary) helps you know how to talk about your innovation without revealing confidential information, regardless of the nature of the invention. The terminology you will use may be broader, and will focus on features, benefits, and the opportunity. It is especially effective to use when pursuing licensees and marketing partners.

The summary may illustrate facts and figures and preliminary market analyses, but it will not reveal specific details. It reflects the CDI benefits, such as those listed in the preceding section. Your invention summary should be concise, professional, and short.

WHY SHOW TESTING?

There is no better way to show market acceptance. What did the consumers (not your family members!) think? Did they know how to use it? Did it work as anticipated? How can an inventor expect a potential licensee to invest hundreds of thousands of dollars on a product that has not had at least minimal testing?

INVENTION (NEW PRODUCT) SUMMARIES

It is best to prepare concise summaries with simple bulleted benefits. Don't use gimmicky or old-fashioned pictures or cartoons. Serious business people will view them as being unprofessional. If you have a patent or patent pending, consider having the summary color copied with a photo image of the invention.

Don't use outdated clip art, cartoons or "hip jargon." Remember, you're trying to attract professional business people, not sell advertising.

With your new product summary complete, you can use it as a guideline when talking on the phone, or you can send it off with cover letters to prospective partners. It can also be faxed to an individual you are talking with on the phone. An example of a new product summary is given on the next page.

Finding and Interviewing Manufacturers and Marketers

Your invention disclosure establishing the date of original conception is written, you have hired an attorney and written a new product summary, and you are now eager to locate your manufacturing and marketing partners. First, make a list of all the potential candidates you would like to start interviewing. Then, begin interviewing these potential associates (preferably on the phone) without disclosing any confidential information and without a confidentiality agreement. Use your new product summary as your guide. If your initial discussions with a candidate garner their interest, follow up with a second conference, either by phone or in person. These follow-up discussions will require a confidentiality agreement to be signed first.

Your initial primary objective is fact finding—that is, confirming the cost of manufacture and verifying marketability. It usually does not matter whether you first interview sales and marketing experts or potential manufacturers. You may start with existing contacts or seek out total strangers. What is important is to begin the process by locating as many viable entities as possible and gathering as much information as you can. You want to find the right fit for your innovation, your team, and you. In a way, you can say your initial discussions are a form of networking in the trade.

When speaking with a prospective partner that may be both the manufacturer and the marketer, it is best to start with marketing first. Since you will be evaluating their ability to manufacture **and** market your invention, you'll also want to talk with their manufacturing head sometime afterward. If it turns out the company is only interested in manufacturing the innovation and not marketing it, make that note in your journal. You may be able to match up the company later on with a qualified marketer. When you talk with manufacturing-based enterprises that make similar products as yours, ask them if they are able to recommend marketing experts in your trade. They may already be working with some. When you are interviewing sales experts, ask them if they know manufacturers. Networking will help you quickly locate several potential manufacturing and sales partners.

In Chapters 11 and 12, you will learn where to find marketing and manufacturing team members and how to evaluate them.

Searching by Phone

Telephone interviewing-networking-prospecting is the most effective means of quickly locating potential partners. It is fast, inexpensive, and easy.

When you are doing telephone interviews with potential marketing partners, ask for the President of the company if it's a small- or medium-sized company; otherwise, ask for the National Sales Manager. If you are licensing your innovation, the president is the

---

INCORPORATE

Independent vendors should consider incorporating to create a professional presentation to companies.
–Michael S. Neustel, U.S. Patent Attorney

SOME INDUSTRIES REQUIRE WORKING THROUGH OTHERS

Although the highly visible invention promotion companies do little to support the true goals of the independent inventor, some industries require the inventor to work through well known creative groups and representatives, which are sometimes referred to as brokers. The toy industry is a good example of this, as they will not work directly with unknown independent inventors. Instead, they provide independent inventors with the names of companies and individuals, with proven track records in the industry, through whom they should work.
–Paul Lapidus, CEO, NewFuntiers

New Product Summary

YCN *Your Company Name* company address, city, phone and fax numbers, website

New Product Summary—M2K Square Bottom Bag

Product Background

Paper bags have been the most popular style of bag to carry fast food and food service items in high-volume outlets in North America and much of Europe and Australia. Paper bags are the packaging of choice due primarily to their ability to square out and stand up.

In some fast food outlets thin-gauged plastic T-shirt bags using space-consuming racks to support the bags for loading are used. Generally speaking, T-shirt bags do not have the aesthetics sought by the fast food trade, are difficult to load, and don't stand up after being packed.

Since plastic T-shirt bag racks require too much space, have poor aesthetics, are not cost effective in smaller sizes, and do not cooperate with condensation problems, they are not a suitable paper bag replacement. Thus, paper bags dominate the fast food industry until a suitable alternative has been developed.

Product Summary

M2K bags combine the best attributes of both paper and plastic into one single product. They can be loaded faster than paper and stand up just like paper after loaded. Using M2K bags makes it a simple task to check order accuracy and speed customers through the drive-through lanes.

M2K bags have also successfully eliminated the condensation issue associated with plastic. The proprietary "Automatic Ventilating System" (AVS) automatically opens and vents condensation after the bags are loaded.

The proprietary hook dispensing system eliminates the need for space consuming racks. The hooks allow different bag sizes to be placed neatly and conveniently along the side of a counter top. The bags can be quickly loaded, are self-opening when dispensed, and stand up afterward.

Product Benefits

There are many benefits for supermarkets. Here is just a partial list:

❏ Lower overall bag costs
❏ Lower labor costs
❏ Improved productivity
❏ Convenient, easy to use
❏ Little training required
❏ No space-consuming racks required
❏ Bags automatically ventilate allowing condensation to escape
❏ Per unit cost is 20% less than paper

Patent Status

Seven U.S. patents are pending in the U.S. with international rights reserved.

The Opportunity

A limited number of licensees in the U.S. and internationally are sought.

EVEN IF . . .

Even if you have patent pending, you should still attempt to utilize Confidentiality Agreements when disclosing your invention. There will be several other confidential matters you'll need to discuss later on.
–Michael S. Neustel,
U.S. Patent Attorney

WHAT ARE THE ODDS?

Maybe there is a 1% chance that an invention will be stolen if disclosed to a company. However, everyone knows there is a 100% chance that an invention will never be licensed if not disclosed at all. Logic dictates that successful inventors must disclose their invention to companies!
–Michael S. Neustel,
U.S. Patent Attorney

PRACTICE ON WORST CANDIDATES FIRST

When starting out to call prospects on the phone, try practicing on some local companies first. Next, call some mega-giants and get beat up a little. Last, start calling the more qualified candidates. This can be an especially good way to get comfortable with phoning prospects if you are new at this approach.

one who must ultimately sign your license agreement anyway. Besides, you will most likely be dealing with small, medium, and large companies in which the presidents tend to be more accessible.

In any case, if you cannot get through to the president, try to talk to the Vice President of Marketing or the National Sales Manager. Never present your new ideas and inventions to the head of R&D or New Products departments without first going through the president or sales manager first. The head of R&D is not paid to evaluate the ideas of outsiders. Besides, R&D people don't usually decide what products are going to be manufactured and sold.

If it is a company's policy to run all new ideas through R&D before being reviewed by sales, this is probably not a company for you. This is most likely a company with a NIH attitude. Move on to other possibilities.

If the candidate on the phone is interested, you can fax or email a new product summary while you talk. Your objective is to have this person ask for more information, which means getting a confidentiality agreement signed and then discussing the more pertinent details.

Using Letters of Inquiry

Even though it is usually best to establish initial contacts by phone, you may sometimes do it via mail or email. Via mail or email you should send a letter of inquiry accompanied by a new product summary. With email put both documents in one single attachment.

Letters of inquiry should be concise and professionally prepared. Address a letter of inquiry to the president of the company or in certain cases the National Sales Manager. You'll need to do some research to get these names ahead of time. You can usually do this via phone or by looking it up on the company's website.

After sending the letter and summary, you should follow up by phone afterward. When you call, ask for the individual by name. If your call is screened by someone else, tell the person who you are and that you want to make sure he received your letter. In any event, if you can't get through, try to leave a voicemail. Be pleasantly persistent in trying to get through, which is a trait that most company heads appreciate. In fact, this same persistence will send the important message that you are committed to the success of your idea.

Sample letters of inquiry to domestic marketing companies, domestic manufacturers, and potential importers are provided on the following pages.

Either the indented or block style format illustrated is acceptable with 1-inch margins on both sides and the bottom. Make sure your letters are typed using a similar professional format with clean, crisp copy. Position your copy on the page so it is well-balanced. Do not use handwritten letters or summaries. Send your letters via Priority Mail in one of the red, white, and blue post office envelopes. It sends just the right message and is affordable. Don't use first class . . . it shows no sense of urgency or importance.

Oh yes . . . and never say things in your letters (or in your conversations for that matter) such as, "I have always wanted to be an inventor and hope to find someone to license my ideas." You must appear professional as an expert in your field. This statement conveys just the opposite.

Using the Internet

In many respects this is similar to letter writing, but certainly much faster. However, stop and think awhile. When is the last time you received any form of important documents or proposals via the Internet? Yeah right . . . email is really more akin to junk mail. Keep in mind also that any heading you have in an Internet email such as "New Product Opportunity" sounds just like junk mail!

For this reason, soliciting partners via the Internet blindly is not usually a good idea, unless you can locate a specific person within the company and only if you can write a very effective tactful email. The one exception to this rule would be when you are contacting overseas partners. Then, email is not only preferred, but it sends the right message. In such a case, you'll still want to try to locate the right individual to whom to address your message.

All this doesn't mean you can't use email to send documents, and in fact, it is preferred over fax and snail mail. However, it is best used after you have made the initial contact by either phone or regular mail. After the individual has given you his/her email address, you can forward the new product summary and perhaps other data as needed.

An example of an email message follows. But keep in mind, it will be far less effective than any other means if it is a blind mailing.

A Sample Email Message

Interviewing with Confidentiality Agreements

After the initial contact, either by phone or letter, some parties will want to meet to discuss the opportunity. If the plant is nearby, you can visit it in order to evaluate its abilities. If the manufacturer or marketer is out of the area, there are alternatives other than jumping on an airplane to fly to the prospect's facility.

One of the best ways to meet out-of-town prospects is by asking the national sales manager when he/she will be visiting your region next. National sales managers travel a

---

USE LETTERHEAD

Send letters on your own letterhead. If you are an independent developer just starting out, it is easy to make your own using the computer. You can quickly model your letterhead after any style you like. Legally you can use your own name and add "Company" to it without creating any problems with mail delivery, business licenses, bank accounts and so on. For instance, if your name is Ralph Smith, call your company the Ralph Smith Company and add your home address. You can use an existing bank account.

PHONE, LETTER OR EMAIL?

Some inventors may prefer to send out letters of inquiry, others may prefer to call on the phone, use the Internet or meet in person. Generally speaking, calling on the phone is more personal and far more effective. In this age of swift communications, the telephone is the preferred method. Let's not forget, we are building relationships and there is no better way to do this than by hearing a real voice. You can even leave voice mail messages, which usually get a return call. Even if you have a tough time connecting with the president, at least you will be remembered if you follow up with a letter and new product summary afterward.

Letter of Inquiry to a Domestic Marketing Company

YCN **Your Company Name** company address, city, phone and fax numbers, website

April 24, 2004

Mr. President (or National Sales Manager)
Company name
123 Wonderful Street
New York, FL 45678

Subject: Marketing Opportunity—Square Bottom Bag

Dear Mr. National Sales Manager:

Your company has been identified as a potential marketing partner for the new M2K plastic square bottom bag we have developed.

This new bag is quite unique and addresses many issues that existing plastic or paper square bottom bags don't. For instance, these new bags breathe to allow condensation to escape; they take up one-sixth the space of paper; they are made from standard high density, FDA-approved polyethylene film; they automatically dispense; and best of all, they cost 20% less than paper. The initial test responses show substantial market potential in the fast food and grocery trade.

We are interested in talking to you about this opportunity and for your evaluation and possible partnering. If you are interested in hearing more, please feel free to call anytime.

Sincerely,

Your Name

Letter of Inquiry to Domestic Manufacturer

YCN *Your Company Name* *company address, city, phone and fax numbers, website*

April 24, 2004

Mr. President (preferably the person's name)
Domestic Manufacturer
123 Wonderful Street
New York, FL 45678

Subject: Seeking Manufacturers

Dear President:

Your company has been identified as a potential manufacturing partner for the new M2K plastic square bottom bag we have developed.

This new bag is quite unique and addresses many issues that existing plastic or paper square bottom bags don't. For instance, the new bags breathe to allow condensation to escape; they take up one-sixth the space of paper; they are made from standard high density, FDA-approved polyethylene film; they automatically dispense; and best of all, they cost 20% less than paper.

We are interested in locating a dedicated manufacturer to produce these bags. It would require co-extrusion capability with up to a 60" web, six-color print capacity, and high speed, bottom seal bag equipment, with a cut-out table capable of handling up to 4 lanes of production. An investment on your behalf will be required in order to modify your existing machinery.

Should this opportunity appeal to you, we are interested in talking to you and discussing the details.

Sincerely,

Your Name

Letter of Inquiry to Potential Importers

YCN *Your Company Name* company address, city, phone and fax numbers, website

April 24, 2004

Chinese Culture Center (or any other country)
832 Stockton St.
San Francisco, CA 45678

Subject: Seeking Exporters

Dear Consulate:

We are searching for Chinese plastic film and bag manufacturers as manufacturing partners for the new M2K plastic square bottom bag we have developed. This new bag is quite unique and addresses many issues that existing plastic or paper square bottom bags don't.

We are interested in locating low-cost manufacturers to produce these bags for export to the United States. It would require co-extrusion capability with up to a 60" web, six-color print capacity, and bottom seal bag equipment with the ability to make cut-outs. A modest investment on behalf of the manufacturer will be required in order to modify existing machinery and make the appropriate dies.

Would you please forward this request to the parties you believe would be interested. They may contact me directly, preferably via email.

Sincerely,

Your Name

lot so it's usually not too difficult to schedule a time to meet. Another excellent approach is to plan to meet prospects at an upcoming major trade show in your region, or preferably at a major national trade show. After making several initial contacts with different individuals, you can schedule several meetings. You'll also be able to meet other potential manufacturing and marketing partners you may not have known about at the trade show.

During these more serious discussions, you will be disclosing confidential details of your invention. Before you do, have a confidentiality agreement (sometimes referred to as a nondisclosure agreement) signed by the heads of the companies you are considering. Use the technique illustrated in the next section and you'll have a high degree of success.

Asking Others to Sign Your Confidentiality Agreements

You will find that in your initial contacts, those prospects you've talked to that are interested in learning more about your innovation will usually sign your confidentiality agreements with no or few modifications. Follow this model and use these guidelines and you'll have a high degree of success.

1. In your initial contact with prospects, you will have discussed the potential marketability, manufacturablity, or both. In those discussions you will have talked about finding the right fit and will have used important words such as you are looking for the right "marketing partner" and "having a team effort." You want these conversations to be friendly so the other party is receptive to taking the next step.

2. When an individual expresses an interest to meet with you, try to arrange a meeting as previously suggested, for instance, on their next visit to your city. When you make the appointment to meet, tell them that you will fax them your standard confidentiality agreement. If there is any question as to why one is needed, tell the individual that your legal counsel requires it. In this case, your legal counsel may be an internal company attorney, outside counsel, or your patent attorney. Whoever it is, they'll agree.

3. Regardless of whether you will be meeting in a couple of days or a couple of weeks try to do this when the discussion is fresh in your mind and theirs. Neatly fill out a confidentiality agreement and fax it as soon as possible. It is far less intimidating to do it this way than it is to bring it with you to the meeting and ask them to sign it. Better yet is to use a form you can keep in your computer and fill in the company and individual's name.

4. Frequently the recipient will sign and return the agreement right away. If you have not received it a week prior to the meeting date, call the person. In that conversation confirm the meeting time and place and ask if they've had the opportunity to return the confidentiality agreement. If there is resistance, this is usually when you'll hear it. If necessary, remind them that your legal counsel will not allow you to divulge any information or show the innovation until it is signed. If you want to add a bit of humor, tell the other party that your attorney requires everyone to sign your confidentiality agreements . . . including your mother!

5. If you still have not received a signed agreement a couple days prior to the meeting, call again as a reminder. If their attorney has made any changes, it must be faxed to you right away so you can review it. When you receive the faxed copy with the changes, you may have to consult with your attorney.

FRIENDS AND ASSOCIATES SIGN AGREEMENTS TOO

Everyone should sign a confidentiality agreement, even those who consider themselves good friends and think it is not necessary. If there is resistance, tell him/her that it is something your attorney requires with everyone . . . including your mother! That'll get a laugh and a willing signor.

UNWILLING TO SIGN?

The most common reasons companies won't sign confidentiality agreements are: 1) They are megagiants that are really not interested; 2) It was a blind mailing . . . you did not talk to them in advance; 3) They are too paranoid they will "miss out on the opportunity if they sign it;" or 4) They think you are a "flaky inventor." Use the method in the following pages and you'll get results.

6. Be prepared to cancel the meeting if they do not return the signed agreement prior to the meeting or there are major changes. You will look inexperienced and vulnerable if you don't cancel or postpone it. This company is probably not the right fit and you can be assured that there will be a multitude of other changes and modifications along the way that can slow down and complicate the effort.

Remember that using confidentiality agreements is something your legal counsel will advise you to do as it is important to the future of your project. Don't sacrifice or jeopardize your proprietary position by acting irresponsibly. In the next section you'll learn more about the specifics of confidentiality agreements and how they should be legally used and applied.

Confidentiality Agreements

Confidentiality agreements cover a wide spectrum of information and materials that you may wish to keep confidential. In summary, a confidentiality agreement is between two parties, the disclosor (you) and the disclosee (the second party). It includes all patent-, patent pending-, and invention-related information, marketing data and testing, manufacturing data, drawings and processes, plus much more. The confidentiality agreements you use must be tailored for invention and patent development. They cannot be a "standard agreement" with multipurpose usage. When you read the following agreements, you'll see the breadth of the scope and the specific added protection for inventors and entrepreneurial companies.

You want to use confidentiality agreements with any entity to whom you may reveal confidential material or who will occupy a supporting role in your efforts, regardless of whether you are already doing business with them. This pertains to subcontractors, parts suppliers, engineers, advertisers, and so on.

PAPER TRAIL

Signed confidentiality agreements and letters of rejection are excellent to include in your paper trail . . . keep them in your files!

There are three basic types of confidentiality agreements used with product development. The most common is a broad-based nondisclosure agreement that can be used with just about anyone associated with the project. The second type is a simplified version you can use with plant visitors, parts suppliers, and those who will not be an integral part of development but may be asked to supply certain technical expertise. The third type of agreement is a independent contractor's agreement for those who will be doing certain specific work for you, such as a prototyper.

If you are a company, you should also have an employee trade secret program in place. This means each company employee signs a confidential agreement and promises to keep all aspects of the companies business, including its trade secrets, confidential. This type of program should include the required assignment of all patentable matter to the company. This is standard procedure with most corporations and something you'll want to have in place if you don't already. You can read more about this in the PQM section in Chapter 6.

Confidentiality agreements protect you and show a serious professional intention. For a few hundred dollars, you can ask your attorney to pull one out of his/her computer and tailor it to your needs or you can use or adapt the agreement shown on the following pages. You can also purchase these editable forms, which are complied in the Resource Guide on our website at http://www.frompatenttoprofit.com.

Regardless of which agreement you use, make sure you are the one being protected. If the other company says it has one you can sign, keep in mind that it will probably protect them, not you. You may have to sign one of theirs if they have proprietary information they'll be sharing with you. But make sure they sign one of yours too, so that you are protected on your proprietary data.

Broad-Based Confidentiality Agreements

This type of agreement is used in a primary role with those individuals and companies that are an integral part of your development. It is essential you have these entities sign one before revealing any information pertinent to your inventive subject matter or the innovation. While not 100% foolproof, a broad-based confidentiality agreement plays an important part in your invention-related activities and reinforces your first-to-invent rights and reduction to practice. Without question, it is far more valuable to use than not to use.

A broad-based agreement signed by both parties states that you, the disclosor, will be disclosing certain confidential information to a recipient, which you request kept confidential. It should be signed by an officer of the disclosee's company.

The Confidentiality Agreement on the following pages is a well-written, broad, and universal document that covers most applications related to the protection of an inventor's proprietary information. It has been refined over many years of product development and reviewed, perfected, and approved by patent attorney Michael Neustel (see appendix for more information). The key points it covers are:

1. Disclosee promises to keep the information confidential and will not disclose the information to any third party.

2. Disclosee will not disclose to others in disclosee's company any information they do not need to know.

3. Great harm may come to disclosor if information is leaked by disclosee and that disclosor can and will hold disclosee liable.

4. Disclosor requests that disclosee have a confidentiality program or an Employee Patent Program to protect disclosor's proprietary information and property.

5. Any innovations or improvements to disclosor's invention that are developed with the disclosee's assistance will become the sole property of the disclosor.

6. Disclosee will not compete against disclosor.

7. Disclosee will not try to design around disclosor.

Usually the only problem you may encounter when using this agreement is that the disclosee may sometimes want his/her attorney to review it first. The attorney will invariably want to make certain changes. If this happens, read the changes carefully. They are frequently innocuous. For instance, they may want to insert language to indicate that the agreement does not pertain to information the disclosee already knows or that is publicly known. This statement is obvious, because the agreement only refers to new confidential information anyway. Generally, this agreement will be signed, unaltered, 80%–90% of the time if you use the preceding approach and are sending it to small-, medium- or large-sized companies.

TOO TOUGH

Many companies do not want to deal with inventors because they believe them to be too difficult to work with. The typical inventor is expected to buck corporate policy, be difficult and unsophisticated. It's OK to be that way if you are financing your own production. If you want to play in the leagues of a Fortune 500 manufacturer, clean up your act. Your first step should be to take "inventor" off your business card and put "Product Developer" or "Product Designer" in its place. Big companies hate dealing with inventors, but they'll deal with product developers.

Basic Rules of Using Confidentiality Agreements

CONFIDENTIALITY AGREEMENTS WITH CORPORATIONS

Most giant corporations will not sign the confidentiality agreements of others. Being afraid of being sued, they may ask you to sign theirs instead. If you elect to do this, talk to your attorney first. Invariably their agreement protects them a lot more than you . . . if at all.

Be careful, some confidentiality agreements have clauses that say something like, "if any of the subject matter discussed was or is now being worked on by them, all proprietary rights will be theirs." The problem with this clause is the burden of proof needs to be clarified. You don't want to give them the opportunity to say "we already knew about it" and that's the end of the subject. You must have some means of verifying their claim so you will not be pitted against their multimillion dollar law firm.

There are some legal guidelines you should follow when using Confidentiality Agreements. Keep them in mind:

1. A confidentiality agreement is a contract of sorts. Therefore, always give a copy of the signed agreement to the disclosee else it will not be legally in force.

2. Do not disclose any technical details that will jeopardize your inventive matter until it is signed with a signed copy in your possession. Any prior disclosure may be considered public domain.

3. If you don't understand changes that are being requested by the second party, talk to your attorney. Usually the changes are simple ones and may not require your attorney's involvement, but be careful.

4. If there are state laws governing the use of these agreements, make sure they are a part of it.

5. Make your agreements clean and neatly drafted. Better yet, print the form from your computer and insert the company information and the officer's name.

6. Make the city and state in which the agreement is executed yours. If there is ever a challenge to the document, it must then be addressed in your area, instead of in the disclosee's, which might be 3000 miles away.

7. Make sure the agreement is signed by an officer of the company with the authority to do so.

Tailoring a Confidentiality Agreement to Your Needs

The confidentiality agreement is easy to modify to fit your specific needs. The obvious changes you may want to make are:

1. Instead of DISCLOSOR, use your or your company's name.

2. Instead of DISCLOSEE, insert the person's name.

3. In the second paragraph, insert the field in which the proprietary property relates. For instance, "thermoplastic square bottom bags and related bagging systems" or "tortilla manufacturing and related machinery used for cutting and cooking." You must be specific. You cannot expect someone to sign an agreement that is exceptionally broad with wording such as "plastic bags" or "tortilla manufacturing."

4. In paragraph 4, in the section where DISCLOSEE and DISCLOSOR further agree, insert the name of your state, even if the DISCLOSEE is out of state.

If you have any questions as to this document's legality or acceptability in your trade, consult your legal counsel.

Simplified Nondisclosure Agreements

A simple one-page confidentiality agreement (nondisclosure agreement, or NDA) may be used with those not integral to product development. It may be used for visitors, including those who may be servicing equipment in your plant, marketing personnel, agencies that are interested in selling the innovation, advertising agencies and so on. A

Sample Confidentiality Agreement

CONFIDENTIAL INFORMATION AGREEMENT

DISCLOSOR wishes to disclose to DISCLOSEE certain confidential, proprietary property, which relates to _____ (INVENTIONS) for the purpose of permitting DISCLOSEE to possibly assist DISCLOSOR in the development and marketing of the "INVENTIONS."

DISCLOSOR wishes to maintain the confidentiality of the material disclosed to DISCLOSEE and to preserve to himself the commercial benefits from utilization of such material, except as may hereafter be specifically agreed in writing between the parties.

DISCLOSEE desires to evaluate the feasibility of performing certain tasks and procedures concerning DISCLOSOR's proprietary property, to bid on performing such tasks, and/or to perform such tasks as may be agreed upon between the parties hereto.

THEREFOR the parties agree as follows:

"Confidential Material" shall include all information relating to business programs, inventions, trademarks, copyrighted material, trade secrets, existing products, potential products, applications, systems, components, technologies, pending/abandoned patent applications, and business topics. Without limiting the generality of the foregoing, Confidential Information includes, but is not limited to, the following types of information and materials, whether or not reduced to a tangible medium or still in development: human or machine readable documents, audio tapes, video tapes, computer discs, data, source code, object code, documentation, research, development plans, designs, drawings, specifications, engineering prints, machines, prototypes, articles of manufacture, processes, procedures, client lists, customer information, market research data, marketing techniques, marketing material, timetables, strategies, pricing policies, financial information, prospective trade names or trademarks, and other related information. DISCLOSEE expressly understands and agrees that Confidential Information shall also include all information received by DISCLOSEE prior to the signing of this Agreement that would otherwise qualify as Confidential Information.

Unless DISCLOSEE specifically identifies with written consent of DISCLOSOR that certain material is not encompassed by this agreement, all material disclosed by DISCLOSOR to DISCLOSEE relating to INVENTIONS will be presumed to be confidential and will be so regarded by DISCLOSEE unless such materials are publicly available.

DISCLOSEE agrees:

(1) That it will maintain the confidentiality of DISCLOSOR's confidential material and of existence of same except with the written permission of DISCLOSOR;

(2) That it will direct its employees to maintain such confidentialities and will limit access to confidential information to the minimum number of employees necessary to complete DISCLOSEE's tasks, all of which employees shall be identified in writing to DISCLOSOR upon his request;

(3) That it will not disclose to any third party, including subcontractors of DISCLOSEE, without written authorization from DISCLOSOR any of DISCLOSOR's confidential material;

(4) That it will use DISCLOSOR's confidential material solely to perform or determine the feasibility of performing certain tasks to be explicitly specified by DISCLOSOR;

(5) That it will not use for its own benefit or the benefit of any third party any of DISCLOSOR's confidential material;

(6) That it will not contract or negotiate with customers of DISCLOSOR for DISCLOSEE to provide to such customers products manufactured by, or caused to be manufactured by, DISCLOSEE which incorporate or utilize any confidential material of DISCLOSOR, and;

Sample Confidentiality Agreement (continued)

(7) That, except as may be further directed or requested by DISCLOSOR, it will not sell, other than to DIS-CLOSOR, any products manufactured from tooling or molds provided by DISCLOSOR, or developed in accordance with or in response to DISCLOSOR's confidential material.

(8) That upon the termination of the relationship between the parties, which may be accomplished via fif-teen (15)-day written notice by either party with or without cause, at the stage of negotiation, DISCLOSEE shall return any and all documents of any nature, originals and copies, to DISCLOSOR within ten (10) days of such notice. Furthermore, any information, technical or engineering procedure devised for concept which is developed at any stage during these negotiations or other contractual relationship between the parties shall be the sole property and for the sole benefit of DISCLOSOR (except as may be specifically agreed in writing hereafter) and shall not be used for any other purpose by the DISCLOSEE, its agents, or representatives.

(9) The DISCLOSEE agrees to indemnify the DISCLOSOR, heirs, successors, assigns and legal representatives for lia-bility incurred to persons who are injured as a consequence of DISCLOSEE's use of any Confidential Materials, equipment, products, and prototypes, or as a consequence of any defects in the invention.

DISCLOSOR and DISCLOSEE further agree:

(1) That should this agreement be breached, money damages would be inadequate compensation, and there-fore any court of competent jurisdiction may also enjoin the breaching party from disclosing or utilizing confi-dential material encompassed by this agreement;

(2) The prevailing party shall be entitled to reasonable attorney fees in addition to any other amounts awarded as damages;

(3) Laws of the State of _____ shall govern this agreement and it shall be deemed executed in the city of _____ , state of _____ ;

(4) Both parties to this Agreement consent to the exclusive jurisdiction and venue of the state and federal courts located in the State of _____ , United States of America. All parties agree that this choice of forum is convenient and waive any objection to the submission of such jurisdiction; and,

(5) This Agreement sets forth all of the covenants, promises, agreements, conditions and understandings between the parties and there are no covenants, promises, agreements or conditions, either oral or written, between them other than herein set forth. No subsequent alteration, amendment, change or addition to this Agreement shall be binding upon either party unless reduced in writing and signed by them. This Agreement may be executed in parts that, when taken together, shall be deemed to constitute one instrument.

Both undersigned parties hereby represent that they have authority as agents or representatives of the respec-tive parties to bind the parties to this agreement.

Executed by the parties this _____ day of _____ , year _____ .

_____   _____
DISCLOSOR   DISCLOSEE (COMPANY NAME)

_____   _____
NAME AND TITLE   NAME AND TITLE

good rule of thumb would be to use it with those who do not represent a threat to trying to design around your invention. A sample simplified agreement is provided on the following page.

Independent Contractor's Agreement

If you will be employing prototypers, engineers, or other experts that may play a part in the invention's design and engineering, you can use an independent contractor's agreement. This form is similar to the broad-based agreement, but it refers more specifically to the fact that the recipient will be doing certain work for you related to the product engineering and design. The important provision is that any discoveries or improvements made by the contractor are owned by you.

Prototypers, engineers, and designers see these types of agreements all the time and readily sign them. The *From Patent to Profit* Resource Guide has one developed by patent attorney Michael Neustel; you may copy and use it as often as you like.

Employee Patent Program

Every entrepreneurial company or start-up should have this program in place. It is the foundation of a patent quality management system. Such a program essentially gives notice to the company that it regards all of its operations as confidential. It instructs employees not to discuss plant activities with others outside of the plant. It also informs employees that if they make a discovery (or invention) related to the business matters of the company, that invention will be the property of the company. Further, it says that employees will cooperate in filing and assigning to the company, any and all documents, if patent applications are filed and pursued.

Sometimes referred to as a Trade Secret Program, this program is usually revealed in a two-step process.

- First, the president of the company announces the commencement of the Employee Patent Program in a memo that discusses its importance.

- Second, the employee is asked to sign a nondisclosure agreement, which includes an agreement to assign to the company all relevant inventions and patents in which he may participate or develop.

For the most part, the interpretation is the same as the broad-based agreement and relates to current and future discussions. The confidential material included is essentially the same as the broad-based agreement. The primary difference is that it does not have the stringent provisions or the detailed summary of inventive matter discussed in the prior agreement. It is a much easier agreement to read and should not require the signee's attorney's review. Your attorney should prepare this form.

Inexperienced or unaware employees can inadvertently leak information when they socialize with others from competing plants. The Employee Patent Program explains the sensitivity of confidential matters within the plant and the importance of maintaining all information relating to the company's products, processes and know-how confidential. Without it, employees really have no way of knowing what is confidential and what's not. If you are licensing your innovation, your licensee should also have this form of program in place (as noted in the broad-based confidentiality agreement).

RESOURCE GUIDE

Tailoring confidentiality agreements shows a professional image for a company. It's easy to do. You can find all these documents in a Word format for easy modification in the *From Patent to Profit* Resource Guide. Visit our website or see the appendix.

THE RIGHT AGREEMENT

Confidentiality Agreements have been an important part of business for many years. If you use one in your efforts, which almost all inventors do use from time to time, make sure it is one that is tailored for new product development. When revealing your new product to others, don't use a general form such as a plant visitor agreement or a universal confidentiality agreement. These forms will lack substance when referring to manufacturing and marketing.

Simple Agreement

CONFIDENTIAL INFORMATION AGREEMENT

The below named "Disclosor" wishes to disclose to the below named "Disclosee" certain "Confidential Information" that is directly related to _____ ("Inventions") that has not been publicly disclosed by Disclosor or a third party. Disclosor wishes to disclose this Confidential Information to Disclosee for the sole purpose of permitting Disclosee to consider the development and marketing of the Inventions. The "Confidential Information" includes all of Disclosor's trade secrets, drawings, documents and materials whether written or oral, computer files and data, tapes, prototypes, data, tangible or intangible, and information relating to the Inventions provided by Disclosor not known prior by Disclosee or the public. Disclosor requires Disclosee to maintain the confidentiality of all Confidential Information disclosed in order to preserve Disclosor's rights to it.

Unless Disclosee identifies in writing to Disclosor that certain Confidential Information is already known by it within thirty (30) days of receiving the Confidential Information, and is not therefore encompassed by this agreement, all material disclosed by Disclosor relating to Inventions will be presumed to be confidential until such material is publicly available.

By signing this agreement, Disclosee agrees to (1) maintain the confidentiality of Disclosor's Confidential Information and acknowledges existence of same, (2) use Disclosor's confidential material solely to determine its interest in licensing and marketing said Inventions, and (3) not use for its own benefit or the benefit of any third party any of Disclosor's Confidential Information unless expressly allowed in writing by Disclosor.

This relationship may be terminated via fifteen (15)-day written notice by either party with or without cause. Within ten (10) days of receiving such notice by either party, Disclosee shall return any and all documents of any nature, originals and copies, to Disclosor. Furthermore, any prototypes, information, intellectual property rights, technical or engineering procedure devised for concept, which are developed at any stage during these negotiations or other contractual relationship between the parties, shall be the sole property of, and for the sole benefit of Disclosor and shall be promptly delivered to Disclosor and not be used for any other purpose by anyone. Any termination of this agreement does not release Disclosee's obligation to maintain confidentiality of the Confidential Information.

Disclosor and Disclosee further agree that if there is any dispute regarding this agreement or its contents that the Laws of the State of _____, USA, shall govern this agreement and it shall be deemed executed in the city of _____, state of _____; and all amendments or exceptions to this agreement must be in writing.

Both undersigned parties hereby represent that they agree with this agreement.

Signed, this _____ day of _____, 20_____

_____   _____

Business Plans and Prospectuses

We all know what business plans are and how they relate to a start-up business and a company looking for loans or attracting investors. If you are launching a start-up company, it is an essential element. Many major corporations also use business plans when launching new products. These plans are specific to the new product and business unit.

Business plans are rarely used with small established companies that intend to launch a new product that is an adjunct to an existing product line. They are also rarely used with businesses and independent developers who intend to license their technologies. In such a case, a business plan would be a waste of time as it could not be used by any company that plans to license your innovation. A licensee must write its own business plan. It must do its own independent research and verify market potential. It is not going to rely on your business plan, nor would that be a responsible thing to do.

If you fall under this category don't waste you time writing a full-fledged business plan, since it simply will not be useful. There is a smart alternative you can prepare however, that's a lot less cumbersome, yet may be used by any potential licensee or marketing partner to write its own business plan. It is called a prospectus.

The Invention Prospectus

Invention prospectuses are becoming more and more important. It is a means to prove that you know what you're talking about and have qualified potential marketability. Compiling a prospectus is most important if you are not familiar with your potential marketplace. It is the perfect means to prepare for future discussions with qualified partners and licensees. With a prospectus you'll be able to anticipate their questions and have qualified answers. Without a prospectus, you may appear as an amateur and not be taken seriously. Even experienced product developers can benefit from writing a prospectus.

Invention prospectuses are almost always well-received by potential partners. It allows the interested party to fairly quickly understand the nature of the invention and the potential market. It shows you know what you're talking about and that you've qualified the opportunity. If your innovation is a major opportunity, you must have an invention prospectus to substantiate it. If it is a not-so-major opportunity, you'd also be wise to prepare a prospectus. It will almost invariably improve your odds of attracting marketing partners or licensees.

In your development activities you've compiled a substantial amount of information demonstrating the marketability and patentability of your innovation. With a little effort, you can compile this information into an organized, easy-to-read format commonly referred to as a prospectus.

In essence, a prospectus is a simplified form of a business plan and includes pertinent information about the innovation, industry analysis, target market, and start-up requirements. Usually a prospectus is drafted on standard $8\frac{1}{2}$" x 11" paper and may be from six to ten pages long, depending on the complexity of the technology. A licensee or manufacturing or marketing partner may use some of the information in its business plan should it elect to prepare one.

A prospectus is never given out to a prospect in the initial stages, and only after a confidentiality agreement has been signed.

AN INVENTION PROSPECTUS IS IMPORTANT

This is becoming an increasingly important part of the invention-licensing process. A well-written prospectus is the inventor's version of a business plan. I don't think there is any doubt about it. Give a prospective licensee a prospectus and you are greatly increasing your ability to be successful .
–*Stephen Gnass*

A prospectus may be prepared in various formats. The prospectus on the following pages exemplifies a fairly simple one. You may follow its guidelines to prepare yours.

Title. The title refers to a new product, not an invention. Include a confidentiality notice under the title. Note that throughout the copy you will see the words "new product," instead of "invention."

Field of the Product. This one paragraph describes the field of the new product, much like the first words you used in your new product summary sheet and even in the first paragraph of your patent application.

Background of the Product. This is a summary of the present products used in the industry. Some of this information is what you also may have included in your new product summary sheet and patent application.

Description of the Present State of the Art. In greater detail you describe the various types of competitive products being used, their deficiencies, and current suppliers. This is important because this section will also determine what competitive responses may be once the innovation is introduced. Depending on the nature of your invention, you'll want your discussion to reflect the content. That is, if your invention is not highly technical, don't use technical terminology and vice versa. This section will show your partners exactly how much you know about current market conditions. If you can't compile this write-up, you are not well prepared and have not done sufficient research.

Product Summary. Remember in your patent application you discussed prior art products and their deficiencies and then in your summary of the invention, you discussed how your invention solved those problems. Now, in this section of the prospectus, you will explain in layperson terms—not technical patent terminology—how your invention overcomes the shortcomings of the prior art products.

These paragraphs should reflect in substantial detail how and why your product works. You may include some drawings to illustrate your claims. You may even be able to use some of the same drawings you used in your patent application or sketched in your journal, depending upon their quality.

At times, this summary may be accompanied by actual samples or a prototype. If a prototype is not available, it may be an important reason you are pursuing development partners. You may even ask in the prospectus for assistance in prototyping.

Test Results. Concrete facts supporting the need for innovation and its marketability are most helpful in getting interest from potential licensees. Financial decisions are usually based on verifiable facts, not speculation. If you do not have test results, you should ask to pursue testing with a potential partner.

Patent Status. Summarize your patent status and summarize what it covers, but do so in layperson terms. Remember, patent attorneys are not reading this, but potential manufacturing and marketing partners are. Never falsely claim that there is a patent pending when there's not. A potential partner wants to know that it is going to be protected with broad patent protection and this is where you must make the statement. If you cannot give an honest, thorough analysis of the patent protection you expect to achieve, you may be in trouble. In this section, you should also say who your legal counsel is and per-

YCN **Your Company Name** *company address, city, phone and fax numbers, website*

NEW PRODUCT OPPORTUNITY SUMMARY
(The following confidential information is not to be copied
without prior written permission from Your Company Name)

Field of the Product

This new product opportunity relates to bags commonly used in fast food, food service, supermarket, retail and other related trades—to carry food or merchandise from a store to home. More specifically, this new product relates to a flat bottom, stand-up style plastic bag and system that can be used to substantially improve packing efficiency and customer through-put. This new product with its related innovations is called the *M2K™* bag.

Background of Bags Used in Food Applications

Paper bags have been and still are the most popular style of bag to carry fast food and food service items in high-volume outlets in North America and much of Europe and Australia. In these high-volume outlets, paper bags are used primarily due to their ability to be carried in a user's hand and to square out and stand up after they are loaded. Paper bags also breathe, which means that hot french fries and hot foods will not become soggy.

In a few fast food outlets, thin gauged plastic T-shirt bag systems using racks to support the bags for packing are used. Generally speaking these thin-gauged plastic bags do not carry the valuable aesthetics for the fast food trade; are not easy to load or to stand up after being packed, unless permanent, stationary racks are used; and are not cost effective for smaller bags. Their use with hot foods does not allow condensation to escape, thus limiting their use to primarily cooler-temperature foods.

Being able to quickly load the right-sized bag with the right-sized order is important in high-volume outlets in order to ensure food quality. Since larger plastic T-shirt bag racks require too much space, have poor aesthetics, are not cost effective in smaller sizes, and do not cooperate with condensation problems, they will not accomplish this objective. Thus, paper bags still dominate the fast food industry until a suitable alternative has been developed. This has been accomplished with the M2K bag and system.

Description of the Present State of the Art

The most popular paper bag used today are brown kraft bags made from 30 lb. paper (except the $1/_7$ Bbl, which is about a 52 lb. paper), typical sizes are #6, #8, #12, #420 and a $1/_7$ Bbl handled bag. The largest U.S. supplier is Stone/Gaylord. Others include Duro, Willamette, Port Townsend, and Ronpak. These paper bags are usually put up in bundles of 500 to 1000 bags. Print copy is normally two or three colors with promotional bags being printed up to 6 or even 8 colors (3c/2s or 4c/2s). These promotional bags have high quality prints that are frequently printed on a partially bleached white kraft paper, which is a paper that is usually made with a larger percentage of recycled newsprint.

Of the less popular plastic sack variety, the largest supplier is most likely Sonoco Products Company plus a few others. These bags are typically in sizes of $1/_{10}$th to $1/_6$th Bbl and are of the traditional plastic sack system as seen in supermarkets. Some of them are vented with "C-shaped" flap holes, which assist in venting condensation from french fries and other hot foods. They are typically printed with up to four colors (total on two sides), and are almost never considered for promotional bags, which would require the higher quality six and eight color printing.

One of the greatest problems associated with prior art paper bags is that when drive-through orders are packed, it is not easy to see into the bags to verify if an order has been correctly filled. This is confirmed by the fact that this is the number one customer complaint in the fast-food industry. Paper is also vulnerable to water damage, unsightly grease spots, bug and pest infestation, and potential paper dust food contamination. Other problems associated with prior art paper bags are the high cost of inventory, freight, and handling. It is a well-known fact—paper bags are far and away the **number one expense item** in a store and rising expenses is the **number one complaint of franchisees**.

The prior art plastic sacks may overcome the objective of being able to see into a bag to see if the order was correctly filled. However, in addition to the previous problems discussed, contents loaded in thin-gauged plastic T-shirt sacks cannot be easily seen and the bags cannot be easily carried by the users filling or completing an order. Also, condensation from hot foods cannot escape due to the flimsy film collapsing around the hot contents shutting off an escape route. Last, plastic T-shirt bags in sizes smaller than $1/_{10}$ Bbl require longer handles and are difficult to manufacture cost effectively on domestic bag machinery. The result is that plastic is rarely used in fast food stores in the small and medium sizes.

Summary of the M2K™ Bag

M2K bags combine the best attributes of both paper and plastic into one single product. They can be loaded faster than paper and stand up just like paper after loaded. Using M2K bags makes it a simple task to check order accuracy and speed customer through-put in the drive through lanes.

M2K bags have also successfully eliminated the condensation issue associated with plastic. The proprietary "Automatic Ventilating System" (AVS) automatically opens and vents condensation when the bags are loaded. Unlike a traditional C-shaped vent that must be forced open, the AVS system automatically opens up when loaded. The greater the load, the more the vents will open.

But M2K bags don't stop there. The proprietary hook dispensing system eliminates the need for space consuming racks. The hooks allow many different bag sizes to be placed neatly and conveniently along a counter top's side. The bags can be quickly loaded, dispensed, and they will stand up afterward. Best of all, the loading and dispensing is faster than with paper!

M2K bags require far less cube than paper. Hence, the inventory, freight and handling are lessened by 60%–70%. M2K bags are also grease-resistant, bug- and pest- resistant and food contamination-resistant. M2K bags have proprietary, patented die-cut handles that are strong and improve zero-defect manufacturing. M2K bags are also aesthetically pleasing since they look like a high quality, sanitary fast food bag! And of course, M2K bags are also cost effective compared to paper.

The proprietary processes developed by Your Company Name in Your City, Your State, provide reliable stand-up bag performance and consistent, high-volume, high quality output. The production process *produces one million bags a day* or more! It would be extremely difficult for T-shirt bags to compete with the quality and cost effectiveness of the M2K bag and its processes.

Unlike many other prior art attempts at making stand-up style plastic bags, which utilize heat sealing on 45-degree angles, gluing bag bottoms, or folding and gluing already-made bags, the unique proprietary processes developed by Your Company Name eliminate the slow, tedious manufacturing processes associated with 45-degree angle seals. Prior art, plastic stand-up style bags are typically priced anywhere from 20%–300% higher than the M2K bag and will not approach its overall performance level.

Test Results of the M2K Bag

In testing of the M2K bag, much success has already been achieved. Confidential details may be available to certain parties. Tests include dozens of lab applications for food quality, health and safety concerns, in-store user and surveys of customers. The tests were conducted by major and minor restaurant chains . . . names of which are commonly known throughout the U.S. In every test, the M2K bag has met or exceeded expectations, with a minor, single-incident, exception.

For the most part, all those chains that have tested the M2K bag have expanded its use and some would like to begin national rollouts.

Patent Status

The M2K bag has a bundle of six patents issued and three pending that accomplish these impressive objectives. The granted patents are on its proprietary die cut handle, U.S. No. 5,338,118; the square bottom bag,

U.S. Patent 6,095,687; the apparatus and process of producing the cold seal, U.S. Patent 6,319,184; the self-opening system, U.S. Patent 6,171,226; the automatic ventilating system, U.S. Patent 6,113,269; and the high output manufacturing process, U.S. Patent 6,186,933.

The handle style is an important contributing factor to the self-opening dispensing. Without it, the bag will perform poorly. Of the additional U.S. patents pending, the two dominant ones to issue soon have the broadest possible scope. The overall scope of the patents and patent applications are considered extremely broad and covers the bag, the means of making it square out and stand-up, at least two manufacturing processes, the Automatic Ventilating System (AVS), and the system (or method) of loading, dispensing, and using. The patents do not infringe or cause the infringement of any known existing patents anywhere in the world. PCT patents have been filed that blanket most of the above subject matter.

The patent counsel for the 10 patents pending is the well-known IP firm of Townsend, Townsend and Crew in San Francisco, CA, which has successfully prosecuted and defended the client's patents in years past. They are known for their aggressive litigation and patent defense ability.

The Marketing Opportunity

The M2K bag is now a tested, proven bag and system with all the right operational and economic attributes. It overcomes the real problems associated with fast food and food service applications and many supermarket applications. Sales of the M2K bag can also be made with the confidence of knowing that the existing major paper bag producers will most likely not have any immediate competitive response. Large T-shirt bag producers will also not be able to respond with T-shirt bags . . . or by converting their machinery to make similiar problem-solving products. This would be especially true in light of the solid patent protection with a scope broad enough to keep out clones and copycats for years ahead.

The Licensing Opportunity

Presently there is an opportunity to license a few key manufacturers/marketers in the U.S. However, in no condition whatsoever is this Marketing Summary an offer to license the proprietary property.

Barriers to Entry

There are four barriers to entry for a new licensee. They are:

- **The right equipment.** Bag machinery, which has been exclusively developed by Your Partner Company's Name is necessary. Upon the signing of a confidential agreement with Your Partner Company's Name, a prospective manufacturer may learn some of the unique manufacturing aspects of making the M2K bag, the cost of the machinery, and the anticipated one million bags a day output. The Your Partner Company's Name bag machines are turnkey and ready to run upon delivery. Your Partner Company's Name will ensure a quick run-down on the learning curve. Higher quality printing is also a serious consideration for a manufacturer.

- **The right manufacturing approach.** Does your manufacturing team apply TQM principles to produce high quality products? There is no doubt that the bag-making process developed by Your Partner Company's Name is ideal for this high quality approach to manufacturing. The result of establishing high-quality output first will clearly result in the million bag a day high-volume output shortly after production begins.

- **A license.** Those seeking to enter the M2K bag market will need to have a license agreement in place prior to beginning production. Interested parties should contact Your Name at Your Company Name for a term sheet or to review a license agreement.

- **Your commitment.** The production of high quality M2K bags means a manufacturer must make a substantial commitment . . . one that is not taken lightly. New product launchings require some flexibility in the ramp-up phase, and as part of the start-up team and opportunity, a manufacturer will be asked to be highly responsive to the customer needs. Are you up to it?

MARKETING PLAN

Market Description

The target market for the M2K bag includes fast food, food service retailers, supermarkets and retail outlets. Research has revealed an additional opportunity with retail applications such as supermarket (bakery, take-out and front end use) and discount store (larger bags) and video store applications for small- and medium-sized bags. Thus far, the approach has been only to major fast food chains that are interested in cutting costs and improving productivity and the freshness factor of the foods they make.

Market Size and Trends

The target market is potentially huge. In other words, you could say that most every retailer wants to cut costs and improve food freshness attributes. Your focus would most likely initially be to only the most progressive top chains that stand to gain the most. Or, with certain retailers, the focus would be on those that have a desire to improve the image of their stores and eliminate unsanitary paper bags or the inefficiencies of T-shirt bags and racks.

From this viewpoint, we believe the M2K bag will capture the lion's share of the target market in the coming years. The food-related supermarket and paper replacement potential market looks like this:

Fast food	19.5 billion bags	$350 million
Food service	9.0 billion bags	$160 million
Front end applications (10% conversion)	6.0 billion bags	$ 75 million
Others	6.0 billion bags	$100 million
Total	40.5 billion bags	$685 million

These figures represent a total potential target market in which capturing even a reasonably small percentage, for example 10%, represents a major opportunity. We believe this estimate to be conservative. For instance, the largest paper bag supplier, Stone Container's estimated kraft bag sales (excluding the S&G handle sack) is about $350 to $400 million annually.

Other Considerations about the Target Market

It is possible that the right marketing effort may make an impact on the current plastic T-shirt bag market used in supermarket front ends and fast foods, by introducing a large, cost-effective, 1/6 Bbl M2K front-end bag. One major fast food retailer using T-shirt bags has expressed a desire to convert to the M2K bag, and another large entity is evaluating it now. The retailer reports many food quality complaints from the use of T-shirt bags with its products. These companies have identified the root problems associated with the current, T-shirt bag systems, hence allowing the entry of a higher quality, high-performance alternative. Some replacement of T-shirt bags in supermarket applications would also be inevitable. In heavier gauges, the M2K bag may replace the S&G style of handled paper bag estimated at over $200 million in annual sales.

Market Readiness

Soaring paper bag costs have caused fast food bag prices to rise dramatically. Franchisees are crying out for ways to improve their profits by cutting expenses. Environmentalists, realizing the harmful glut of paper that is filling our landfills, are also promoting the use of plastic bags more than ever. This has only helped the industry to have greater interest in a plastic alternative. A number of fast food chains will listen to those who offer solutions that can be verified. The M2K bag is the only product that meets the high-quality standards, with the operational, high productivity requirements of the fast food industry, that is available in a full range of sizes and can cut bag costs substantially.

Marketing Assistance

Some marketing aids have already been developed and may be available from Your Company's Name.

haps you may include a brief biographical statement about yourself. You may include information that is relevant to your innovation and new product development or patent activities. You want to be perceived as a specialist in your particular niche.

The Marketing Opportunity. Use simple, straightforward language that substantiates the product's potential and viability and summarizes the marketing opportunity. If your invention is not considered a breakthrough, don't call it one. Also, you do not want to use hype as it will send the wrong impression—in fact, it'll suggest just the opposite.

The Licensing Opportunity. Summarize the basis of the licensing arrangement you are pursuing. Be somewhat specific, but leave the details open for discussion. You do not want to use this section as a term sheet. Nor is this the place to say, "I want a 10% royalty and an up-front deposit of $500,000." Those details are rarely part of a prospectus and should be revealed later on in a license term sheet or a copy of the license agreement itself.

Barriers to Entry. What resistance will there be? If you can't answer this question, you have not done enough research. In this section you want to make a list of the requirements needed to get your project launched.

You are the team captain and if you can't define the barriers now, you will have to find out. They may include purchasing equipment such as in the example, or perhaps modifying existing equipment. Other barriers may be training aids, brochures, advertising, and so on that you believe would be necessary to succeed. Make one of the barriers to entry "your commitment" like that shown in the example.

You'll need to know all this up front else it could kill the opportunity.

Marketing Plan. This is a simple marketing plan that discloses various elements of how you see the product being sold. Don't worry about how a potential partner views this as they may have their own ideas of which markets are best to approach. They may even identify other marketing applications and find other uses for your product.

In actuality your marketing plan discusses a long-term opportunity. However, it is up to the reviewing partner to agree or disagree.

Market Description. Clearly identify and describe the various markets in which your product may be applicable. Don't use pie-in-the-sky examples, but try to qualify them based upon solid research, interviews, or otherwise.

Market Size and Trends. Be accurate in identifying your market and its size. Notice that in the model, the market size in any one segment is listed with actual numbers and supported with a corresponding dollar volume. It's important you have verifiable facts in this section, as you may have to produce them for a prospective partner. Remember, overzealous projections will affect your credibility.

Other Considerations. In this section you can talk about speculative areas that may also be developed. You may also list other untraditional approaches to sales that may not be listed in the previous section. This is where you can make comments about why your market may be larger than anticipated, but note that you are not relying on it as part of your analysis.

THE WORST APPROACH

Experts agree that the worst way to launch a new product is to develop it completely and then tell the marketers what it is. Right behind that is to let engineers develop what they think may sell and have them launch the product. From this perspective, the partnership concept is validated. An inventor needs both engineering and marketing department input to make a new product successful. Your primary objective is to qualify the concept, but you should also be able to get your manufacturing and marketing partners involved as early on as possible so they may provide vital input into the design and marketing.

Market Readiness. How is the timing? Let the reader know why you believe the timing is good. Try to cite concrete reasons.

Marketing Plan Assistance. You can state your willingness to share additional information or support for their marketing effort. If you're unsure, make a list.

Generally speaking things such as sales projections are not necessary. This is what your partner must prepare, not you. Typically *sales projections* in a prospectus would be used to illustrate ramp-up potential and to give a manufacturing-only partner an idea of what to expect. Sales projections should be for about three years.

Start Up Costs. Last, you may also include a final section on start-up costs. You would generally want to include this in any innovation that is targeted for quick and easy conversion. In such a case include a simple summary of the costs required to launch the project. Remember that most companies will expect to invest money in new projects, so do not be afraid to ask for it here. In this section you may include die and mold costs, packaging, and a license agreement fee.

Formalizing Your Prospectus

Once you have written your prospectus, it is best to neatly package it in some form of binding or otherwise. Even if it is only five or six pages, a bound set with a title page and front and back cover will send the right message.

Use attractive, businesslike colors such as blues and greens for more technically oriented prospectuses or perhaps flashier colors and some dazzling art for those innovations that have real visual or emotional appeal. Colors such as chartreuse, lavender, and pink should be used cautiously. Remember that the reviewing audience usually consists of conservative executives, production managers, and sales professionals.

Send a brief cover letter with each prospectus. You may use language such as, "I have enjoyed talking to you these past few days," and "I look forward to discussing your interest on the project in the enclosed prospectus." And again, never send out a prospectus unless the recipient has first signed your confidentiality agreement.

When sending a prospectus use Priority Mail or an overnight delivery service. Do not fax it. In some circumstances you can email it, if it has been attractively drafted.

Benefits of Preparing a Prospectus

The benefits are numerous. For those who are not experts in the field—your innovation's niche—a prospectus is the perfect way to get there fast! The more you become the pre-eminent expert in your field, the more success you'll have developing new products and patenting them.

By preparing an invention prospectus, you are in essence qualifying and assessing your invention at the same time. The research you conduct and the information you find will also be invaluable when you meet with prospective partners. You really should know this information before you begin pursuing manufacturing or marketing partners and licensees. Without this knowledge base, you'll be a misfit. With it, you'll be credible and people will listen.

In the next two chapters, we'll discuss how to apply some of this knowledge to find, interview, and evaluate manufacturing and marketing partners.

MANUFACTURING PARTNERS

*It starts with quality . . . the right price . . . dazzling service! . . .
and a special kind of partnership*

Manufacturing a new product can be a challenge. Seemingly simple products can have fatal flaws if an inexperienced manufacturer uses the wrong manufacturing process or wrong materials. Customers can be forever lost if a single shipment arrives unknowingly defective or late, or the product may never reach its potential if the price is too high.

Either your company is going to manufacture your innovation or an outside subcontractor or licensee is going to make it for you. Regardless of your approach, this can be a costly experience if you don't know what to look for in advance.

This chapter explains how to avoid tricky issues that can destroy your innovation's potential and shows you how to make it cost effectively with the right quality and with a minimum investment of your capital. After all, your manufacturing partner has a lot to gain from a successful launch. Thus, it should be willing to invest its capital and its expertise to make sure the product is produced with the right quality, at the right price, and delivered on time.

If you are not going to produce the innovation yourself, you must locate manufacturers with the ability to get the job done right from the onset. It takes a lot of manufacturing savvy to accomplish this. The right manufacturing partner should be willing to help with prototyping and should be willing to gear up for production quickly, as well as absorb most, if not all, of the cost of doing so.

As discussed in Chapter 2, the right manufacturer is an integral part of your team. In exchange for a manufacturer or subcontractor's participation and start-up investment, you can give it a supply contract. Even if you plan to license several producers nonexclusively, you'll most likely select a favorite producer you can work with early on. You can consider granting this company "favored nation status," with a jump on the marketplace and perhaps a short-term exclusive in exchange for their prototyping assistance and manufacture of working models for tests and trials. If you plan to assemble the innovation in house and use subcontractors for some or all of the components, your subcontractors can also help in the same manner.

The key is finding manufacturing partners who are motivated, smart people who can solve start-up problems and are willing to make a time and financial investment.

Your Manufacturing Objective

With the right mindset from the beginning, your primary objective is then to locate willing manufacturers with all the right attributes that can mass produce your innovation. It is most important that you find a producer able to make your invention the way you want it made—not the other way around. You simply cannot allow a manufacturer to dictate how your innovation will be designed or how it will work.

Invariably there is some give and take, based on current state-of-the-art processes. The manufacturer can provide input, but if you modify your invention to make it more and more generic as most manufacturing entities would prefer (It's easier to make a simple, generic product.), you may destroy your invention's uniqueness. This could be a disaster. Keep in mind that you are customer driven and your invention must not lose its uniqueness based on generic manufacturing processes. If you allow a production-driven mentality to dictate design and performance, you'll be in trouble. To have a successful product launch means applying CDI and not kowtowing to the mediocrity of manufacturers who prefer generic production processes.

Your manufacturing approach is not related to the marketing department other than its qualification of the production models. Separate a company's ability to market from its ability to manufacture. When evaluating manufacturing abilities, use manufacturing-related questions. As you read this chapter, make a list of the crucial needs and attributes of your invention . . . and make sure the manufacturer can meet them.

With all this said, it really comes down to finding the right manufacturer no matter where it is located. Be patient and keep networking and searching and you will find the right fit.

> **A BIG EXPENSE FOR YOU IS A SMALL EXPENSE FOR THEM**
>
> What could cost you a small fortune for R&D and prototyping, the right manufacturer can do at a fraction of the cost and in a fraction of the time. If you are not a manufacturer, you need a close association with one that is flexible, future thinking, and real smart.

Reliable, Cost-Effective Manufacturing

Countless numbers of inexperienced inventors and small businesses start out by relying upon local machine shops, brothers-in-law, and neighbors to "make their invention at the right price." Now if this were the case with all successful innovations, don't you think that store shelves would be filled up with products made by all the local businesses, with very little room left for those that are mass produced?

The truth is this is not the case. Instead of looking at local producers, or friends and relatives, you would be wise to search everywhere you can on this earth to find the right manufacturing partner, one that'll give you the quality, price, and service you require. If that means going overseas, then that's where you'll have to go. But that doesn't mean you have to jump on a plane, fly there, and start your search. Fortunately, because of the Internet, you'll be able to do most of your searching from the comfort of your office or home.

From here on, this chapter will show you how to find, interview, and evaluate manufacturers and how to get what you want from them. What you want is a highly marketable innovation made the way your want it, made with tremendous customer appeal. Best of all, you'll end up with a dedicated production partner who will offset a substantial amount of start-up costs.

The Wrong Approach

Before zeroing in on the right approach, let's look at a somewhat comical approach with which many inexperienced companies and inventors approach manufacturing a new product. They start out by talking to a relative, let's say a brother-in-law, who owns a local machine shop. He says, "This is just what I've been looking for. I'll make the prototype at cost, sell you the necessary molds and tooling at cost, and work with you to develop the final product. I'll even assemble them for you if you want. All I want is to be the exclusive manufacturer when the orders start coming in. The best part is 'You can trust me.'"

Oh boy, are you asking for it! Here's what's wrong with the above scenario:

1. What do the words "at cost" mean?

2. He is talking in terms of "the prototype," when it will more likely take several before it's right.

3. After prototyping, there is no mention of working models to be used for testing.

4. How much are the molds and tooling going to cost? Who'll own them?

5. How do you know his manufacturing costs will be competitive with those of much larger firms or importers? (Or course, they won't be.)

6. If he is going to assemble it, will he be cost competitive?

7. Who's paying for the cost of package design and the packaging supplies? You? . . . up front?

8. Who pays shipping? Does he already do a lot of shipping to potential customers?

COST-TO-SELL RATIO

Per the pricing schedules in Chapter 10, we know a 4 to 1 minimum ratio is usually needed. However, in this example the approach is all wrong. There is no qualification of the market price. This project is doomed from the onset.

So without qualifying all the above, here's what the novice inventor does next. He says, "OK, why don't you check out pricing and then tell me how much it is going to cost to make." After a few days the brother-in-law gets back to the inventor and says, "I've got a price. They cost about $4.25, each but that doesn't include packaging because I don't know how much that'll be for the final assembly. I'll have to have some contract labor come in and do it. But together it shouldn't cost more than $.15 cents per box and the labor to assemble and pack them should be about $.25 cents each. Oh yes, I can make the prototype for $2,000." So the inventor adds the manufactured cost to the "other costs," such as pallets and shipping cartons and shipping costs to the distributor. He estimates that it'll be close to $5 by the time it is delivered to the high-volume distributor or retailer for his retail innovation.

The inexperienced inventor then says, "Well, that makes it easy then. It will retail for $19.95 (4 times cost). The distributor will pay about $8–$10, and I'll get the other $3 to $5 profit."

Unfortunately, this approach is not only backwards, but it will invariably create some ill will with the brother-in-law when the invention doesn't sell. After a few months of working together, the situation looks like this:

1. They are now on their fourth prototype. The third one was worse than the previous. The brother-in-law blames the inventor. The inventor says the brother-in-law should have known better.

2. The brother-in-law keeps asking for more money for the work he is doing "at cost" because the inventor keeps changing the product. The inventor is becoming frustrated with the brother-in-law because he just can't "get it right." He doesn't understand why he has to keep paying the guy for getting it wrong. After all, he is the one who is supposed to be the engineering expert. "I trusted him!" him the inventor says!

3. The inventor is now out $6,000 and disillusioned. The brother-in-law is becoming irritated and won't let go of the project because now he is more determined than ever to "get his 'just due money' out of the project."

4. The inventor still doesn't have a properly functioning prototype and still has to keep in mind that he has to pay for the upcoming mold costs that were estimated at about $12,000. The brother-in-law then tells the inventor that because of all the changes the inventor has made, the molds are going to cost closer to $14,000.

5. The inventor's heart sinks when he realizes that he doesn't even have the packaging designed yet. His neighbor agreed to do him a favor. With his discount from the graphic design company where he works, it'll only cost about $4,000. He asks himself, "How much are the printed cartons going to cost anyway?" He is shocked when he is told it'll be about $8,000 for a short-run. He considers hiring an amateur student to do the artwork because "that is all he can afford" and save a few thousand.

6. It finally sinks in and the inventor realizes that most of the other competitive products on the market retail for about $10. Then he finds out that the product can be imported for about $1.80 and the mold costs are only about $3,000. The brother-in-law assures the inventor that, "That doesn't matter, because ours is much higher quality and made in the U.S.A. and everyone will buy it because of that."

This scenario can go and on. It usually gets to a point where it is a no-win situation for both. The inventor is out thousands of dollars, not counting the other costs he has for patent applications and so on, and the brother-in-law is becoming impatient, pressuring his sister to divorce her inventor husband. Well, hopefully it's not that bad.

The biggest problem with what has happened thus far is that the inventor is blindly going into business, with no qualified research. The inventor is creating an overpriced product with doubtful performance and amateur, inadequate packaging. Further, he has no marketing plan and no financing to build inventory for product ramp up. There are also many other factors to consider, such as bar codes, legal warning messages, and so on. This inventor needs one thing very badly: a manufacturing expert on his team who has made similar products before, one who can give him some guidance on all those matters.

LICENSING CHANGES PRICING STRUCTURE

When partnering with licensees, the cost structure becomes that of the company manufacturing and marketing the innovation, which is quite a bit different from your company being the supplier. In such a case, you are taking a royalty and the cost analysis is the responsibility of your licensee. However, you should have a good idea of price structure up front, because if you don't, you'll not be able to qualify the market sell prices, nor convey profitability to a licensee. Keep in mind, a licensee is licensing "money earning potential." If you can't show them the money, they'll not be interested.

The Right Approach

There are some basic tenets you can follow to make sure you avoid the previous fiasco. First of all, we know that overpriced products don't sell well. Don't fall into the trap of thinking that all you have to do is mark up by four or five times whatever the cost of your invention is and that then it is automatically going to sell. You need to find out exactly what unbiased customers are willing to pay, not what your brother-in-law estimates the cost or what your wife, mother, or friend say they would pay.

Basic Competitive Research Needed

Before you get too far along in pursuit of a manufacturer and make a lot of assumptions, you need to do some basic competitive research to assess the opportunity. Without the upfront research, you really won't know specifically how to qualify the manufacture of your innovation. Frankly, you should already have a good idea of what customers will pay because you should know what competitive products sell for. In the case of the previous inventor who took the wrong approach, the market price should be just under $10.00. Until you can determine this and be assured of a corresponding quality level, how can you pursue the manufacturing of your innovation?

With a brief, but detailed, summary of competitive products, your competitive research will determine a few important points that will guide your development. They are:

1. List competitive products and their sell prices. Determine your product's price level.

2. Describe the unique position of competitive products. Describe your product's unique position, the attributes it must have and the required quality level.

3. Determine cost based on the pricing model in Chapter 9.

Going back to your customer-driven innovation approach, you have determined that your innovation most likely will have as good or better quality than similiar products, be more attractive than the others, and have all the new design attributes you have invented. So with the previous example, you target your price level to be no more than $9.95 retail. Your challenge is to manufacture your innovation at a competitive cost based on the targeted cost in your pricing model.

In our example, your distributor cost should be about $4.50 wholesale (55% discount off list price is standard.). Thus, you're looking for a cost of about $2.24, or less if at all possible.

Obviously, a cost of $5.00 isn't going to work if the innovator plans to go into business and be the supplier. However, if you elect to license the innovation, there's profit enough to pay a royalty. However, the bigger problem arises when large orders are placed. The brother-in-law's job shop will never be able to handle them. How is inventory going to be built up? Where will it be stored? Who's going to finance it? Who's going to pay for product liability insurance? Will a company such as Target or Walmart have trust in a small job shop to deliver and support its requirements? Of course, all of these problems are deal killers. In fact, you'll never get the first order based on being an inadequate supplier.

Common Manufacturing Processes

Obviously there are many types of manufacturing processes. You will need to find the one that fits your invention the best. Since most new innovations are relatively simple,

NO PRECEDENT?

If you invented the TV, it could be said that there is no precedent, but there is. You could ask people how much would they be willing to pay for a "radio in the home that also has pictures, like the movies." Then you can use this data to determine your target market and see if it can be made cost effectively. If TV was introduced at $5000 apiece some 60 years ago, you can see the market would be extremely small. Fortunately, with the lower cost of volume production and the invention of the assembly line (thanks to Henry Ford), the price was substantially less, allowing TV to be mass manufactured and marketed.

this should not be a difficult task. Most new products tend to use materials and components that are readily available anyway. These include everything from raw material to screws, wire harnesses, computer chips, plastic sheeting, moldings, and so on.

Your innovations may use one or more of the manufacturing processes in Table 11.1. If there is a combination of two or more and minor assembly is required, you'll most likely want to have your primary producer do it. For instance, if your innovation is primarily made from wire with two screws and a simple patented injection-molded plastic cap on the end, have the company forming the wire assemble the final product for you. Thus, no extra step or extra shipping cost is required for assembly. If the same wire bender can also package the finished product in retail bags and then pack them 24 per case, you'll be even further ahead. These are some of the manufacturing issues you will want to address up front when you seek and interview manufacturers.

Don't Know How Your Innovation Will Be Made?

If you are unaware of how your innovation may be made, you're going to need some expert help. Hopefully your quandary is only that of being unfamiliar with manufacturing processes and not a problem of unavailable technology.

There are a few ways you can do some research to find the best process for mass manufacturing your innovation. First of all, try asking anyone who is experienced in the manufacturing and assembly trade. Many will have a broad-based understanding of how most products are manufactured and assembled. If assembly is minor, then it is usually done by the main manufacturer anyway.

If you are employing a prototyper, he/she will almost always know the best manufacturing process as well. Make sure you cover this with him/her when you discuss your invention. You don't want to have a prototype made that will result in a product that is not cost-effective to manufacture. You want to gather as much information as possible so you will have the best understanding of how your invention is to be made.

The last, but perhaps best, alternative is your local SBDC. Without question, its manufacturing consultants will know the best manufacturing processes for your invention. If not, they'll be able to find out. They may even have a list of qualified local producers. But keep in mind, it is still wise to look everywhere in the world to find the very best fit. You don't want to settle on a domestic supplier because "he's a nice guy" or a "friend of mine." Spending money with nice guys and friends is OK as long as they can be competitive and you can capture the intended mass market. But if they turn out to be a high-cost hindrance instead, as illustrated in the previous example, you may end up derailing your invention's development and product launch.

Simple Inventions

If your invention is relatively simple, it will be a lot easier to get it made and prototyped and a lot easier to market. Finding marketing partners should not be too difficult to help you qualify marketability. Simple innovations are a lot easier for existing companies to take on as an improvement to an existing product or as a line extension item.

If you, a marketing partner or licensee, is going to be importing, that too is greatly simplified. The less sophisticated the invention, the less assembly and the less you have to rely on others to manufacture the various parts or assemble them. With relatively simple inventions, the marketers will also find that promotion and advertising is a lot easier.

WHO ASSEMBLES?

If your innovation is something that requires assembly of a few parts—nothing massive—it is almost always best that you have the prime manufacturer, the one providing the costliest component, be the assembler. They can buy the other inexpensive parts with little mark-up. But if your prime manufacturer and assembler has to buy the costliest component, it would have to charge a much larger mark-up on the expensive component.

Table 11.1 Basic Manufacturing Processes

Manufacturing Process	Description
Injection molding	Used for small plastic products and parts such as hooks, holders, caps, small containers, and a multitude of parts used in assembled products. Molds can be expensive; the larger the part, the costlier the mold.
Vacuum forming	A sheet of flexible plastic is heated, drawn and pressed into a two-part mold. Typical products include containers, dishes, and parts that do not require substantial thickness. Molds are fairly inexpensive compared with injection molding.
Blow molding	Used when making large containers such as trash bins and buckets. Molds are relatively inexpensive. The plastic material is "blown" into the mold, thus filling it out and creating the container.
Die cutting	Most thin plastic sheeting, leather, cardboard, and paper products may be die-cut into pre-determined shapes. Some common die cut products include plastic grocery sacks, gaskets, cardboard cartons, and shoes. Dies are fairly inexpensive to make.
Computerized numerical control (CNC) manufacturing	Parts made on CNC machines may be virtually any material, but they are usually some form of steel, aluminum, or metal. The exact dimensions and specifications are fed into a computer which, in turn, directs the CNC machine to cut, hone, drill, and so on, until the final part is formed. CNC machines are commonly used in machine shops to make parts for industry.
Metal stamping	A relatively simple process using a high pressure machine. Sheets of metal are placed beneath a die, registered if required, and cut into the desired shape. Large pieces of sheet metal may also be cut and formed in a similar manner. Items like dogtags are made using metal stamping.
Metal forging and casting	In this process, which is not too common any longer, forged and cast products are formed when hot or poured into molds to form the desired shape. Metals typically include iron, brass, and bronze.
Wire bending and forming	Commonly used to bend round or flat wire stock from a thickness of $1/32$" up to $1/2$" and more into various forms and shapes. This process is inexpensive to set up and to prototype. Common products you've seen using this technology are magazine racks and hair clips.
Sewing	Many products are commonly sewn, such as purses, canvas bags, backpacks, and so on. Most contract work is done outside the U.S. in Latin American and Asian countries.
Woodworking	Some products may be predominately wood and may be mass produced in woodworking facilities. Certainly this is the case with furniture.
General assembly	Some products are nothing more than a combination of pre-existing and/or new parts that are assembled in a certain configuration that gives the desired outcome. A simple example would be a lamp. The wiring, base, and lampshade are made from existing components. A complex example would be the steering system in a boat, which is a combination of an existing pulley system, cables, steering wheel, rudder, and so on.
Handmade	In smaller numbers of units, some products are handmade. Prices will be substantially higher than mass-produced products, but the uniqueness and style of individual quality usually stand out.

Complex, Sophisticated Inventions

If your invention is relatively sophisticated and costly to develop, you may want to consider subcontracting or licensing as your approach. For instance, your invention may be a new electronic personal alarm device that warns you when someone has entered your home or your car. This is not going to be a simple device to develop and will require quite a bit of expertise and technology. It will also require a substantial financial investment—perhaps hundreds of thousands of dollars, even millions, by the time you have built up a minimum inventory.

Keep in mind that with such a device there is the cost for the plastic housing, the electronics for the device, the electronics for the home, those for the car, the microwave technology, plus extensive packaging and manual writing. Thus, it will require several parts suppliers and one that must assemble it. It will require extensive testing in a large variety of automobiles and homes. It may also require meeting certain government regulations, such as UL approval and perhaps others. An investment of a few hundred thousand dollars is not unreasonable.

In such a case you will want to try to seek out a U.S.-based company that is active in the sales of either alarm devices, home security, or perhaps car theft devices. You will either want to license them and work with them as you prosecute your patents or you may want to have a contract with them for a certain time period where you are paid as a project leader on a monthly or hourly basis, or a flat fee.

You also have another development avenue to consider. Can you invent a means to streamline the product so that it will be much simpler to make? For instance, how about a simple device that connects to a car's ignition wire that automatically calls your cell phone if the wire is cut or the ignition is tampered with? Or a similar-type motion sensor in your home that dials your cellular phone from your home phone when triggered.

No matter whether your inventions are simple, complex, or somewhere in between, or how you decide to go to market, you'd be wise to have or find development partners that can help defray costs. Whether you go into a buy/sell arrangement or if you license your inventions, the principles are the same. You want your inventions to be dazzling, high-impact innovations that are highly marketable. The Strategic Guide in the appendix will guide you as you do this.

Choosing Domestic or Foreign Partners

You basically have two approaches to consider. You can either have your innovation made in the United States or you can import it. But keep in mind, whatever your choice, the best way you are going to get competitive pricing for the quality you desire is by matching up your invention to the right manufacturer, regardless of where it is located.

Windy Wheels: US Patent 4,735,429 is for those who are tired of pedaling a bicycle but like fresh air. You can harness the power of the wind and sail like an ancient mariner on your wacky and wonderful Windy Wheels. Hop on the bike and head down the road. If there is sufficient wind, the self-adjusting sail speeds you on your merry way. Going upwind? Sailors tack from side to side but that can be risky on a busy street or narrow sidewalk. Need to stop quickly during a strong gust of wind? Find something big and ram it. All kidding aside, this is not a simple invention to pursue, nor easy to test, nor easy to get insured! You'll need deep pockets to pursue it and sales would probably by fairly limited. Looks like something Da Vinci would have passed on centuries ago.© Totally Absurd Inventions, www.totallyabsurd.com.

Furthermore, you want to establish a special partnership with your manufacturer. That is, you want a partner that is going to invest its time and money to prove it can make your innovation with the right quality and will then work with you to build inventory. In return, it could become your dedicated supplier, providing it is able to maintain proper quality and keep up with demand.

Again, look everywhere in the world to find the most cost-effective manufacturer that can give you what you want . . . what your customers must have! The four basic— and most important—elements you want from a manufacturer are:

1. **Quality.** Can it provide the right quality your customers will expect?

2. **Service.** Can it ship on time?

3. **Price.** If 1 and 2 are OK, is its price competitive?

4. **Partnership.** Is it willing to invest its time and money to be your primary manufacturing partner?

For now, these four elements will be the central focus of your efforts regardless of whether or not you'll be subcontracting the manufacture for your company, authorizing a producer for a licensee, or licensing the patents outright. When you find a prospective manufacturer, there will be much more to qualify. In the following sections, you will read about six additional topics you need to consider when choosing your manufacturing partner.

NON-EXCLUSIVE LICENSING

If your innovation is one that will be licensed on a nonexclusive basis, then it will probably be up to the licensee to determine where it is manufactured. But nevertheless, you should still do your research to qualify manufacturers—overseas or domestic—so that you can match one to a qualified licensee if necessary.

Domestic Suppliers vs. Importing

There are both benefits and negatives involved in importing an innovation or having it made domestically. Whatever fits your innovation best, you'd be wise to evaluate as many viable producers as possible so you'll have competitive information and will be able to make an intelligent decision. Table 11.2 compares the factors involved in using domestic and import suppliers.

Generally speaking, importers will have lower costs than domestic suppliers. This is principally due to lower labor costs and lower start-up costs. However, it may have higher shipping costs and a long delivery time. Typically with importing you always have to have a container "on the water," or there can be a serious break in supply. Of course, this could spell disaster to the introduction and continued sales of any innovation.

In contrast, domestic suppliers may be able to ship faster with more economical shipping rates, but if the cost of the product is too high, or the cost of start up—molds, dies and so on—too great, it may be prohibitive to produce in North America.

All in all, if you can find a U.S. supplier who can make your invention at the right price, it is probably the best approach to take. But before you just jump in with an open checkbook, make sure you can strike the right kind of partnership.

Importing Your Innovations

There are several considerations when importing. Most important is that you'll want to be overly careful and overly thorough when delineating your requirements. One error can spell disaster:

1. If your innovation is relatively small-sized, there is probably going to be a substantial cost savings by going overseas. We don't make small things in America very well. Labor on a per-unit basis is usually quite high.

Table 11.2 Comparing Domestic and Import Suppliers

Domestic Suppliers	Import Suppliers
They speak English	May need an English-speaking agent
Shipping lead times are usually pretty short—2 to 3 weeks	Lead times may be 6 to 8 weeks or longer
Can usually warehouse product so re-orders can be immediate	If it can warehouse product, re-orders take 3–4 weeks; otherwise it'll still be 6–8 weeks.
Quality is generally excellent	Quality is generally pretty good to excellent
Overall start-up time is short	Start-up can easily be twice as long
Coordinating packaging is easy	May take extra time forwarding art overseas
Defects and returns are easy to take care of	Must have a working relationship for credits; large problems can be a complete disaster
Customers can easily audit the plant	Difficult for customers to audit
Labor strikes may disrupt service	Labor strikes unlikely
There are no political threats to disrupt supply	Political strife may have a major effect on supply
Shipping strikes may have a minor effect on supply	Shipping strikes may have a major effect on supply
It is easy to get credit	Difficult to get credit; you usually have to commit up front with a letter of credit
Can use the "Made in U.S.A." logo	Must mark with country of origin
Prices are usually higher than imports and may not be competitive	Pricing is usually excellent
Start-up cost for molds and dies tend to be expensive	Molds and dies are inexpensive, perhaps one-fifth the cost of domestic
May not have a steady capacity to take on an expanding business	Can usually increase capacity fairly easily

GLOBAL PRIORITY MAIL

One of the best secrets in shipping internationally is the U.S. Postal Service's Global Priority Mail. If you have documents, drawings, or small parts that can be put into a "large envelope," it costs only about $9 to ship most places in the world. A large envelope is the 9" x 12" size.

2. With relatively large, more expensive products, importing may not be the best choice. It is sometimes easier to absorb U.S. labor costs in larger costlier products than to ship from overseas.

3. Initially you do not have to travel overseas to meet prospective importers. Most correspondence can be done via the Internet. Samples can be sent via UPS, Federal Express, or Express Mail to most countries. It may take up to a week to arrive, so allow for the extra time. It's best to send drawings and artwork via email.

4. Early on, establish a working relationship with one key employee who speaks English. Message the contact regularly—at least a few times a week—to see how he/she responds to your requests. Does this person do so promptly or does he/she take days to get back to you? Are you informed when it may take a few days to get an answer? If the employee or company is not very communicative, see this as a red flag.

5. You have to be overly concerned about quality control. You need to have an agreement on what acceptable quality is and how defective products may be credited. It is doubtful you'll return defective merchandise. Usually it's just destroyed or discounted if it is still viable.

6. If your product is going to be packaged in a plastic bag, you want to see samples. There are many different kinds of plastic bags—high density (hazy), low density (clearer), and cellophane (clearest . . . but weakest). You want the right one with the right quality.

7. If your product is going to be in a printed box, you want to see representative samples so you can select the board quality that fits your innovation best. You'll also want representative samples of print quality and a sample of a carton in the exact size of your innovation. It is usually best to have artwork prepared in the United States and then send it overseas via email.

8. No shipment should be made until a representative sample case is sent to you. In other words, if there are 12 products per case, you want to examine a complete case. Only after you have signed off on the sample case should the shipment be released.

9. Your initial payment terms should not be Letter of Credit (LOC) "payment in advance." Initially it should be LOC "payment upon arrival" to the U. S. port of your choice. Ideally, you'll even want to extend those terms to make payment upon delivery to your facility or even 10 days afterward.

10. You will have to make sure that follow-up orders are placed right away since it generally take three or four weeks manufacturing and three or four weeks shipping time, or six to eight weeks overall. Thus, you need to maintain a steady stream, a continuous flow of product coming in.

11. Once you have established acceptable quality and have a better grasp on sales, you'll want the overseas producer to maintain a working inventory at all times. Thus, subsequent deliveries should only take three or four weeks.

12. Keep in mind that Chinese New Year comes mid- to late-January to early February every year and factories shut down in most of southeast Asia for two weeks. Prepare ahead of time and place orders well in advance.

After working with your overseas supplier for while, you can schedule a visit to the plant. If you do it right, it should not be costly. With the savings you'll have on the up-front mold or die costs, you'll be able to afford it. Keep in mind that traveling to most parts of southeast Asia will cost about $1,000 round trip if purchased in advance and about $500 to $800 to Mexico and South America. Hotels will generally be about the same price as in the U.S.

When importing, don't be fooled by the extremely low prices you encounter. There are hidden costs as well. Here is a typical cost model you can use:

Cost to manufacture, package, and containerize	$ 2.00 each
Shipping cost (8%)	.16
Duty (3%)	.06
Import agent's fee (5%)	.10
Document processing costs ($100 or so)	.02
Inspection fee (1%)	.02
Drop off and delivery of container (usually $300)	.05
LOC cost (2%)	.04
Total cost	$ 2.45

VISITING HONG KONG

Kelly Masterson says, "With the guidance of *From Patent to Profit*, I learned how to reach key people in corporations on the Internet. At the company's website, I searched until I found the names of key contacts and printed the page. After all was said and done, this company paid my expenses for two trips to Hong Kong, enabling me to bring the first 200 Bubble Puffs home for consumer testing. The people in Hong Kong arranged my hotel . . . in the best location for us to conduct business. They picked me up at the airport and were very giving and gracious to me on both trips. My Hong Kong manufacturers were rewarded with their first order of Bubble Puffs— 111,000 pieces . . . nearly $200,000.

The extra charges are usually about 20% to 25%. Duty can vary depending on the type of product you are importing. Inspection fees may occur once in every ten shipments, but they may be a bit costly. Letter of Credit costs may also vary; talk to your banker. Last, import agent's fees may be as great as 8%.

On top of all those costs are your subsequent costs to palletize, warehouse, process and ship future orders. So, if you use a specialized U.S. freight-forwarding service that offers palletizing and re-shipment, you can take the above scenario and add the following:

Total import cost	$ 2.45
Cost to palletize (3%)	.07
Warehousing cost (1%)	.02
Order processing fee (1%)	.02
Shipping costs (8%)	.20
Delivered cost to your customer	$ 2.76

IMPORT HARD COSTS

Usually you can add on about 22% in hard costs to bring an import into the U.S. If duty from a particular country or particular product group is higher, then that must be added on.

The alternative course is to have the merchandise shipped directly to your facility, in which you would then recalculate your costs to palletize, process, and so on. Of course, on top of these costs you will want to have sufficient profit in order to pay administrative costs, office, telephone, travel, insurance, sales commissions, and so on. As a simple rule of thumb; use a 35% minimum margin (divide the delivered cost by .65 to get your 35% margin).

So with the above example, your sell price should be about $4.25. If this were your example for the previous example with the retail price target of $9.95, you're right where you want to be.

Import Repackaging and Reshipping

There are several freight companies that service the United States and specialize in handling your incoming shipments, palletizing, and warehousing them and shipping merchandise out as required. They are usually located near ports. They will usually have a palletizing charge, a warehousing charge, and an order processing fee to ship out subsequent shipments. Most of these fees are very reasonable as shown above.

The main reason these companies have low-cost incoming fees is that they are more interested in your out-going freight, or in other words, shipping out goods to your customers after they have arrived at their warehouse. Even those shipping rates tend to be discounted because they will be shipping a lot of other merchandise from their warehouses as well. Your new product will be considered an add-on and represent additional profit in their normal scheme of shipping.

Some of these companies specialize in large-volume shipments, others in low-volume, less-than-pallet sales. Take some time and do some research in all major cities near the key incoming ports—Los Angeles, Oakland, Portland, Seattle, Houston, Miami, Jacksonville, New York, and sometimes Chicago. They're not difficult to find. Some of them will have multiple warehouses throughout the U.S. and may be able to warehouse at the various points once volume is higher.

Import Agents

Throughout the United States there are import companies that serve as agents to overseas manufacturers, such as those in China. Most are located near major seaports. Some may serve only as your agent to process your order through customs at the port and arrange for subsequent shipping. Usually the fee for doing this is about 5%, but it could be a bit more, depending on how much the company does.

Some companies take on much broader responsibilities. They may be in a position to make all the arrangements, including quoting prices from overseas suppliers, arranging for shipment, and processing through customs. Some of these import agents will even warehouse for you. Of course, whatever they do, whatever job they perform, they'll build that into their profit margin.

If this type of import agency fits your needs, treat it as if it were the supplier and compare its costs to other quotations you receive. Just make sure that all facets (duties, container charges, and so on) are included when comparing costs.

Some chief advantages to a well-established, full-service import agency are:

1. They can deal directly with the overseas manufacturer. Thus, it may eliminate your need to hire or use one of your employees.

2. They can track shipments and, with their experience, give you a fairly accurate account of when the container is going to arrive. At times, they can even ask the shipping company for favorable handling of the container if there is a critical deadline for you to meet. This would mean the company would off-load and process your shipment ahead of others.

3. They arrange payment of all duties, processing fees, and inspections at the port of entry.

4. They will arrange pick up of the container at the port, drop off at a warehouse, and pick-up again after it is unloaded.

5. They arrange for the transfer of all monies on the LOC.

6. Some have warehousing, which may be ideal for smaller volume sales. However, large volume sales are usually best going through major freight companies.

7. Some may be able to extend credit, if they have established working relationships with the overseas producer.

A qualified import agent is almost like having an employee on your team who is doing all the paperwork, taking all the phone calls, processing shipments, and so on. This does not mean that you never talk to the manufacturer overseas, but it can help tremendously as you take care of other matters in your busy schedule.

Domestic Production

It is obviously easier to calculate domestic production costs. There are rarely add-on costs other than freight, perhaps. It is also easy to establish a return policy if there is defective product. But most important, it is a lot easier to work with a manufacturer that can quickly ship working models for testing and the first article in a production run.

Usually the biggest drawback with domestic suppliers is cost. Higher labor and operating costs tend to make domestic production higher priced than production overseas. However, this is not necessarily the case if your product is large, bulky, high priced or not extremely high volume. Larger products may cost less with domestic production because labor cost becomes less of a factor and shipping is cheaper.

FINDING EXPORTERS AND IMPORTERS

Inventor Kelly Masterson says, "Just starting out, not really knowing where to start and due to limited finances, I took the least expensive approach. It turned to be the best most effective approach too . . . I used the Internet. I searched for makers of the products I needed and put together a very basic "Help Wanted" list with the required materials and processes. Every day for a week I sent ten emails to different makers. I also worked with the Trade Councils of different countries and had them post the Help Wanted Lists for me. A company in Hong Kong was just one of the makers who responded and we have now been working together for three years. *Kelly is the inventor of the Bubble Puff™.*

Seeking prices from domestic producers is an easy process. The chief negotiating points become:

1. **Terms.** What are the terms? Can you get extended terms such as 60 days?

2. **Warehousing.** Can it warehouse for you?

3. **Blanket orders.** Can you give a blanket order for a lower cost and draw product as required?

4. **Start-up costs.** Will it be willing to absorb the start-up costs if you give it a large blanket order?

5. **Test product and samples.** Will it be willing to provide adequate samples for testing and proving product performance and quality?

Generally speaking, all these things should be something a domestic producer should be willing to do for you since they are things that importers generally can't. If they can't do these kinds of things, then where is the incentive to buy? You'd be much better off importing an item at a much lower cost.

Dedicated or Multiple Suppliers

At times it is advantageous to have multiple suppliers. One or more may be domestic, and one or more may be overseas. In such a case, you will still have one main supplier with others augmenting its production. The main supplier may come from either the overseas entity or the domestic producer.

If you plan on licensing a company to market your innovation, then all potential manufacturers, domestic or foreign, are potential suppliers. The marketer will determine which ones it wants to partner with. In such a case, you will want to make sure that the suppliers your licensee selects have written authorization from you to manufacture your patented products. This written authorization may be in the form of a "Manufacturing Only" license, which is discussed in greater detail in Chapter 14.

Some chief reasons for seeking multiple suppliers are:

1. Purchasers of the product may require two or more suppliers.

2. There may be two or more marketing licensees, each of which will invariably want their own separate supplier.

3. The volume is too great for one supplier to handle.

4. It is beneficial to have regional suppliers because of high shipping costs.

5. Two or more suppliers help keep costs competitive.

6. An import supplier augmenting domestic production can help keep costs down by dollar averaging.

7. A domestic supplier can help keep a steady flow of product in the marketplace in the event there is a disruption in the import supply, a rush order to fill, or a sudden increase in supply required.

It's not difficult to find and interview multiple suppliers. Just use the format discussed later in this chapter and you'll see how it's done. All you have to do it keep your options open and strike the best deal.

CONFIDENTIAL RELATIONSHIPS

You want confidential working relationship with your manufacturers. You want them to work solely with you and not use any of the technology or trade secrets you bring to them to the benefit of competitors or others. Use the same confidentiality agreement with your manufacturing partners as you do with prototypers.

Do this in advance and you'll be setting the stage for a long-term confidential partnership . . . one that benefits both parties.

Finding, Interviewing, and Evaluating Manufacturers

Finding, interviewing, and evaluating potential manufacturing partners is not a simple task, but with some perseverance, you'll have several good alternatives. If you are unfamiliar with manufacturers in the field of your invention and are looking for manufacturers to serve as subcontractors or suppliers to marketing partners or licensees, you follow the same process essentially.

Your objective is to find at least two good candidates you can compare to determine which one has the greatest interest in partnering or licensing. From this perspective, you want to make all the right decisions based on what is best for your innovation. You must be analytical in your approach.

Finding Manufacturers

Begin by networking. Your initial objective is a fact-finding mission, nothing more or less. You do not want to be making any commitments until you've located several good options.

There are many good avenues to search. As you search, keep an open mind and look for every potential manufacturing candidate in the entire world. Do not limit yourself to your metropolitan area or just the U.S., Canada, and Mexico.

If you are licensing your invention and subsequent patents, remember that the manufacturing partner may be a company separate from the licensee or may be the licensee's manufacturing department. Regardless, you still want to go through this exercise to qualify the cost and quality of your invention. Besides, a prospective licensee may decide that it would be best to have the product made by one of your subcontractors anyway.

If you are exclusively licensing your invention and patents, the decision to settle for one or more suppliers will be that of your licensee, not you. By presenting to the licensee your list of qualified producers, you're greatly improving your ability to license. So be prepared.

Here's a list that should provide a never-ending list of potential contacts:

1. **Companies you already know.** If you work in the field of your invention, you may already know some qualified candidates. Start with them first.

2. **Personal recommendations.** Network with business associates and others in the field of your invention—for instance, sales reps and association members.

3. **Trade shows.** Go to trade shows as a visitor, not as an exhibitor. You can effectively network with dozens of experts in your industry in a short period of time. Most decision makers attend shows to visit with customers and to see what competitors are doing. You will also be able to meet privately with many of them as well as with those to whom you sent a new product summary or have signed confidentiality agreements. This is an exceptionally cost effective way to meet decision makers.

4. **The Thomas Register.** Many companies have a multivolume set of these valuable resource books. So do most major public libraries. You can also check out the Thomas Register on the Internet at www.thomasregister.com (or see the appendix), which has a slimmed down version with enough detail to help you contact potential

---

TALKING TO PARTNERS

Remember that when you are talking to potential manufacturing and marketing partners and answering their questions, you have the stage. This means you should take the time to frame the answer to any question the way you want. Thus, think about what you're going to say before you say it. Respond in a way that fills your needs. For example, if a potential manufacturer asks you, "How much do you have budgeted for molds?" don't give a dollar amount, but position your response something like this, "We have anticipated the first order to be about $8,000—that is, of course, once we have signed off on the quality and performance. After that, we anticipate a monthly volume of about $10,000 for the next four or five months, then steadily increasing. You should amortize the costs of the molds into the first five to six months' purchases. Is that going to be a problem for you?" By framing it this way, you'll find out how willing of a partner they'll be.

285

manufacturers. The Thomas Register lists millions of U.S. companies in virtually every imaginable field and includes the company's size, phone numbers, names of contact people, etc.

5. **Trade journals.** Looking through trade journals may produce some good results. It is, however, different from looking in journals for marketing experts. When searching for manufacturers, you should look in journals in the field of your invention and in the field of the manufacturing process. For instance, if your innovation is to be made from a blow molding process for the recycling industry, look in journals targeted towards both trades, blow molding and recycling. Many manufacturers place ads in the journals seeking new product opportunities. Some may also be marketers in the field, which is what you'll want if you're looking for licensees.

6. **Yellow pages.** If your invention is relatively simple and can be made by a simple process such as injection molding or wire forming, and you live near a major metropolitan area, try the yellow pages. There may be vast differences in pricing based on the type of injection molded product you have, so you'll need to be thorough. However, with some processes, such as wire forming and blow molding, the price differential is going to be minimal.

7. **Trade associations.** Almost every manufacturer in every field in industry, every product line, belongs to some form of association. They tend to be specific to an industry, such as electrical supplies, specific food products, medical equipment, and so on. Track down the associations related to products in your field and you'll be able to get a list of members. You may have to join the association, but fees are generally low, up to $100 or a little more. There's an added benefit to being a member. Not only does it give you some credibility but it also provides information from its newsletters alone that may be invaluable in your invention development.

8. **Recommendations from your marketing partner.** If your marketers are your own in-house sales people, or you have an outside marketing partner already interested, ask them for their suggestions. Many marketing companies and potential licensees already have established relationships with qualified manufacturers.

9. **The Internet.** Search the many search engines, preferably google.com and yahoo.com, and see what you find. Search deep into the search categories since your partners will most likely not be major national companies, but one of the smaller companies in the field. There is almost no end to the possibilities you can find on the Internet.

10. **Product designers and prototypers.** If you have employed a product designer or prototyper, they may know of some candidates.

11. **Importers and agents.** There are four ways to locate importers and agents. First, you can try the local telephone directories if you are near a major seaport. The problem with the listings in telephone directories is that they do not usually define the importer's category. Second, you can use the Internet search engines. Third, you can visit an International Trade Center located in many major cities, such as New York and Los Angeles. Fourth, you can contact the Trade Development Councils of the various countries you believe would be appropriate for your product. They are usually located in major U.S. cities or you can contact them via the Internet. Virtually every embassy with every country has some form of trade development council and

---

DON'T EXHIBIT

Purchasing a tradeshow booth and exhibiting your new product could easily cost $10,000 or more. In fact, it is not uncommon to spend $25,000 and up on booths at prominent trade shows. And all you would do is anger prospective buyers anyway. Why? Attendees at trade shows like to BUY, not just visit. If you don't have inventory ready to ship, your product will get a big black eye. Instead, make a deal with a manufacturer and/or marketing company and let them handle those details later on when the product is ready for ramp up.

can provide a list of manufacturers in their country that are interested in exporting to the United States. The lists can be faxed or emailed to you in a matter of hours.

12. **The government.** Yes! The U.S. and many state governments have a myriad of agencies available to help you find the kinds of manufacturers you seek. The best starting point is the SBDC. Your local SBDC should be able to connect you to various state and county agencies that assist in manufacturing—for instance, in California, there is the California Manufacturing Technology Center (CMTC).

13. **Manufacturers that cannot make your product.** During the networking/interviewing process, you will have several conversations with producers that cannot make your innovation, but may know a company that can. Every time you talk to a manufacturer that says it can't make your new invention, ask, "Do you know a company that can?"

Once you've located potential manufacturers, the introduction and the interview process begins.

Interviewing Manufacturers

With a list of potential manufacturers, you can start contacting them by phone, mail, or the Internet if you elect, and by faxing, emailing or snail mailing a New Product Summary. If you have not prepared this instrument, you're not ready to proceed. If you are licensing your innovation, you'd be wise to have your prospectus ready as well.

When you begin contacting manufacturers, keep your options open. You want at least two or more viable suppliers if the manufacturing company is a subcontractor or supplier to your company or to one of your future licensees. You will be evaluating many facets, but perhaps most important is a company's willingness to partner and use its own resources to be a part of the ramp up and product launch. This includes evaluating pricing, mold and die costs, delivery times, credit terms, and the other intangibles such as their willingness to participate in prototyping and providing the original working models for testing.

There are two key points you'll want to convey in your discussions. They are:

1. Tell potential manufacturers that you are evaluating possible suppliers of your new product.

2. Tell them your selection will be based on your making a smart business decision . . . that is . . . "in the best interests of your product."

Once you are comfortable with your results, you should select one manufacturer to start out. With the promise of being the initial dedicated supplier to launch the project, the right manufacturer will be more willing to commit its time and monetary investment up front.

> **The primary, dedicated supplier will be the one who can give
> the right quality at the right price with reliable shipping and is
> willing to work as a partner in the venture**

Hopefully your marketing department or partner is waiting in the wings, ready to go and sell a lot of product! If not, don't worry right now. However, you'll want to focus on securing one soon. If you anticipate having a marketing expert on your team imminently,

JOB SHOP MENTALITY

If a manufacturer has a "job shop" mentality, you're asking for problems. Job shops typically take a job at a given price, produce it, and then ship it. If they get the order, fine . . . if not, fine. This may be OK for initial prototyping, but this in not a healthy mentality for a long-term supply relationship. You need a manufacturer that is experienced in supplying products to industry . . . not just bidding on a job and shipping it when finished.

TOOTHPASTE

Toothpaste is as old as the wheel. Almost 5000 years ago ancient Egyptians were cleaning their teeth using a recipe of powdered ashes, myrrh, powdered eggshells, and pumice. The paste was applied to their teeth with their fingers. By 1000 BC, the Persians had further refined the toothpaste recipes to include crushed oyster shells and burnt gypsum. As early as the eighteenth century, ingredients such as ground brick powder and china were commonly found in toothpaste. It was not until 1873 that the Colgate Company introduced an innovation called Colgate Dental Cream. It came in a jar, smelled much better than the other products on the market, and did not contain soap. It became so popular that it established the basis for virtually all toothpaste formulae since then and founded the multi-billion dollar Colgate-Palmolive Co.

have some friendly up-front conversation. Here's how you may approach your discussions (to qualify a few important points) leading to a manufacturing partnership. Read the section Qualifying a Working Partnership on the subsequent pages for a complete list of partnership qualifications.

We (the team) are seeking a manufacturer who's interested in being our manufacturing partner on a new product we are developing. We're looking for the right fit. We want to partner with a company that can give us the quality our customers' expect, has excellent service, ships on time, and can produce our product at the right price. We intend to make the right business decision and seek a company that is large enough to ramp up to $1 million in annualized sales by the end of year one. Does this sound like it is of interest to you?

To start out, we are focusing our efforts and the major part of our budget on marketing to make sure we get off to a good start. After all, sales is what we are all looking for. Since you're interested in being a supplier, let me ask you some things to see if there might be a fit. Some things that are important to us are: 1) We would like to have the cost of the molds amortized into our first year's sales. Is this something you can do? To give you an idea . . . and providing all goes well, we estimate our first twelve months sales at about $250,000. The first order we give to you will be about $10,000. By the end of the first year, the annualized sales should hit the $1 million mark, as I said before.

As I said, a lot of our budget is dedicated to marketing, and we believe it is going to be a challenge to fill the pipeline and guarantee a continuous supply (especially true with importing). What sort of terms do you think you can arrange for us? I understand that the industry standard is 1% 10 days, net 30, but frankly, we are looking more towards 60- to 90-day terms to get started so we can use our cash flow to ramp up and launch and maintain a strong advertising campaign. Would this be acceptable to you? (If not, ask the manufacturer if it will ask its raw material supplier for a special consideration on your purchases since this represents a new opportunity. When importing, it will be difficult to get extended terms, but try. And of course, if you are asking for them, make sure you have excellent credit.)

I would like to ask a few more things that are important to us. Quality is tantamount to our product and success. Our product is positioned like the 'Lexus' in the industry, not the 'Chevrolet.' We are dealing with several service-oriented distributors and we need to verify the quality and service you can provide us. Let me expand on this. We need a simple straight-forward guarantee policy that states that within one year, if a product is defective, it may be returned and may be immediately replaced at no cost to the customer . . . no questions asked.

This also brings us to another question. We—and our customers—need to be assured that your company's quality is going to be acceptable before ramp-up begins. I see the products you are making. They look quite good and appear to be acceptable, but our customer's don't know you and can't visit your plant. We will need to assure our customers of our product's performance and quality, and we will work with you to attain those goals. Our customers will be asking us to provide samples in order to prove your quality, and frankly, to prove your service too. There will be some focus groups (or in-house tests if indus-trial/commercial) used to test the product's quality. We would ask you to provide these working models—samples—at no charge. Just so you know, it may take as many as 100 units to get your quality qualified.

One last matter I'd like to talk about. We strongly believe in using a team effort to make this project a big success. We are talking to you about being a part of this team effort. If we decide to proceed as partners in this team effort, you'll soon be talking more with our marketing partners (or marketing depart-ment) than with me. Nevertheless, I'd like to ask you something that could make our jobs a whole lot easier. Are you in a position to assign a project engineer to oversee your operations to fine tune the product and make sure the ramp up goes smoothly? This would take a big burden off your back if there was someone who you could rely on to help make this happen. It would be someone that I could talk to from time to time so that you wouldn't have to be burdened by the little details. He or she could just get it done instead. Do you have someone like this?

If you get all this . . . and there's no reason why you can't, you'll be set to really make things happen. Just remember, if there are some things the potential manufactur-ing partner can't or is not willing to do, then you'll have to take that into consideration. But after talking to several good candidates, you'll have a much better understanding of who can do what and at what price level. And without question, you'll find some who will be far more willing to work with you than others. What's key to your effort is to find a manufacturing partner that is really committed to helping you launch your prod-uct and be part of a team effort.

Remember, if a domestic manufacturer says the molds are far too costly to amortize over a year, find out how much they'll be. If they are going to cost $20,000–$30,000 or more, you may want to consider importing instead. Most of the cost of a mold is labor. Since overseas labor is so inexpensive, mold-making costs far less than in the United States. Couple that with the fact that import pricing is so low, it is usually fairly easy to get the overseas company to absorb the cost of the mold in the first order (or better yet, the first year's commitment).

After all, wouldn't you rather be paying $20,000 up front for several thousand sell-able units instead of paying $20,000 for a mold and still have nothing to sell?

It is also easy for an overseas company to start out with an inexpensive one- or two-cavity mold. They do not need the more expensive multi-cavity molds domestic pro-ducers would require in order to be cost competitive due to high U.S. labor rates. The importer will just make more molds later when volume increases.

And don't forget that every time you hear a "no," you are that much closer to hear-ing a "yes." Just don't make the mistake of thinking that the first company you talk to is going to be "it" . . . regardless of how nice the guy is. You have to make "a smart busi-ness decision." That concept should pervade your discussions. Anytime you are asked about who else you're talking to, let them know there are others you are talking to or will be contacting. You don't have to say who they are.

One of the best reasons to have a marketing expert/partner on your team early on, is that he/she (the head person usually) has probably had a lot of experience in talking to manufacturers addressing all the above matters.

SMALLER IS BETTER

Don't forget that smaller companies here in the U.S. tend to make better partners. They are much more flexible and can get things done faster than large, committee-driven companies. Overseas, it is usually true as well. In either case, it is also best to steer clear of those that are too small, since they may not have sufficient capital to ramp up your project.

CHAMPIONS TO THE CAUSE

The value of having at least one "champion to the cause" is immeasurable. This person can help cut through red tape, track progress and orders, ensure excellent quality and timely shipments . . . and really help propel your efforts forward fast. Always be friendly and keep in mind that innovating is a team effort. Try to find as many champions of your cause as you can, whether they work for the manufacturer, a marketing partner, or at times even a customer.

Evaluating a Manufacturer

It is best to base your decision on selecting the right manufacturer based on the company's ability to perform, not on how much you like the executive with whom you are dealing. While you may like a particular individual, he/she must have the right support team.

When evaluating manufacturers, evaluate them based on these primary considerations:

1. **Quality.** The quality of your invention will make or break its success. A manufacturer absolutely must be able to give you and more importantly, the customers, the quality they expect. A manufacturer's track record must prove it can do exactly that. After reviewing some of the products it has made and evaluating and qualifying its production capability, continue your evaluation.

   If a manufacturer cannot deliver the expected quality now, it could take years for it to do so. It is extremely difficult to convert a so-so quality company into one that makes excellent quality. Chevrolet is not going to compete with Mercedes in the near future, nor is Timex going to compete with Rolex. If a potential candidate cannot give you the quality you need, do not waste your time. Keep searching.

   If a manufacturer can produce the required quality, you must also determine if they can do it consistently. A modern quality management system and philosophy will most likely give you that result. If a manufacturer has been trying to improve its quality due to quality problems in the past, be cautious. It could take two or three years before the new companywide philosophy is securely in place and the quality of its output is consistent.

2. **Price.** A producer must be able to manufacture your new product at competitive prices. You can get a good idea if it can by comparing the prices of some of its other products. If a producer talks about how profitable it is, beware. It may be cherry picking only super high-quality, high-profit products. On the other hand, if it is the low-cost producer in the industry, ask why. Is it because it has generally mediocre quality and a generic approach to manufacturing? This, too, is not a good sign.

   You want to locate companies that are not low-cost generic producers, but those that focus on cost-effective, quality manufacturing. Competitive companies within an industry usually have similar pricing. You need to know that your manufacturer is "lean and mean" and not loaded with excessive profit demands.

3. **Service.** Are they known for excellent customer service? Do they return phone calls promptly? Do they have an 800 phone number? Are they polite and courteous? Do they do what they say they will do and when they say they'll do it? Are their deliveries on time? A marketing partner, retailer, or industrial/commercial customer must feel good about purchasing a new product, especially if it is from a new, untried supplier. Do not settle for anything less than dazzling customer service from your manufacturer.

4. **Capacity.** If a producer does not have the capacity to take on additional output, you are probably wasting your time. If it cannot give customers the quantity they require with a timely delivery, consistent orders will not follow. If the potential manufacturer says it can expand quickly in its present location or some other plant location, get all the specifics. At the very least, make sure your manufacturing expert has suf-

ficient capacity to handle the start-up production requirements, while expansion is being completed.

5. **Finances.** What is the present financial condition of the manufacturer? This is particularly important if a producer must expand to accommodate your requirements. It's also extremely important for its role as a partner, since you'll be asking it to contribute some of its financial resources during ramp-up and perhaps afterward. If necessary, request a profit and loss (P&L) statement or an annual report. If they are reluctant to supply this, explain that "it is something your accounting department requires." They'll do it. Another strategy is to ask others in the industry about the company's financial reputation and history. Believe it or not, some companies have a bad reputation for bankruptcy filings.

6. **Reputation.** Research the producer's reputation to verify it does what it says it can do and has satisfied customers. An unreliable manufacturer may make your new customers angry and kill your innovation's launch. Try to find out if the company goes through frequent management changes. If so, you could be in for a roller coaster ride. A new president may not see the potential of your product in the same light as did the old president.

   What is its reputation regarding patents and licenses? Is it a licensee for other products? Have they ever knowingly infringed on a patent? When talking to them about patents, listen carefully. Ask around to learn what their attitude has been toward other intellectual property. If they have infringed on others, they will probably look for a way to get at you.

7. **Trust and philosophy.** Can you trust the owners and management? During the interview, look the executives straight in the eye. If they cannot look back at you, you'll have a tough time ahead.

   A company's business philosophy must be similar to yours. Are they individuals with whom you would do business? Remember that they will be your partners for a long, long time. The last thing you want is a company that is late with royalty payments and then pesters you to death with excuses for their inability to perform.

8. **Innovative or generic approach.** Is this company innovative or is it just another old-fashioned, generic producer? Most innovative companies can quickly gear up and make new products. Does the company have good problem solvers and do they tend to be responsive to customer's needs? If a producer has a generic attitude, it would be difficult to suddenly change it. Evaluate other products the company has produced, see the production hands-on, and feel confident they are capable of providing you and your potential customers with high-quality products.

9. **Commitment.** Above all, is the manufacturer willing to make a *total, absolute commitment* to your project? The only way anything will happen in a timely manner is with their total commitment to your project. You must ask for this and they must give you the right, aggressive response. Your success depends on it!

Ideally, you want manufacturing partners that correctly answer all these questions. Sometimes, however, we do not have this luxury; it is not a perfect world. If a candidate does not fulfill most of the preceding attributes, keep looking. Remember: if a manufacturer does not meet your standards, but wants your new invention badly enough, it will be willing to make necessary change. Make sure that they prove it to you first.

LOOK AT YOUR PRODUCT CONCEPT OBJECTIVELY

Is it an adaptation or a brand new innovation? Is your product based on well-known and available components and/or product features, ones that are easy for manufacturers to find? Or, is it a product based on a new material, technology and/or methods? This impacts your product's ability to be manufactured as well as protected, licensed, and even successfully sold to consumers in the future.
—*Paul Lapidus, CEO, NewFuntiers*

With these basic qualifiers verified, you'll next want to make a list and qualify all the important partnership considerations.

Qualifying for a Working Partnership

Once you've identified some candidates, you'll want to whittle them down to see who will make the best partner based on your needs. Here are several partnership factors you may want to consider:

1. If a domestic supplier, does the producer have warehouse space available? If not, you'll have to find it. This may be important.

2. You want to be able to give your manufacturer blanket purchase orders and be able to draw against them. For instance, you give it an order for 1,000 units with the agreement that you'll draw 250 at a time. If domestic, you want orders shipped within 24 hours of receipt. If overseas, you want repeat orders shipped within 48 hours.

3. What are your domestic supplier's payment terms? They're going to have to be pretty good since the import prices are going to be so much better. With good credit, you or your marketing partner—depending on who is going to take the billing—should ask for 60- to 90-day terms. That's a lot, but it's not impossible to get.

4. If your new invention represents a fairly high volume of raw material usage, such as plastic resin, ask the manufacturer to ask for special terms from its raw materials supplier in order to offset the favorable terms you are requesting. Frequently they are able to get it. This is commonly done by raw material suppliers in the plastics trade that may be interested in capturing emerging markets, such as Dow Chemical or Exxon or Lyondel.

5. You may want to consider having your manufacturer bill extremely large orders direct. In such a case, make an arrangement in which the manufacturer either pays you the difference between your cost and the sell price, or, as in a more common approach, you split the difference.

6. Ask your supplier if it will be willing to appoint a dedicated engineer or project manager to the project. He may even be given the title "Project Engineer." This special person should be someone who is knowledgeable, ambitious, eager to get ahead, and a good problem solver with good people skills. This person is typically someone who will be a "champion to your cause." This person will want to have a vested interest in your success. He/she can be invaluable to you by following up on orders, and ensuring quality is correct, shipments are on time, and so on. Build a relationship with this individual and it will be like having your own employee as a liaison to the manufacturer.

7. Find out who does the package design with some of the company's existing products. Would your marketing partners want to work with them?

8. Most important of all . . . who is going to pay for die and mold costs? Your manufacturing partner should be willing to accept a large blanket purchase order with the cost of dies and molds amortized into it.

You may ask, what are the qualifications a manufacturing partner would want in order to make all these commitments. That's a fair question. If you have your marketing partner in place, that partner's commitment is about all that'll be required. In other words, if your marketing partner is your own internal marketing department or manufacturer's representative's agency, then its sales commitment and your subsequent purchase order is usually all that will be required. Obviously your company's credit plays an important part as well.

If your marketing partner will be taking the billing from the manufacturer, such as may be the case with licensing or private label, then that entity is going to have to make the commitment and it will have to have its credit in line as well.

Without a marketing partner committing to substantial sales, you can only rely on your own credit worthiness. However, in such a scenario you have to seriously ask yourself if you've got the cart before the horse. How can you even consider making such a commitment with having your sales/marketing team lined up?

Nevertheless, and this is important, should you not have a marketing partner on your team, you still should be qualifying manufacturing partners so that when you talk to potential marketers, you'll be able to qualify materials, costs, delivery times, and so on. It is irrelevant if they ultimately prefer some other manufacturer to make the product for them. What's important is that you'll have all the p's and q's qualified for those discussions with prospective marketers.

Requesting a Quotation

With a good understanding of your product's requirements, you'll want to have a firm price so you can proceed. This is usually done with a written "Request for Quotation" (RFQ).

A Request for Quotation should include the following elements:

1. The due date to submit a price quote. In other words you want the quotation to arrive to you no later than _____.

2. A complete description of the item being quoted

3. Product and shipping specifications, which may include:
   A. Raw materials specifications
   B. Complete packaging specifications
   C. Shipping containers (master cartons? pallets?)
   D. Terms (or you may try requesting a certain term such as 60–90 days)

4. Freight preference (For instance, free on board (FOB) delivered in the United States on orders of 250 units or more.)

5. Lead time for the initial order (or in other words, a turnaround time saying that upon receipt of the initial purchase order, how long will it take to ship)

6. Warehousing terms and requested turnaround time for subsequent purchase orders (or draws against the original, if that is your choice)

7. A time duration for the quotation. (You should accept a minimum of 30 days, but you should ask for price guarantees of at least 90 to 120 days, or even up to a year.

An acceptable price quotation from a potential supplier should be on the letterhead of the manufacturing company, include responses to all the above, and also include:

1. A date

2. A price-per-unit of the item being quoted

V-CHIP

Canadian Tim Collings is an electronic control device consultant and professor. In 1995 he became concerned about the impact of violence on television on children. As a result, he developed a decoding device that parents can use to program or block out various levels of violent programming. Broadcasters agreed to code their programming to indicate the level of violence and the V-chip was born. The letter V stands for Viewer, not Violence.

3. Any additional charges if applicable: artwork, printing plates, dies, and so on

4. A time duration for the validity of the quotation (Accept a minimum of 30 days, but try for more.)

5. A signature from an authorized officer of the company.

If anything is missing from the price quote, ask the manufacturer to provide that information in the form of a memo.

Even though an inventor who is planning to partner or license his/her invention would not normally be securing price quotations or submitting purchase orders, this is a valuable document to have. When talking to prospective partners and licensees, make sure your homework is done. It can significantly improve a licensee's ability to make a decision in your favor and allow it to go into production quickly. If nothing else, you can use the information as a competitive quote. Don't worry about hurt feelings . . . that is the way business is done at times.

Submitting a Purchase Order

If you are purchasing the product for sales and distribution through your company, you'll want to submit a purchase order once you have received an acceptable price quotation. If your marketing expert or a marketing partner is going to be doing this, then it will submit a purchase order.

A purchase order is a document that outlines the products, terms, and conditions of purchase and is usually based on the terms and conditions previously outlined in your RFQ. It must be thorough, for the oversight of any term or condition would not be a binding part of the purchase order.

Once it is signed by you (or the issuing party) and accepted by a manufacturer (or receiving party), it is a legal document obligating the signor to purchase the products as outlined. Acceptance by a manufacturer is usually in the form of a written acknowledgement or confirmation. If this is not common practice with your supplier, request confirmation in writing in the body of the purchase order.

Prior to shipment by a U.S. manufacturer you should try to visit the plant to sign off on the first production articles (referred to as "first articles"). If that is not possible, the manufacturer should send them overnight to you. Or, prior to shipment by an overseas company, make sure you have signed off on a representative sample overnighted to you as previously outlined.

Working With Manufacturers

During the initial stages of development, there will be a lot of questions as your innovation metamorphoses from being a prototype to a working model to a marketable product. Thus, all the preceding tenets regarding prototypers also apply to manufacturers. The exception may be with overseas manufacturers, who will solely be following your instructions. Any minor changes they may want to consider should be sent to you for your pre-approval.

Be determined to solve any manufacturing problems associated with the design. Regardless of how your new invention is going to be produced, you cannot accept the mediocrity of everyday generic mass-manufacturing processes. You must maintain your vision and insist that it be fulfilled. Challenge your manufacturing partner to use its

---

PACKAGING INCLUDED

Frequently price quotations will include the cost of packaging. This is common with products that are imported. It can also include domestic quotations. In either case, you have to qualify the type of packaging, the quality of the bag, chipboard box, shipping containers, and so on. In how many colors will it be printed? Who does the artwork and who pays for the printing plates? Be careful that your quotation on packaging is not for a basic "one-color" print. Read more on packaging in Chapter 10.

ingenuity to help solve these problems. Believe it . . . there'll be a lot of unforeseen start-up problems.

Again, you'd be wise to have a single inside engineer appointed the project manager to oversee the intricacies of your project. With smaller companies that individual may be the owner of the company. However, larger companies will usually be able to designate one of their better employees as the project manager or engineer.

Fulfilling Legal Requirements and Certifications

Many products have legal requirements or must have certification. The more common ones are UL approval, law labels, and FDA or FCC approval.

UL approval is commonly required for anything electrical or electronic. (For more information, visit http://www.ul.com/.) Law labels are legally required on products such as pillows, blankets, and all fabric-related products like shirts, pants, and so on. (For more info visit http://www.abflo.org/about.html.) FDA approval is required on most food-related packaging and on drug products. (The FDA list can be found at http://www.fda.gov/opacom/laws/.) FCC approval will be needed on electronics that use any form of radio frequency for transmission, such as cell phones. (Visit http://www.fcc.gov/aboutus.html for additional information.)

There may be other legal requirements or certifications as well. For instance, if you are manufacturing a product for the automotive trade, there will be certain standards set by an automaker that must be met for durability, performance, and safety. Most industrial products and chemicals may also require some form of certification based on industry standards. Many products require warning labels for use, such as thin-gauged plastic bags or concrete products that require warning labels for their Portland cement content. Even shipping cartons may have to meet certain requirements by a customer.

The good news is that your manufacturing partner will usually know what legal requirements will need to be fulfilled. Hopefully, you do too. You can also consult with your marketers who should have plenty of experience as well. Since the responsibility usually rests on the manufacturer, they'll usually know best. Another good way to find out is to see what legal warning labels and markings are used on competitive products.

Should you be importing product, you'll need to supply all the proper language and print requirements to the overseas company. Some inexperienced overseas producers may not be aware of certain legal requirements, law labels, weights and measures markings and so on. If you're not sure, start asking.

Working with Multiple or Dedicated Suppliers

Frequently you will require multiple suppliers based on industry demands or the fact that a single supplier would be inadequate to handle your particular needs. Also, if your approach is licensing, then you'll want to determine your approach—either exclusive or nonexclusive licensing. The answers to these concerns will be determined by your marketing or licensing approach. You'll read about this more extensively in Chapters 12, and 14–16.

Here are some advantages of employing multiple suppliers should your particular needs dictate:

1. You'll be able to compare pricing. Two suppliers tend to keep each other competitive.

DR. PATRICIA BATH

An ophthalmologist from New York was living in Los Angeles when she received her patent, patent No. 4,744,360, on a method for removing cataract lenses. It was the first patented medical invention by an African-American. Her discovery transformed eye surgery by using a laser device making the procedure safer and more accurate. Today laser surgery for eye vision correction and other abnormalities is standard procedure.

2. If there is a shipping strike, bad weather conditions, or internal financial problems with one supplier, then the other can take up the slack.

3. Having suppliers in different regions may result in lower shipping costs. For example, an East Coast supplier may ship east of the Rocky Mountains and a West Coast supplier may ship west of the Rockies.

4. If one supplier is domestic and one is an import entity, you may be able to lower overall costs through dollar cost averaging.

In contrast to this scenario, there's good reason to employ only one dedicated supplier. Those reasons are:

1. It's easier to deal with only one source when placing purchase orders, tracking shipments, and monitoring inventory.

2. A dedicated supplier that will go out of its way to help improve the product provides a big incentive.

3. A single supplier may (or may not depending on the nature of the product) have a lower cost than multiple suppliers because a larger volume means longer production runs and may mean lower raw material costs.

4. A single supplier may be more dedicated to your product and your company.

Regardless of the number of suppliers your innovation requires, you'll always want to be ready to fulfill sales requirements quickly.

Chapter 12

MARKETING PARTNERS

Nothing happens until your product gets sold.
Get the right sales expert on your team from the outset.

Regardless of your approach to commercialization, you're going to need a marketing partner on your team. Remember, nothing happens until something gets sold. Who's going to do this? Your marketing partner is usually one of the following:

- Your company's internal marketing department, which typically has a national sales manager and support staff. These individuals are usually salaried employees with the company with various forms of sales override and commission incentives.

- Outside commissioned sales reps such as manufacturer's representatives. They work on a commission-only basis. This may be from 5% to 10%, sometimes more and sometimes less, depending on volume and profitability. These people invariably get paid after shipment is made or payment is received.

- A separate marketing company that you will supply under its name, typically as a private label or OEM production. This means it pays you for the merchandise.

- Licensees . . . exclusive or nonexclusive. The licensees pay royalties on sales.

You can see that with the exception of your own marketing department, your marketing partners do not incur any out-of-pocket expense on your behalf. Even at that, an internal marketing department really doesn't incur any new sales expense either unless you're a start-up company. If the marketing companies you've been pursuing want to charge you money, they're clearly not the kind of marketing partner you want.

The ultimate responsibility of your marketing partner is to generate sales . . . nothing more or less. If it can't get the job done, you don't want it as your partner. Your marketing partner must give you these guarantees in order to be considered. It makes no sense at all partnering with any sales/marketing entity that says, "We'll try." You want qualified marketing experts who are already active in the field of your innovation and have sufficient experience to qualify its marketability.

Getting a marketing partner is the single most important requirement for a successful invention. However, it is usually the last aspect most inexperienced inventors consider. Frequently it's not until long after a lot of time and money has been spent on patenting and prototyping that they even think about marketing and sales. While technically different, sales and marketing are interrelated because a successful marketing strategy results in sales.

Your sales/marketing partner typically consists of several individuals or a company. It's not a single individual. Granted, you'll be dealing with the head of the entity, such as your sales manager, the national sales manager of a private label company, the president of a manufacturer's rep agency, or perhaps the national sales manager of a licensee. But you'll have the support of an entire sales team behind you.

In an ideal world, your marketing partner is on your team from the outset. If not, you'll want to get him/her in place at the earliest possible moment. There's no need to repeat the importance of your marketing expert. It's threaded throughout the book. From this point on in this chapter, we'll be expanding on this.

Your Sales/Marketing Objectives

Your marketing partner will be invaluable in the following ways:

- **Qualifying the inventive attributes.** The right marketing expert with experience in a given industry can usually tell you if your innovation's attributes are marketable. After all, this is what you've been looking for all along—verification of marketability.

- **Recommending improvements and modifications.** They can also give you feedback about how your innovation may be improved. One thing is for certain—you do not want to waste your time and money on further developing an innovation with attributes that do not have solid marketability. Keep improving and creating until you get the outcome you desire.

- **Helping to determine the target market.** Being perfectly positioned with existing sales contacts, your sales expert will also be perfectly positioned to help determine the target market for your new invention and advise you about which target markets to avoid due to competitive responses or barriers to entry. If your potential marketing partner does not have existing sales contacts in the field of your invention, you're talking to the wrong person!

- **Determining price.** Your objective is to enter the marketplace with a target price that will maximize sales. An experienced sales expert in your field can help determine the target price in a very short time frame.

- **Determining the quality required.** The right marketing expert can help determine the exact quality a new invention should have and may also know a producer to make it.

- **Setting up test marketing and evaluation of the innovation.** A marketing expert can arrange for and coordinate testing of your new invention. Having him/her on your team early on can help get your innovation's potential evaluated by retailers, distributors, and end users—and test marketed, too.

AN INVENTOR'S BEST FRIEND

Your marketing partner is your best friend. Sometimes we get caught up in our creations and lose objectivity. One thing for certain, you can't have tunnel vision and see your innovation's sales only in your perspective. By listening carefully to smart marketing experts in the field of your invention, you can learn all you need to know in order to change, modify, or improve the invention and transform it into a high-volume success. If the marketing experts you talk to just don't buy into your invention, you had better find out why or put the invention on the backburner until you find out.

- **Helping to locate a manufacturer.** Your marketing expert is likely to have several good contacts.

With a well-tested prototype that is suitable for sales presentations, the right marketing expert can use it to start lining up sales. This may be done even though you have not made a commitment to a manufacturer. With orders or commitments, the manufacturers you speak to will be a lot more interested in partnering. They will be much more willing to invest their own time and money, build up inventory, and absorb or amortize start-up costs if your marketers have solid commitments from major retailers, distributors or customers. In many cases, your marketing partner may not need commitments from its customers, but would be willing to commit to a manufacturing entity based on its experience and know-how in the marketplace. You best believe that these kinds of start-up sales scenarios happen all the time, and you'll be giant steps ahead if you can get the same commitment from your team. Frankly, if you've got a great idea, these kinds of commitments are not hard to get.

By beginning your sales activities before you have contracted with a producer, you will be developing orders and commitments that will put you in the enviable position of knowing what your cash flow is going to be. You'll be in a much better bargaining position to a seal a deal with a manufacturer. You'll also significantly improve your negotiating position and that will drive manufacturers to action quickly.

Last, keep in mind that your sales/marketing experts are not distributors, but they will arrange distribution for you. They are not graphics artists, but they can work with graphic artists to help prepare packaging, sales aids, and promotional brochures. And your marketing experts are almost *never* someone or some company that is not established and experienced in the field of your innovation, like an invention promotion company. They are experienced professionals who have all the right contacts to make sales . . . and make them now! They'll get the right appointments with the right accounts and save you a huge amount of time and money.

If you have been talking to companies that cannot do this, you're talking to the wrong ones! Your marketing partner absolutely has to be in a position to get your innovation to market and secure sales.

Marketing Strategies

There are four basic marketing strategies, three of which may apply to innovation marketing. It's wise to know these strategies because it can also help you identify potential marketing partners or qualify new hires based upon their experience.

The first strategy is called **defensive marketing**. This is almost exclusively the strategy used by companies that are the high-volume producers in an industry and are considered the dominant suppliers. Their approach is to defend their present market position. They are rarely innovative and rarely oriented toward market creation. Generally speaking they do not make good marketing partners with small companies and independent developers. Their marketing approach just doesn't fit.

The second strategy is an **offensive strategy**. In many ways, you can say this is the marketing strategy of the second tier companies that "try harder" to "make it your way." Of course, their objective is to become a dominant supplier and to move into a defen-

UNABLE TO SEE THE LIGHT

In the late 1970s and early 1980s, GM did not see the direction in which the auto industry was headed. Resisting the transition to smaller, higher-quality cars, Chevrolet lost its sales leadership. To illustrate how backward their vision was, in early 1997, Chevrolet finally announced that it would no longer make large cars!

sive marketing position with occasional offensive sales bursts in attempts to capture additional market share. While this might sound appealing for an innovation, it rarely happens until the later stages. That is because the focus is not on market creation and new product opportunities, but on trying to capture market share—really two entirely different objectives. Offensive marketing is akin to deploying major national sales campaigns, or at the very least, broad regional pushes.

The third strategy is referred to as **outflanking maneuvers**. This fits well in the earlier stages of new niche product sales and market creation. With some established sales and depending on the size of the sales team, a new product may be sold to markets, regions, or customers where the dominant competitors would not normally go. Doing so builds volume and lowers manufacturing costs. Thus, it puts a company in a position to do two things: to solidify its present niche and customer base and to build up a war chest of inventory and money so it may go on its own market offensive. If this strategy is used from the onset of a new product launch, it may be considered aggressive. In such a case, beware of competitive responses.

The fourth strategy is commonly referred to as **guerilla warfare** and is the most commonly used sales approach to ramp up new niche products. Just like its name suggests, it's perfect for smaller companies and those marketing partners that are striving to get a foothold in the marketplace. Guerilla tactics usually start with a small, aggressive sales effort with a narrowly defined customer and in a narrowly defined region, one that's considered easy to protect. From the initial penetration, sales are expanded to other narrowly defined targets. Sooner or later, sales may increase to a level sufficient to expand sales through some broader based outflanking maneuvers.

In a ramp-up application, you want your marketers to be a company that is skilled in the last categories—guerilla warfare and outflanking maneuvers. That's what will get your innovation established. Ideally, a marketing partner with skills to perform offensive maneuvers will also serve to take your innovation well into the future.

When you are interviewing your marketing partners, you must be talking the same language, else there's not going to be a good match. Usually you can find this out in the early stages of conversation.

Choosing a Partner

Can They Really Sell?

It is important to find the right team players to develop the sales and open up the distribution of new product placements. Many good ideas and inventions have failed because of the lack of a committed or properly trained sales team. Great new inventions and products go nowhere if sales teams do not want to sell them or lack the knowledge and ability to sell them in new emerging niche markets.

There are two classic examples in history of a lack of commitment from sales teams that created opportunities for their innovative competitors. First, accountants in the American auto industry figured more money was made selling large cars than smaller, higher-quality cars. This backwards thinking almost bankrupted the entire U.S. car industry and let in massive competition from overseas. It took years for the U.S. auto industry to only partially recover.

DO YOUR HOMEWORK

The willingness of a company to develop and market a new product depends, in large part, on how secure its patent protection is. Does the patent provide broad coverage of the technology or is it possible to circumvent the patent with a minor design modification?
–*Don Kelly, Intellectual Asset Management Associates, LLC*

WALK SOFTLY

"Be careful who you 'partner' with. They may be your partner today and your competitor tomorrow."
–*Michael S. Neustel, U.S. Patent Attorney*

The second example was in the mid 1970s, when the IBM sales team was informed that the future of computers would be the PC (not IBM's super computer) and was offered the blueprint for developing and mass marketing PCs five years before Apple launched its first PC. However, IBM's marketing team directed its future away from PCs because they did not offer the large commissions available from the sale of its super computers. Could you imagine if IBM had embraced the PC at that time? Could you imagine if it had spent its R&D war chest on software instead? It would be the largest corporation in the world and probably the most influential one ever!

In both examples, the industrial leaders should have had the foresight and the guts to "attack themselves" to retain their leadership position. In hindsight, both have recovered. The auto industry got back on the road to profitability, but with a lesser market share. IBM with a massive makeover in the 1990s refocused its efforts toward invention and innovation. Today, IBM receives more U.S. patents than any other corporation in the world, with licensing income at $3 billion a year.

You want sales experts who are not afraid to attack themselves, not complacent with the past, and not committed to selling only the existing product line. You want those that have the ability to introduce and sell niche-oriented, customer-driven innovations. After all, by the year 2010 or so, almost all products sold in the United States will be niche products in one form or another and will be patented and perfected based on CDI principles. Sales experts who do not address and satisfy true customer needs and desires simply will not be around.

Regardless of whether your sales partner is your internal sales department, manufacturer's reps, private label companies, or licensees, it must have the right ingredients to ramp up your innovation.

Two Types of Salespeople

There is another way to look at it. Generally speaking, there are two types of salespeople. The first wants to take orders, maintain the status quo, and look for additional market share. The second type likes to sell new products, create new markets, and be entrepreneurial. This is the kind of sales expert you want on your team. These sales experts are the market creators, can sell into the future, and are not content with just taking orders. They enjoy selling value-added products instead of products based on lower prices.

Beware of a prospective partner that may call itself a market creator when it is really just looking for added market share. This type of sales organization can rarely follow through on selling a value-added niche product and will invariably fold to price pressures. Their justification for failed sales is almost always, "The price is too high."

Most salespeople do not have the patience to persevere as well as the ability to be persuasive in order to convince a customer to try new value-added products and to experience the benefits for themselves. For instance, they may not be able to persuade a customer that a bottom-line savings is far more important than a per-unit price. If a sales associate cannot show this in concrete terms, then he/she lacks the savvy and knowledge to sell a new innovation.

Market creation sales take considerably longer at the beginning. But they result in loyal customers and sales that are easier to maintain over the long term. Companies that are good at market creation sales know this and when they speak, they'll be speaking the right language. You'll instantly know you have a viable potential partner.

YOUR BEST TEAM PLAYERS

Your marketing expert is usually the easiest team player with whom to work. He/she wants to do exactly what you need—that is, to get your invention sold and producing income! You will find that having a trusted marketing expert on your team will be an incredible asset to your invention's development and lead to its ultimate success.

THIS GUY CAN SELL!

Early last century, Murray Spangler worked in a department store selling carpets, but he developed an allergy to carpet dust. Since cleaning carpets was done by hanging them over a line and then whacking them with a dust paddle, he realized his problem might worsen to a degree that he might have to quit—unless he could find a solution. Thus, he invented the vacuum cleaner. He had a difficult time trying to sell them due to the high cost. But when his cousin, Hoover, saw the vacuum work, he immediately saw the light and said, "I can sell that!" To overcome the high price objection, he cut out the middlemen and sold them direct to the users in a door-to-door (home-to-home) sales campaign. The rest is history.

One of the Best Kept Sales Secrets on the Planet

Often new or small businesses and product developers think in terms of hiring sales people and paying them a salary, plus expenses. They look at this as some form of inevitable expense.

However, hiring a national sales manager to head up the sales/marketing of a new innovation is frequently not a wise choice for a start-up company. The money can frequently be spent much more wisely on product development and promotion, leaving the sales responsibility up to one or more manufacturer's representative groups instead.

Many independent manufacturers' representatives are interested in working with small companies, start-up or established, and independent inventors. More importantly, a high percentage of manufacturer's reps have the all the right attributes. Why is this? First, most are self-employed and are naturally entrepreneurial. Second, many understand what it takes to get a new niche product to market. Manufacturer's reps usually represent many lines of products, most of which are from smaller companies or are niche brands. Unlike large, established companies with in-house, defense-minded sales teams, manufacturers' reps tend to understand the strategies and tactics required to launch new products. They also tend to know how to protect these smaller niches once they're established.

At one time or other, most experienced reps have been hired by major companies to build up sales and were subsequently fired once the job was complete. It is not uncommon for a major national company to hire experienced sales reps to get immediate market penetration and then fire them when new management takes over (usually a future cost reduction move). If the reps can work with an inventor and help him/her get his/her new products launched, they can establish a relationship with long-term security.

Manufacturer's reps work on a commission basis, which essentially means that you do not pay them until something has been sold. Typically commissions range from 5% to 10%, sometimes more and sometimes less; they can be determined by doing some research in your industry. It's better to ask other companies what they pay their manufacturer's reps than to ask the reps themselves. Commissions are usually paid on a monthly basis and include all product either shipped or paid for in the previous calendar month.

You will discover in Part Four how you can license an exceptional manufacturers' representative organization. Not all of them may be interested, but when you find one that is, it can be very rewarding for you and for them. Just keep in mind that when you find an aggressive, successful manufacturers' rep organization, they will be your best friends!

Finding Marketing Partners

If you do not have a sales partner or licensee, finding one is probably the most important step you'll be taking in the coming months. Very little will happen without one, and final development of your innovation will most likely never occur. There are several ways to locate sales/marketing partners and licensees. The best ones are:

1. **Recommendations through existing business or personal contacts.** Talk to friends, acquaintances, and business associates about your innovation. Ask them about companies they may know that market products in your field.

2. **Recommendations from the manufacturers you interview.** When you interview potential manufacturers, ask them for recommendations on some of the companies

DON'T CONFUSE DISTRIBUTORS

Distributors are not specialists in any given field. Salespeople for distributors generally sell thousands of products and are excellent order takers. They "resell" small amounts of many products to many retailers or users.

In contrast, the sales partner you want on your team will handle a small number of specialized products and sell large amounts to a smaller number of distributors or large end users.

PRODUCT SUMMARIES AND PROSPECTUSES

When contacting potential sales partners and more importantly potential licensees, your New Product Summary and Prospectus will become important tools. How are you going to convey the opportunity to prospective partners if you can't talk to them in simple concrete terms? How are you going to convey market readiness and market potential if you do not have solid, concrete estimates? You never say silly words like, "Everyone can use it," or "It is useful in 12 million households." You have no idea that's true and rest assured these comments will only show your inexperience and lack of knowledge. If you are unclear about these facts go back to Chapters 10 and 11 and get better prepared.

they presently supply that are in the field of your invention. Having a marketing partner that is already being supplied by your manufacturing partner is a big plus!

3. **Trade shows.** This is one of the smartest things you can do early on. Go to a major national trade show as a visitor, not an exhibitor. Network with companies in the field of your invention, their national sales managers, presidents, manufacturer's representatives, and so on. Large numbers of independent manufacturers' representatives attend these shows and work in the booths of manufacturers they represent. Seek them out! The value of a trade show is enormous. It alone should give you a never-ending stream of sales/marketing contacts to pursue.

4. **Trade journals.** There are several options to pursue. First, you can place your innovation in the "New Products" section of the publication. If you do, make sure your innovation is well-developed and protected. Get an idea of what you'll need to provide in order to have them publish your innovation in a new products section or perhaps in a feature article. If you are looking for manufacturer's reps, look in the classified section in the back and see if any are advertising for new lines. You can also place an inexpensive classified advertisement seeking sales people. To find out about trade journals in your industry, ask a retail store manager or a wholesale products plant manager for an old issue. Get one or two of the most popular publications and then call the 800 number and ask for a complimentary copy. On the Internet, search google.com for trade journals in your field.

5. **Trade associations.** Every type of product, service, or field of business has some form of related trade organization it may serve. Not surprisingly, they tend to be very helpful and informative and can help locate active marketers (and manufacturers) and suppliers in a given field. The best way to locate associations in the field of your invention is once again google.com. Or, you can always visit your local library and read its publications.

6. **Walking the mall.** One of the best places to get a quick overview of suppliers and marketers in a given trade is to visit stores that carry the type of product representative of your innovation. For instance, if your innovation is a tool, you can visit Home Depot, Lowe's, your local Ace Hardware and True Value stores, Sear's, Target, K-Mart, Walmart, contractor's supply outlets, dedicated tool suppliers, automotive supply and even some drug stores and supermarkets. If you picked up just one new name at each of these outlets, you'll have a list of 13 names! Not a bad start.

7. **Asking a future customer.** If you have a certain customer in mind that you believe would sell your new product, contact the person in charge of marketing or merchandising if it's a retail product, or contact the head of operations if it's an industrial/commercial product. Ask the individual about companies that supply similar types of products or manufacturers' representatives who call on them. If you cannot get these names on a local level, try calling the district office and ask a district manager. For instance, if your innovation is a folding Christmas tree, by contacting Target Stores (or other similar mass retailers), you may be able to locate several companies they purchase their holiday merchandise from. Some may be importers; some may be domestic. The good news is that they are already suppliers to, or sales

FEATURE ARTICLES

If your invention is well developed and has some exciting potential, try contacting an editor with a trade magazine and have a feature article written about it. But be ready, because the response you'll get may be from those who want to purchase and make things happen now!

AVOID SCAMS

Inventors must avoid the marketing scams—those late-night television ads that ask: "Do you have an idea?" They're a national disgrace. Some $200 million will be spent by independent inventors on these bogus marketing scams this year—and they get nothing in return.
–*Don Kelly, Intellectual Asset Management Associates, LLC*

reps calling on, the retailer and if they can partner the marketing with your innovation, they're already in position.

8. **Asking distributors.** While you will not normally focus your efforts on seeking distributors, you may ask them who some of the better manufacturers' representatives are and who some of the more innovative suppliers are. Frequently distributors can also give you good inside information that'll help you sell your innovation to their customers. In other words, they can tell you what'll be required for sales presentations and incentives, ads, and so on.

9. **Yellow pages.** This is probably the least likely place to look, unless your innovation has a very local market. At times, you may locate manufacturers' reps from yellow page directories in major urban centers; however, the problem is that the directories do not list field of expertise.

10. **Internet search.** Progressive marketing companies and suppliers have Internet websites. The better manufacturers' reps do as well. Search through google.com and yahoo.com and start the review process. One of the best aspects of Internet searching is that you can quickly see a company's product line, review its marketplace, and locate all the key executives.

11. **Dedicated Internet websites.** There are several dedicated websites listed in the appendix that have some marketing ability in specific areas. However, these are only a fraction of what's really available. While these contacts may provide you with a starting point, you'll find there are ten others for everyone listed here.

12. **Thomas Register.** As always, you can look up various types of companies in the Thomas Register. The Register tends to be more manufacturing-oriented than marketing-oriented, so you'll have to take that into consideration. It's really an incredible resource and helps you pinpoint the executives names. On the Internet it's thomasregister.com and is free.

13. **Invention trade shows.** Many local inventor organizations have trade shows targeted to exposing their members' inventions. Generally speaking, they can provide exposure but are usually not visited by the decision makers of major suppliers or the retailers in a trade. If you are in business and plan to supply a general regional market, you may find some contacts. If you're seeking licensees or supply partners, this should be considered only one of the many means by which you pursue partners. There is one Internet trade show forum called the Invention Convention® that has had some success. Steve Gnass' website is dedicated to inventors and is another resource for inventors seeking licensees and networking. If it is in your budget, it may be worth the investment. In contrast to the above, we are unaware of any commercial success from attending trade shows put on by an invention promotion company.

14. **Small Business Development Centers (SBDC).** The government's best friend to innovators and start-up companies may be able to help, too. This may be particularly true if your product is to be produced locally. Since the SBDC deals with local innovators, it may be able to refer you to some. It may also be able to refer you to other local and state sponsored agencies that may be active in your field. Make sure to talk to your local SBDC consultant (Remember, it is free.) and see how he/she might be able to help.

INTERNET WEBSITE

If you have a patent pending or issued, consider building a website promoting your new innovation. Instead of just offering it for license, try posting some specifications, features, and benefits, plus photos showing how it works and an anticipated sell price. Make sure that interested parties can send you inquiries regarding availability, comments, and questions. This feedback can be valuable. Based upon the number of customer inquiries you get, you will find it a lot easier to sell your manufacturing and marketing partners on taking on the project. A marketing expert just surfing the web may also discover your website.

---

15. When someone says "no." In your search for a marketing partner, every time someone says "no, your invention is not for us" . . . tell them "thanks" and ask them if they could give you the name of some companies they believe may be interested. This alone can supply you with a never-ending list of potential contacts.

Whatever you do, make finding a marketing partner your priority. By doing so, you'll be gaining invaluable information about your innovation and the industry.

Think Creatively

When searching for potential partners and licensees, put on your thinking cap and search creatively to locate potential sales partners and licensees. For instance:

- Companies in related fields that may be interested in expanding sideways

- Companies with distribution in place that have seasonal business and see your innovation as a means to balance out sales

- Companies with cash interested in diversifying their product line

- Start-up companies with an innovative management team with the finances and desire to exploit new opportunities

- Aggressive manufacturer's representatives with sufficient financing to create their own brand and to either open a small production facility or have the innovation subcontracted

While networking, don't be surprised if you get a call from one of your prior contacts who has changed jobs and is now with a company looking for new products like yours to market! It happens. That's why it is important to tell the marketing experts in your field about your invention even though it may not be a fit.

DURA-FLAME

Inventor Podromos Mike Stephanos pursued the licensing of his orange oil-based charcoal lighter fluid and licensed a company whose business was primarily winter-based products—Dura-Flame. This was a perfect opportunity to give it a summer-based product that sold at exact opposite seasons to its winter-based instant light fireplace logs.

The Attributes of a Winning Sales Organization

For a sales partner to have the ability to market and sell your products, it must have all the right ingredients. There are ten factors to evaluate:

1. **Quality.** Is the company currently selling the same type of quality product you have invented? If not, they may not be capable of selling value-added products. Used car salespeople don't usually jump into the new car market selling Mercedes or BMWs with much success. Sales people who sell low-end generic products can't very easily change and start selling value-added products.

2. **Sales philosophy.** Does it have a defensive sales strategy, trying to hold onto existing business and market share? (Bad news.) Is it hungry, offensive, and aggressively looking for new markets to create? (Good news.) Do its leaders talk about the new markets they have created? (Good news.) Do they only talk about "picking up additional market share?" (Bad news.) When referring to your invention, do they say, "I can sell that?" (Good news.) Are they more interested in hearing what you can do for them? (Bad news.) Do they show the positive attitude, eagerness, and enthusiasm needed to launch a new product? (Good news.)

3. **Niche marketers.** Is this potential partner a small company, cutting out niches with its product line and holding onto them? Is its niche your niche as well? If the interviewee is a company that is losing market share and it thinks your new opportunity

ZIPPERS

The zipper was invented and patented in 1893 by Whitcomb L. Judson. He came up with the concept to help out a friend who had a bad back and could not bend over to button up his shoes. Judson was a native of Chicago who already owned over a dozen patents for mechanical items involving motors and railroad braking systems. His concept was to devise a slide fastener that could be opened and closed with one hand. The term "zipper" was coined after the sound the device makes by B. F. Goodrich, whose company started marketing galoshes using the fastener in 1923.

may just be the ticket to regain its share, turn around and run. Invariably, its bad habits will return, price will become the sole issue, and when pressed by customers to match existing price levels, it will not be able to sell your new innovation any longer.

4. **Track record.** What are some *recent* past successes on new product launches it has made? Salespeople and companies with stories about pioneering products ten, fifteen, or twenty years ago will never make it today. Almost anyone could have been successful launching a new product in an emerging marketplace several years ago. The fact that they have not had any recent pioneering successes illustrates their satisfaction with the status quo and inability to ramp up a new product. If the candidate has some recent successes, the chances are that it will be successful again with your innovation.

5. **Innovative attitude.** Is the sales candidate innovative in its marketing efforts? New inventions invariably require innovative approaches to marketing and problem solving. Your sales expert is your first and most important liaison with the distributors, retailers, and end users. You want partners who have the ability to make things happen, whether that be a suggestion to make countertop displays, hold sales competitions with distributor sales people, or get in the back door and bypass large up-front slotting fees (Slotting fees are frequently charged by major retail chains; the fee gives the customer the right to occupy shelf space in its chain.) The right company can address these very important issues.

6. **Ability to promote.** New innovations need a lot of promotion regardless of how wonderful they are. How is anyone going to even hear about it if it's not "told on the mountain"? Is the candidate willing and able to promote? Does it have positions at trade shows? Contacts at retail establishments and distributors? Can it help make sales and training videos? Develop sales brochures? While not all sales organizations have the ability to do this kind of work, many do. If they don't do this in house, will they outsource the job? If they can't, you will have to do it yourself, do it with the manufacturer, or arrange to have it done. Frankly, this is a job that only the marketing partner should do, as it is the expert, not you, and it should know what will make the new product sell best.

7. **Time and people power.** Does your marketing partner have the time and the people power to promote your product right now? If not, when? If it's not in the foreseeable future, forget it.

8. **Well-connected.** Is the potential candidate well connected with your potential customers' decision-making management? This means merchandising managers at mass retailers and sales managers at distributors if your innovation is retail based, or operations and product managers at industrial facilities as well as sales managers at a distributorship if your innovation is industrial/commercial. Decisions to change are made at the top, not in a buyer's office. Buyers only follow instructions from their superiors. If your sales partners only call on buyers, it may be tough trying to get your innovation sold. With industrial and commercial applications, you'll most likely need a company that also has the ability to interact with end users. Technical questions as to use may arise and they must be answered or they may not reorder. In

such a case, is your sales partner able to work with the end user should a problem arise? With industrial/commercial products, your sales partner must be willing to work with them so you can get the valuable feedback you need.

9. **Finances.** Is the marketing partner financially sound? Does it have the resources to do what it says it can do? Remember that innovative efforts require substantial time and money in the early stages to get them launched. Later, your sales experts will have the luxury of having protection because of patents on your innovation.

10. **Total, absolute commitment.** Will your sales/marketing expert make a total, absolute commitment to do a bang-up job with your product and do it now? Just as with manufacturing, you need a totally committed response to do all it takes to make things happen.

This is a tall order, but a marketing partner really should have all these attributes. Chances are, if they're successful niche marketers, they will! If they are short one or two of these attributes, will they still do OK? Probably just that—OK. But it is best to address the deficiencies ahead of time and see if there is something that can be done to fill in the gaps. Find out if they will consider changing their sales approach, dedicate sales reps, or even hire an individual with ramp-up experience to accommodate your innovation's needs. If they do, and they will make the total commitment to the change, you probably have a winner.

The Magic Words

This may sound trite . . . but you can just about be assured of marketing success if you hear "the four magic words." In your presentations to national sales managers and potential licensees, you'll be listening and watching their response. During the interview, if you hear the four magic words, they'll most likely do a very good job selling your innovation.

An inventor can plan, consult, and learn all there is to know about his/her invention through talking to prospective marketing partners. During these discussions, you may hear objections, preferred changes and modifications, and a myriad of reasons why it won't sell. Take it all in and be objective. Make the appropriate changes if they're proven valid. Better yet, if the prospect says the magic words, make them part of the decision making on those changes.

From this perspective, your planning now takes a positive turn based on pure marketing potential, which is exactly what you've been looking for.

The four magic words? Simple, you can take them to the bank. They are:

"I can sell that!"

The only remaining questions are who'll manufacture the innovation and how much can be sold?

Put it in Writing

Once you're satisfied with a prospective sales/marketing partner, you'll need to put some sort of agreement in writing. If you are manufacturing and supplying the products to a sales/marketing partner or an appointed manufacturer or importer is supplying the products, you'll need some sort of written sales agreement or supply agreement with your

GAUGES

There are no guarantees, but if a prospective marketer drives a luxury car like a BMW, Mercedes, or the like, chances are he/she is financially sound. If the expert is driving a 1983 Toyota pick-up, you may want to find out why. You may also get a gauge on how well a person is doing by the shoes he/she wears. If you've got the money to pamper your feet, you'll do it with expensive shoes. Interesting, but true.

sales team. If you are licensing the related patents, technology, and know-how, it'll be in the form of a license agreement.

Sales Agency Agreements

If you are hiring manufacturers' representatives for sales, you'll want to use some form of sales agency agreement. This may be nothing more than a simple memorandum of understanding or a more formal agreement. In either case, it should include certain important clauses.

This agreement is a legal document and establishes several facts. They are:

1. **Confirmation of the appointment.** It will state that the sales partner is being appointed as the sales agency for your innovation.

2. **Territory.** The sales partner has a specific region it covers. It may be a certain region in the U.S., the entire U.S., all of North America, or any other parts of the world.

3. **Field.** It may also include a specified field the sales partner may sell into and omit others—for instance, retail accounts, but not industrial commercial applications.

4. **Accounts.** At times, a sales agency agreement may include only specific accounts within a given territory. Should this be the case, then those accounts need to be listed, or the exclusions (house accounts) will need to be listed.

5. **Duties and responsibilities.** This basically says that the sales agency will do its best to secure credit-worthy customers and may include performance clauses establishing sales quotas. Typically this would be on a monthly, quarterly or semi-annual basis.

6. **No compete clause.** The agreement should include a clause stating that the sales agency cannot sell competitive products.

7. **Commissions.** This includes the amount of commission to be paid and when it will be paid. Usually it is based on net sales after the deduction of discounts, terms, and at times, freight. It also covers credits against commissions should a customer not pay or should there be returns. Commissions are usually calculated on a monthly basis, typically due the 10th of the month following the month in the accounting period. It can be paid on either the date the product was shipped and invoiced or the date the product was paid for.

8. **A time period.** The agreement should cover a certain time period so that if the sales partner is not performing as anticipated, you can cancel the agreement at the end of the period. This time period may be one year, six months or whatever the parties agree on. Likewise, if you or the supplier is not performing and shipping orders on time, the sales agency may wish to terminate the agreement.

9. **A cancellation clause.** Should either party default or decide not to continue the sales agency relationship, there should be an out clause stating the time period required in order to terminate the agreement. Usually this is from 90 to 180 days, but may be less; 30 or 60 days is not unreasonable in many cases.

There are other clauses that can be included, such as what to do in case of a disagreement or which state or country of jurisdiction is a venue for legal interpretation or settlement of disputes. Generally speaking, sales agency agreements are not too lengthy, not complicated, and frankly, they shouldn't be. At times, you will find that some manufacturers' representatives may prefer to work on a handshake agreement.

Buy-Sell or Supply Agreements

In some cases the sales partner may be doing the billing and will enter into a buy-sell or supply agreement with the manufacturer. This would typically apply with a private label arrangement or when a company's sales partner is going to be purchasing the product direct from you, the manufacturer, or an importer.

In such a case, an agreement that outlines your supply commitment, the sales partner's purchase (and sales) commitment, and perhaps some form of exclusive marketing arrangement may be in order.

This form of agreement may be a simple memorandum of understanding or a more complex document outlining the parties' obligations. It may contain many of the clauses in the preceding sales agency agreement and may be terminated should nonperformance be an issue.

The primary difference is that the sales partner in a buy-sell agreement is not relying on commissions, but on your or some other manufacturer's supply of products to it. Thus, there is no commission clause as income is generated through buy-sell profits instead.

LAUNCHING A NEW PRODUCT

*This can be the most difficult process of all. Just keep focused
on the means and you will have the end results you want.*

In a way, the term "inducing a birth" more accurately reflects what you are doing in the invention process than any other description. Just like a newborn baby, you want your invention to be healthy and to realize its potential. And just as you nurture your children and help them grow into well-adjusted adults, you will want to nurture your invention so it becomes a well-received, mature product. A child without proper parenting will not survive well in the world; a new invention or product also cannot survive without proper guidance.

In contrast to the similarity, there is one tremendous difference between raising a child and raising an invention. That difference is of great benefit to you as the inventor. When children mature into adults, they normally continue to cost you money. But a well thought out invention whose birth has been properly induced will earn you money! Nice difference.

Therefore, your mission, as the parent of your invention, is to induce a healthy birth, control its developmental environment, and give it guidance and direction to become a gainful, useful, successful product. Do this and you may be rewarded with wealth and long-term security.

You should always remember that as the parent of your invention, it is your primary responsibility to oversee its development, guidance, and direction. You must never stop guiding its development until it truly is able to continue on its own. Even then, you should continue to improve it.

In the process of inducing a birth, you must depend on many other people and companies to help get the job done. Work with your development partners, stay on top of developments, continually monitor and help direct your progress, and your innovation will become a success. In fact, when your team is all working together and you are focusing on the means, it happens almost magically. It's true . . . all of a sudden, you will see your innovation mass produced and making inroads in the marketplace.

The Launch Is a Three-Phase Process

There are three phases to inducing the birth of a new product and surviving. Trying to jump ahead too fast can spell disaster. The three phases are:

1. **Perfect the prototype.** As discussed in detail in Chapter 2, you develop simple prototypes that evolve into working models and then into production models. Keep in mind that in certain respects, your first sales may also be considered prototypes. This is particularly true with large innovations such as machinery. Your prototype testing begins with small controlled tests in house in preparation for full-scale tests and customer trials. Do not enter into customer trials until you and your marketing experts are very confident the final prototyping phase has been accomplished. Your final prototype should closely resemble the first production models.

2. **Prove it with working models in consumer trials and test marketing.** Typically you will do this through a limited production of working models. Customer trials and tests are conducted with unbiased end users. Start small and then expand sufficiently to prove and repeat results. One of the most important outcomes you should look for during this phase is consistency. Performance must be repeatable throughout the consumer testing process. One of the best ways to ensure this outcome is to oversee customer trials hands-on, if at all possible.

3. **Ramp it up!** Even though this term suggests massive sales, it really means to gradually expand sales instead. If there is some oversight, it should come out in the early stages. Sales can be ramped up within a customer's operations or within a sales region. Once you are confident and repeat orders are flowing, expand nationwide and hopefully internationally after that.

Perhaps the most important key to making it through these three phases is to make sure your innovation has been well tested and that any last minute changes and modifications have been made to satisfy all customer needs. One simple oversight can spell disaster.

Phase One: The Market Approves the Final Prototype

As you know, prototypes begin as crude models to prove the concept and then go through a metamorphosis to the invention's ultimate marketable form. The final prototypes are those that have sustained extensive tests in your lab prior to consumer testing.

In your marketing-related activities, define "testing" as those tests you conduct *before* you allow end users to try the invention on their own. After you have completed a series of obvious tests, you will want to set up hands-on testing using unbiased consumers and in concert with your marketing partners. This testing will probably begin in the lab or at a plant if it's an industrial/commercial innovation and then may continue at a user's site, if necessary.

The importance of thorough testing may be more beneficial with your marketing expert involved, because that expert is the one that has to carry the project forward from here. Consistency is extremely important at this phase and you will want to observe these final tests with your marketer before the innovation goes into regional trials and test markets.

Never introduce marginal-quality products into the marketplace. It will only result in a negative experience being associated with your invention, regardless of whether the problems are subsequently resolved. Remember that first impressions are hard to change.

Phase Two: Consumer Trials and Test Marketing

After you've achieved the desired test results, you will be ready for some customer trials. Use the word "trials" when talking to new end users, not "tests." Most people don't want to be guinea pigs with new products. At this phase, your marketing partner can accurately say that the testing is done and now is the time for customer trials.

Many times consumer trials are actual sales. Thus, you can use your previous final prototypes (or first-production working models) to initiate sales with friendly retailers or industrial/commercial customers. Begin with a small trial first. This could be a single retailer for retail products or a single plant or business for industrial/commercial applications. Initially, you should observe the consumer trials if at all possible. Customers will invariably use products differently than how you and your guinea pigs used them in a lab. Because there are so many unknown contributing factors to consider even at this late hour, it is best to start out cautiously.

If something needs fixing during these broader customer trials, fix it quickly. Communicate directly to the plant and to those who will make the necessary changes. The chances are that anything that needs fixing now may not be of great consequence, but fix it anyway. Do your best to resolve any potential customer complaints before they arise. If you do, your customers will be impressed and will reorder.

During customer trials, be friendly and enthusiastic. Do not hype the new product, but be encouraging to the users if you are in a position to do so. The positive attitude you convey helps get them involved and makes them part of the success equation. They'll be much more likely, eager even, to help solve problems and give you honest feedback.

If your customer is industrial or commercial and has a training department, get them involved as well. You can "train the trainer," so they can take over afterwards.

Last, consumer trials are a great opportunity to have your sales partners observe the results. You can be sure that the experience will whet their appetites! By involving your sales experts in these start-up consumer trials, they are literally being trained to take over and expand sales when the time is right, or to train the trainer should it be required. It is usually during this phase that an innovation either makes it or goes bust.

Striving to Make It Right. Just like good genetic engineering can determine a baby's future, you too want your new invention to be well engineered. It's a fact: your first objective is to consistently make your innovation right so it will perform consistently. Frankly, there are two approaches to accomplishing this. Make it right from the beginning before market release or stumble around trying to make it right later and most likely see the innovation's ultimate demise.

From this perspective, your primary manufacturer must be totally committed to solving

WORDS TO INVENT BY

Measure twice, cut once!
–George Margolin,
"words to invent by . . . "

USING SEMINARS FOR TESTS AND TRIALS

Seminars can be an effective way for you or your marketing expert to introduce a new innovation. You can provide them in industry, at trade shows, or at customer locations as a means to help get the word out or train employees. In some cases you can establish a home-based business and use seminars as way of introducing your timesaving, productivity-enhancing, self-teaching innovations. Not only do you sell them at the seminar, but if you give them at colleges and universities, you can get paid for the seminars as well. The feedback you will get from your participants will be incredibly valuable! It'll virtually ensure your ultimate success.

**WORK YOUR PLAN . . .
AND KEEP FOCUSED**

Once you have a concept
and market defined, target
your market by effectively
making a plan, then work
your plan, and stay with it
'til hell freezes over—or at
least until you make more
money than you know
what to do with. Then
call me!
*–Doug English,
Patent Attorney*

all the start-up problems, while continuously striving to make the perfect product. You had this commitment from them before you started and now they will be living it.

As you work together as a team, you must always remember that you and your marketing partner are the key interpreters of quality. Everyone will look to you to determine if the quality is acceptable. The quality you want is really nothing more than the quality you believe the customers expect. Do not confuse the customer's expected quality with your personal opinion or the opinions of plant employees or a minority of salespeople. What the masses expect is what you want to give them.

How do you know when the quality is right? There are several signs you can read from your customers' responses to determine if your product's quality, performance, and price is acceptable. Here is a list of some of those signals you should look for during tests and trials:

1. It works the way it is supposed to.

2. Consumers know how to use it.

3. Consumers smile when they use it.

4. Your intuition tells you it is right.

5. It raises a positive emotion with the users, the marketers, and those who are associated with the trials. Remember the importance of elicited emotions in new inventions!

6. They ask where they can buy more.

Before you release your new product to the mass market, you must be comfortable that it is made correctly and performs consistently and that your marketers are ready for a major sales effort.

Recognizing Red Flags. More important than knowing when your new invention is made right and performing well is knowing when something is wrong. If you spot problems and difficulties, your innovation is not ready to release to the marketplace. Be sensitive to red flags that tell you something is wrong. Remember how red flags were defined in Chapter 3? These warning signs are more important than ever now that you're conducting consumer trials.

**RED FLAGS =
OPPORTUNITY**

If you have an established
relationship with your
marketing expert, you
know that he/she can
point out many red flags
long before you will ever
discover them on your
own. Never overlook
these wonderful
"blessings in disguise."
When you solve them,
opportunity abounds!

Don't downplay and avoid agonizing little problems. Overcome them and they will become a gift in disguise. Here are some things to look for:

• Consumers find it difficult to use or difficult to figure out how to use.

• It does not do ***everything*** it is supposed to do or does not do it quite right or consistently right.

• It does not look quite right—perhaps the color is wrong.

• It is too difficult to make in order to obtain consistent performance. Do not confuse difficulty to make with "going down the learning curve" associated with manufacturing new products.

• It raises a negative emotion with users of the product.

• It is not being considered a significant enough improvement over existing products . . . "I like it, but I don't know if I would switch."

• There are no reorders after the first trial.

These kinds of problems must be resolved or understood before you move forward.

Most times, just knowing what the problem is and knowing that it can be resolved will set the gears in motion. Being this close to success, your manufacturing and marketing partners will want to get these last-minute problems resolved and prove it can be done.

Another excellent means to get honest answers and feedback on a new product is by conducting a blind survey, in which those surveyed are anonymous.

Conducting a Test Market. There are two primary choices when test marketing. You can either pursue an easy test market or a tough test market. There are benefits to both. Conducting test marketing in both scenarios is beneficial since it'll give you two separate perspectives on your innovation. A tough test market may not be advantageous if it is based on certain local prejudices, so you'll want to determine what type of test market fits your target market best. Likewise, an easy test market may not be advantageous for several reasons. For instance, regional acceptance may be prejudiced or a given region or customer may not handle competitive products that are sold elsewhere (where the bulk of your sales will ultimately be made anyway); thus, they won't give a qualified comparison or evaluation.

Test markets are usually conducted on a regional basis. Your marketing partners and you will want to select those regions and those retailers or industrial/commercial customers that will give you the information you want so long-term decisions can be made towards capturing the entire target market. There are many good test market regions in the country . . . and some not so good regions.

Generally speaking, regions such as northern California and the Northwest are good test-market regions, as is much of Florida and Texas. Consumers in those areas tend to be more receptive to change. In contrast, areas like southern California and New York may be poor test-market regions. Price is king in southern California and in well-established regions such as the Northeast, change is not as popular. However, if your innovation is one that is made of plastic and will be taking business away from existing paper products, you probably would not get a fair test in a region that relies on the paper industry; thus, in this example, you'd most likely not want to test market in Oregon and Washington.

Since your marketing partner will be selling the new products, it should play a major role in deciding where test marketing (or consumer trials) should occur and how extensive it should be. They should select the first retailer or end-user customer, too. Test markets may involve as few as three to four retail outlets in a national or regional chain, or it may be an entire district. Chain store districts tend to be from fifteen to twenty outlets and are run by a district manager and staff.

It's usually a fairly easy task for an experienced sales rep to get a test market in a chain store. Most districts have specified stores they use for new product testing. Others may be willing to test an entire district.

If you have anything to say about it, a test market should be for an absolute minimum of thirty days. If someone wants to only "try a few," don't accept it. Insist on a longer time frame and give a guarantee that any product that doesn't sell can be returned. It's best to ask for a sixty-day time frame and to accept thirty days if need be. This will give you and your sales team ample time to make any necessary adjustments, such as counter displays and so on.

There is an important pyschological factor in establishing a thirty-day test. First, new habits take about three weeks to take effect. So a one-month test market means that a

WHO INVENTED THE CHIP?

Chip? What chip? The potato chip was invented by chef George Crum in 1853. A dinner guest found the chef's French fries too thick for his liking and rejected the order. Crum then cut and fried a thinner batch, but these too met with disapproval. So Crum decided to rile the guest by producing French fries too thin and crisp to eat with a fork. The guest was delighted with the browned, paper-thin potatoes and other diners wanted them too. As for the microchip, it was invented by two inventors living in two different parts of the country. Jack St. Clair Kilby worked for Texas Instruments in Dallas and Robert Noyce worked at Shockley Semiconductor Laboratory in Mountain View, California. U.S. Patent No. 3,138,743 for Miniaturized Electronic Circuits was issued to Jack S. Kilby and Texas Instruments in 1964, whereas U.S. Patent No. 2,981,877 for the silicon-based chip was granted to Robert Noyce in 1961, who later founded INTEL Corp. in 1968.

new habit has only been in place for about 10 days or so. This phenomena is particularly important with commercial/industrial products. You want the end users to become familiar with your product. After thirty days . . . even if it is marginally better, they'll not want to switch back. Plus, this is ample time for you to make some simple improvements so the innovation is "more than 'just marginally' better."

Ideally a test market is conducted with products that are sold to the retailer or industrial/commercial customer. At times, this may not be the case, but usually it is. Make sure you and your marketing partners discuss this in advance, so that there are no surprises when the question comes up, "Who's going to pay for the goods?"

Conducting Consumer Surveys. Consumer surveys can be an invaluable means to uncover problems and opportunities. They may provide an excellent means to discover how consumers and users value your innovation.

When conducting a survey, you may delineate what objectives you'd like to achieve. However, to have a completely unbiased survey, you should have the survey conducted by some form of third party. In the case of an industrial/commercial innovation, the company itself may generate an internal survey addressing those concerns; however, this too may be biased if the individuals preparing the questions have any degree of prejudice on the outcome. I think you can imagine how a particular buyer may slant questions if he/she has been loyal to a competitor for many years.

There are two excellent ways to have relatively safe surveys conducted by unbiased parties. First, there are professional companies that do this all the time. Do a google.com search and you'll find them. Second, contact a local university and ask their business department head if they participate in consumer surveys. Many major universities do this and use graduate students as survey takers. They tend to be very experienced and tend to produce qualified results for a fraction of the cost of professional surveying companies.

Regardless of your choice, you and your marketers must be in accord with the content being surveyed and exactly how the questions are written and exactly how they are asked. It's simply too easy to slant a question by the simple change of one word, or by a simple inflection in a person's voice. With written surveys, it is usually smart to make sure there is ample space for the survey taker to write his/her comments and suggestions. If you see comments that are repeated throughout, you may have missed something in the question section.

Survey results after a test market provide excellent sales material and even testimonials for your marketing partners.

Recognizing Red Flags Unrelated to the Invention. Being sensitive to red flags throughout the development process is important. But all these red flags are usually product related. Now, during the consumer trials and ramp up, you will need to be in tune to warning signs that are not necessarily relevant to your invention's performance. They, too, can kill a product's successful introduction. Examples of unrelated red flags may be:

1. Service issues, such as the manufacturer's lead time is too long when placing orders
2. Shipping issues, such as the warehousing department does not like how the product stacks up in the warehouse

A $12-Million Red Flag

Chicago, January 17, 1991. On a cold winter day in early 1991, I traveled to the Chicago area to oversee a test my associates and I were conducting at Sears. The test was for a new bagging system we were developing and the stakes were high—$12 million in business. During the past week, and on the flight to Chicago, I had a nagging feeling about a "red flag" emerging over the performance of the new system. I couldn't put a finger on it, but I knew something was not quite right.

When I landed in Chicago, it was snowing hard. I knew I had just enough time to rent a car, drive to the Sears test store, see how things were going, and race to the hotel before I got snowed in. In the test store, I carefully viewed the clerks using the new bagging system and tactfully asked them questions regarding its performance and what they did and did not like.

During the store visit, the red flag continued to nag me. While the clerks liked the new system over the other two systems we were competing against, to me it was just not right. Finally, I had to leave the test store and get to the nearby hotel before it was too late.

At the hotel, the snow pelted the window and built up to form a solid white block of ice. I was feeling depressed because the new bagging system was just not quite right and we really wanted this contract. After I walked around the hotel for awhile and then returned to my room, a light suddenly went off.

The red flag was a warning that the use of the system was not natural. It would require some training to use effectively. Then, the videotape in my brain began and I visualized what would be more natural. I visualized a system that dispensed "backwards" in a simple one-step fashion.

We made the changes and after two months of testing against two other competitors, which were larger, more generic-oriented companies, the survey revealed that we were the choice of the employees and management. We got the contract from Sears. We then filed three patent applications to protect our interests.

3. The retailer, distributor, or customer does not like your choice of vender or the sales rep
4. The retailer is not satisfied with its profit margin, or it makes more money on other competitive products
5. The accounting department does not like the payment terms

These kinds of red flags may be more difficult to overcome if there are some personal issues between your marketing partners and distributors or retailers. Nevertheless, try to be supersensitive to the comments you hear. As some may say, "Try to be the bug on the wall" and listen to all the unsolicited comments. Then fix problems whenever you can.

Phase Three: Ramp it Up!

This is what you've been waiting for! After you and your marketers are comfortable with the test marketing (or consumer trials), it's time to expand sales. Expanding sales may begin as soon as one to two weeks after an initial test market begins. It may take as long as two months and at times, as long as six months. It all depends on the retailer or end user.

Depending on your strategy, the expansion of sales may mean increasing sales by adding on additional retailers and/or distributors in the test market region itself, or expanding sales to other regions and other customers. If you started with three to four outlets in a district, your marketer will certainly be asking the district manager to expand sales to the entire district.

Once this objective is achieved, your marketers will expand throughout the chain and will be looking at applying some offensive sales tactics. Even during this expansion phase, continue to be supersensitive to any red flags that may arise.

Expansion of sales is usually slow and methodical . . . slower than what we want. But there is a reason for this. Retailers and consumers are cautious for one. Another reason is that the retailers or industrial/commercial customers may also have to use up an existing inventory of competitive products they've purchased. They may even have to honor a contract that may have several years remaining. Last, ramping up sales must also be coordinated with production, and production needs time to gear up and build inventory. This is not going to happen overnight and must be coordinated with the production facility. Only after sufficient production capacity is available can sales be expanded into larger regions, nationally or internationally. If demand is great, it will only help keep prices and profits up. Fortunately, these are good problems to have and are ones that your marketer will be handling, not you, the product developer.

During this phase, your sales organization may be ready to "blow the top off of sales," but they may need to be restrained until your manufacturer can gear up for added supply. In your piloting position behind the scenes, keep the pressure on the manufacturer to meet production deadlines while you gracefully work with your sales team and keep a focused effort.

Continuous Improvement to Ensure Product Longevity

The term "continuous improvement" was coined as an integral objective of Total Quality Management (TQM) processes. Continuous improvement is also your theme to guarantee continued success. By employing the concept of continuous improvement, you will be reinforcing future product security. By employing the concept, you will stay ahead of the competition and open the doors for other new products as well. Why?

1. Having the end-user in mind and improving your products accordingly will keep consumers buying them. Remember the words, "new and improved." This is right at the heart of continuous improvement.

2. Continuous improvements may not all be patentable, but every time you make your innovation a little better, it can only improve sales. Should you make new discoveries that are patentable, then obviously you are extending the life of your proprietary position. It's another twenty years of patent protection!

3. Even if your first patent does not have as strong coverage as you'd hope for, you will still be light years ahead of competitors. Improvements may result in a breakthrough opportunity that even with a narrow-scope patent will wield a lot of power.

RAMP-UP

An important factor when designating a manufacturer for a new innovation is its ability to supply large quantities quickly. Why? Well, what happens with a successful product launch with large users is that after they test it and they like it, they'll want to ramp it up nationwide. The producer has to be able to provide a reasonable timetable do this. If they can't, then the test will die and maybe the product line. If the manufacturer of your innovation is a small company and you expect your innovation to sell to a potentially huge user, how is this small producer going to supply them? My reputation is on the line. I would never knowingly commit to selling a new innovation to a huge end user without qualifying the manufacturing ability of the producer to ramp up once we have been successful.

–Joe Truckload Bowers, Marketing Expert

4. New improvements also include breakthroughs in production processes, which mean lower costs and putting competitors at a further disadvantage.

5. New improvements build incredible brand loyalty as well. When your customers are accustomed to knowing that you're continually making your product easier for them to use, or more beneficial to them, they will stick by you. Just ask Microsoft . . . about its software . . . with its automatic updates . . . discounted new versions. Get the message?

During the process of expanding the market, you should always be observing the product's performance and acceptance. You should always be looking and listening to customer input and using it to your advantage. Once you're established, you can improve and continually maintain your market edge through the same innovative, inventive process that got you there in the first place—that is, your CDI focus and being supersensitive to customers' and end-users' true needs.

The "continuous improvement cycle" from Chapter 4 pertains to product development. Here it is again as Figure 13.1, but as it pertains to the process of continual improvement of an existing innovation.

STAYING CLOSE TO THE CUSTOMER

The Whirlybird, produced by Ortho created a market as a hand seed spreader. When patents ran out, the trademark was what they relied on. Competitors entered the marketplace with cheap knock-offs, but with Ortho's experience in the business, it countered by increasing the guarantee all the way up to 10 years, while the imitators were too afraid to do so! Ortho maintained marketshare.

Figure 13.1 Continuous Improvement of Innovation

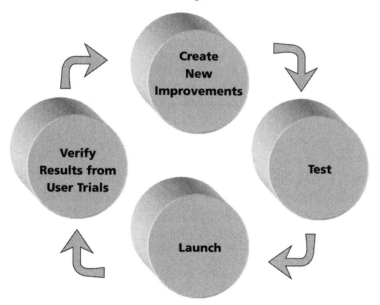

You've heard throughout the book the importance of improvements. They almost always enhance the ability of your invention to become the preferred product in the niche, regardless of competitive responses. Just like the single most important characteristic of successful people is to continually improve themselves, the same can be said of new products—to continually improve their performance and increase their benefits.

Selling Into the Future

After ramp up, your challenge is crystal clear. It is to continually improve and develop new creations so your sales team can continue to expand and improve sales. What you've

The chromium-nickel alloy used for resistance wire was discovered and patented by Albert Marsh in Feb. 1910. This discovery was essential for toasters, as we know them today. It was developed by Hoskins Mfg. Co. and licensed to various toaster manufacturers. It wasn't until Oct. 1915 that James Lamb of L.F. & C. patented a simplified toaster that resulted in a more durable, inexpensive appliance, which was the forerunner of the versions now commonly used. The development and mass production of the toaster in the United States saved time for housewives in the kitchen and led the way to the design and development of the huge number of household kitchen appliances used in the kitchen today.

created is "customer pull through." There's no better way to sustain sales. Sales may be expanded in a number of ways. Here are a few possibilities:

1. Offer more sizes . . . small, medium, and large, not just one

2. Introduce new colors or designs

3. Provide better graphics

4. Initiate a 24-hour delivery policy

5. Provide local warehousing for immediate pick-up or delivery

6. Issue coupons to encourage repeat orders

7. Create high-tech sales presentations with concrete customer feedback (Remember, no one else will be able to do this if you are in the leading position.)

8. Develop co-op advertising campaigns

9. Sponsor golf tournaments for customers or within a region

10. Give local or regional seminars on how to use or sell your innovation

When Abnormally Strong Growth is Anticipated

Highly successful innovations tend to have different problems, the most important being adequate production in a timely manner. If adequate production is not possible with an existing supplier, there are really only three good alternatives to consider:

1. **Subcontract supply.** If your company or manufacturing partner cannot handle the volume, it's most likely written into the contract anyway that you can pursue alternative sources. What's most important is that the quality of the subcontractor is exactly that of the existing quality. You can't skip a beat in this transition. Even the shipping cartons, marking, and so on must be identical. If the augmented supply is from overseas, it's even more challenging.

2. **License or authorize additional producers.** If you've licensed your innovation and your first (present) licensee can't handle the volume, you should be able to do one of two things. First, is there a subcontractor that can fill in for the demand? Do you or your licensee know of a company that can augment supply in order to fulfill demand? Second, you may have to license a competitor to ensure adequate supply. An existing first licensee is probably going to be protective of its business. Perhaps it may suggest a supplemental licensee that is not a threat, such as a company in another region where the present licensee cannot ship cost effectively. Keep your present licensee abreast of your discussions with others.

3. **Hold off sales.** Can you hold off on expansion until the present producer can purchase more equipment? This is not always a good choice. When something's hot, it's hot! However, sometimes a delay only builds up customer demand . . . "We can't keep them in stock." If the delay is too long, consumers will find alternatives, so make sure the delay is acceptable.

Frankly, this is every product developers dream, so if you're burdened with these problems you're successful.

Take Time to Celebrate

Hard-charging innovator-inventors need time for leisure and relaxation. After all, inducing a birth is a lot of work! Take time to enjoy yourself in the process whenever you can. Inducing the birth of a new innovation may take you to new places and you may meet many new people. Enjoy them and enjoy the scenery while you're there. Besides, during rest and relaxation, many new, exciting ideas and discoveries may develop.

PART 4

LICENSING

LICENSING BASICS

*Licensing is the best way to mass produce and
market your ideas without spending a lot of money.*

Licensing appeals to many businesses and inventors because it solves many of the problems associated with product development and the marketing of innovations. Licensing can be a welcome alternative to spending hundreds of thousands of dollars and thousands of hours gearing up to go into business. Licensing is also an excellent means to supplement a company's income by licensing the patent rights in fields outside the company's core business.

Licensing an invention and the related patents can be relatively easy, once an inventor has two qualified prospects willing to partner. Those two prospects are a manufacturing partner and a marketing partner. It's irrelevant if the two entities are two different companies. With a marketing expert eager to exploit sales and a manufacturer eager to produce, licensing your proprietary rights is a relatively straightforward matter.

Businesses and inventors considering the licensing of their inventions should learn all they can about licensing before entering into an agreement. If licensing is your choice, take the time to learn the many nuances of licensing. That will open doors not only for your present innovation but also for future ones as well.

Basics of Licensing

What Is a License?

A license is nothing more than a legal agreement, a contract between two parties. Just as a patent conveys rights to an inventor, a license agreement conveys those rights to a licensee. Simply stated, a licensor (you, the grantor of the rights) gives a licensee (the

IT'S ALL ABOUT MONEY

Licensees don't license ideas—they license MONEY! If you can convince a licensee they will make money from your invention, they really won't care what your invention is. Forget the invention details and focus on making sure you can show the licensee how they will make money.
–Andy Gibbs, CEO, PatentCafe.com

recipient of the rights) the right to manufacture, and/or sell, and/or use your invention and the rights covered under the scope of its patent(s). Either party to a license agreement may be an individual, sole-proprietorship, corporation, or other form of legal entity.

A license is literally a written authorization for a licensee to make, use, or sell products under the scope of your patents and intellectual property. It sets up the terms and conditions with which the licensee must abide in order to exploit the sales of the invention and to keep the license in force.

The key terms and conditions of a license typically include:

1. A list of the intellectual property covered under the agreement

2. The payment of royalties

3. Maintaining manufacturing and sales records

3. Marking of patent number(s) on products

4. Maintaining product liability insurance

5. Provisions in the event of the company's sale

6. Provisions in the event of bankruptcy

7. A time frame

It may also include:

1. Acceptable minimum quality standards (important with trademarks)

2. Minimum sales volumes

3. Minimum royalties

4. A right to subcontract

5. The right to new developments related to the invention (although this is usually not recommended)

In a license agreement, the provisions should be clearly written so they cover your needs and the needs of the licensee and avoid any ambiguity.

What Rights Do Licenses Grant?

Licenses almost always grant some form of proprietary property rights. This is usually in the form of intellectual property such as patents, but it may include trademarks, trade dress, copyrights, trade secrets, customer lists, and know how. If a patent pending is licensed, then the licensee is licensing the rights in anticipation of the appropriate patent rights being issued at some later date.

It is not necessary to have a patent granted to execute a license agreement. In fact, it is usually wise to license your invention during the patent pending phase and not wait until after the patent is granted.

Why Are Licenses Needed?

Frankly, licenses are not always needed. If you are going into business, you'll not require a license since you own the underlying patent rights (or intellectual property) anyway. In certain rare occurrences, licenses may not be necessary, as may be the case between good friends or family members. However, it's still advisable to have some form of written license agreement in place in the event one of the parties ceases business operations or dies.

FIRST OPTION

Licensing is usually the best option for inventors. It is much cheaper to first try to license your invention and then attempt to manufacture instead of vice-versa.
–Michael S. Neustel, U.S. Patent Attorney

THE BEST WAY TO PROFIT

If you think that licensing is not the best way to earn money from an invention, think again. Prolific inventor Jerome Lemelson licensed his patents (over 500!). He was the founder of the Lemelson Foundation, Lemelson National Program in Invention, Innovation and Creativity and the National Collegiate Inventors, and Innovators Alliance.

FACT

It is very difficult to find a willing buyer for an unproven patent. There are too many variables that can affect the patent's value. Besides, if you sell your patents, you are probably selling out for a lot less than if you licensed them.

At times, license agreements are multipage documents negotiated by the parties or their attorneys. This might be the case with the sharing of major technology, such as reactor processing technology to make plastic resins or the "cold draft" brewing process of Sapporo in Japan, which Miller Brewing uses to make Miller's Genuine Draft.

However, licenses do not have to be long, legal documents. Some may be as simple as a two or three pages. What is most important is that the license clearly covers the subject matter to avoid misunderstandings between you and your licensee. Most license agreements for most products are about five to ten pages in length, which should be sufficient to spell out the parties obligations and to avoid any misunderstandings.

Many misunderstandings tend to involve the reporting and payment of royalties. With simple built-in safeguards, this can be avoided. Other misunderstandings may deal with the transfer of a license agreement or the validity of the patents. Your license agreement should contain a simple method of accounting and verification so later you will not think you are being underpaid; this is discussed later. If a company changes management or is sold, you want to make sure that payments and adequate product quality continue. The best way to do this is with a license agreement that contains the proper language. In the event of a licensed patent not being valid, it is usually a matter for your attorneys to address. This may be covered in a license agreement as well.

Types of Licenses

There are four types of licenses you can use, one of which is by far the most common with the licensing of patents. The approach of whether to grant your patent rights exclusively or nonexclusively is your second consideration.

Generally speaking, the four types of licenses are:

1. **A license to manufacture, sell, and use.** This is the most common type of license; it grants all your underlying patent and IP rights to the licensee. It may be used with a single entity you've licensed that will be both manufacturing and marketing your innovation. It may also be used with a marketing partner or a manufacturing partner even though the manufacturer will not be marketing your invention to the trade, but to or through a dedicated marketing partner. In such a scenario, both may be licensed, but with separate responsibilities. With this type of license arrangement, the licensee pays you royalties based on the number of licensed products it sells. If a marketing partner and a manufacturing partner are both licensed, then typically only one pays you royalties. When the manufacturing and marketing partners are the two different companies, there are usually better ways to accomplish this objective, which discussion follows.

2. **A sales-only license.** This type of license may be used when licensing a sales partner that is not going to be producing the innovation, but will be purchasing product from either your company or an authorized manufacturer, or at times, will be an exclusive sales arm of the manufacturer and work on a commission basis. If the sales-only licensee is purchasing product from an authorized manufacturer, you must also issue a manufacturing-only license (or a simple manufacturing authorization) to the producers you've appointed. Either the sales-only licensee or the manufacturing-only licensee would typically be designated to pay the royalties in this type of licensing arrangement. The primary distinction as to which one pays royalties

---

PLYWOOD

Around 3500 BC the Egyptians first thought of sticking several thinner layers of wood together to make one thick layer. It was originally tried during a time when there was a shortage of quality wood. Thus they glued veneers (very thin layers of good wood) over the "not-so-good wood." Plywood as we know of it today was invented by Emmanuel Nobel, the father of Alfred Nobel, the inventor of dynamite and founder of the Nobel Prize. Emmanuel found that several thinner layers of wood bonded together were stronger than one single thick layer of wood. His idea and methods of making plywood are now a standard in the construction industry.

usually rests with the entity that will be billing the merchandise to customers in the trade. For instance, the sales-only licensee pays royalties if it is purchasing licensed product from the authorized manufacturer. The manufacturing-only licensee pays royalties if it is billing the licensed product to the trade customer that is sold by the sales-only licensee; in that case, it would also be responsible to pay commissions to the sales-only licensee.

3. **A manufacturing-only license.** A manufacturing-only licensee is designated only as a supplier to or through a "sales-only" licensee. The one exception to this rule occurs when the manufacturing licensee is billing the product or sales made by the sales-only licensee. In this case, the manufacturer pays the royalties. A manufacturing-only license may also be used as a means to augment the production of another licensed manufacturer. This may also be accomplished as a sublicense arrangement between a licensee and the subcontracting supplier, but it is best you are the one who officially licenses the subcontractor as it provides better safeguards. Technically speaking, a manufacturing-only license still allows the manufacturer to sell the product (to a sales-only licensee), so you can say that a manufacturing-only license allows the licensee to sell product as well. Another way to accomplish the manufacturing-only relationship is to provide the manufacturer supplying product with a sales-only licensee with a written authorization to make product under the scope of your patents (and IP), but only for sales to an authorized licensee. Essentially it is the same thing as a manufacturing-only license, just titled a bit differently and usually not as formal.

4. **A license to use.** This is commonly granted in computer software sales. When you purchase software, you enter into a license agreement upon installation of the software into your computer. This license agreement states that the software is for your personal use only and is not to be copied, manufactured, or sold to anyone else or for the use of any others.

If you are subcontracting the production of your innovation for sales through your company, a license is not usually needed. All you have to do is place orders based on your specifications. At the least, you'll certainly want to have a memorandum of understanding making the producer aware that the product is patented and that no sales may be made to any company other than yours.

Try not to make these various forms of licenses too complicated. Generally speaking, when licensing patents most objectives can be accomplished through the use of the first license type—that is the license to manufacture, sell, and use.

Approaches to Licensing

Exclusive Licenses

Any of the preceding license types may be granted on an exclusive basis. That means that the licensee has the exclusive rights, regardless of the license type. For instance, an exclusive right to manufacture, sell, and use means that there will be no other licensees. Or, an exclusive license for manufacturing-only means no other licensees may manufacture the innovation. For an exclusive sales-only license, there will be no other sales licensee. An exclusive license to use means no other entity may use the patented technology.

Exclusive licenses are generally best for niche market products. It gives the licensee an incentive to invest its time and money to be your partner and protects the licensee's interests and investment. If you are granting an exclusive license, the license must also have built-in protection for you, since you will be relying solely on a single licensee for your income (or production in the case of an exclusive manufacturing-only license).

Exclusive licenses provide the licensee with an incentive to make things happen because the licensee has proprietary patent protection. In return, these licenses usually include performance clauses based on guaranteed sales volume, minimum royalties, and a time frame to get the licensed product to market.

It is common to grant an exclusive license based on different fields. For instance, you may grant an exclusive license under your patent rights for a company to sell and market in the retail trade and an exclusive license to a separate company to sell and market in the industrial/commercial trade. When granting an exclusive license, it is common to include up-front money and have long-term guarantees on sales and income.

It is also common to grant an exclusive license based on different regions. For example, you can have one exclusive licensee for the eleven western states and another for all those east of the Rocky Mountains. This may be necessary with products that have high shipping costs that may negatively affect sales potential.

Last, exclusive licenses may be for a specified time period. This gives the licensee a jump on competitors. This is one way of overcoming a start-up licensee's objections. For a specified time, it will have an advantage in a marketplace that must eventually have two or more nonexclusive licenses as may be demanded by the trade. This type of license agreement is typically referred to as an "Exclusive and Nonexclusive License Agreement." In other words, it is exclusive at the onset and becomes nonexclusive after a certain time period.

In addition to the standard licensing clauses, it is essential that exclusive licenses include:

1. Guaranteed sales
2. Guaranteed royalties
3. Immediate recourse if sales and royalties are not met

Exclusive licenses are not usually advisable for commodity-oriented products. For example, exclusive licenses destroyed Sony's Beta Max and allowed the VCR alternative to dominate the video market. But if you are carving out a specific commodity niche that you want to protect, it may be best to have a single, dedicated exclusive relationship.

Nonexclusive Licenses

If you plan to license several users, you will use a nonexclusive license. Nonexclusive licenses are typically used for commodity-type products, such as an aluminum can's flip top. Take care to license nonexclusive licensees at an "arms-length." Manufacturers tend to be very possessive of their sales and customer base and would frown on your showing favoritism to a particular licensee. Nonexclusive licenses will usually contain:

1. No sales minimums
2. No guaranteed royalties, just a set rate
3. Recourse in the event of nonpayment, infringement, etc.

FLIP TOP CAN

One of the most prolific nonexclusive licensing arrangements came from the invention of the pull-tab aluminum can that we know today. It was invented in 1962 by Mr. Ermal Fraze in Kettering, Ohio. If you recall, with prior art pull-tab cans, the pull tab became detached upon opening and was usually discarded. Environmentalists roared when problems arose. This included litter in our national parks, animals eating the small aluminum rings, or the rings getting caught in the beaks of birds. The pull tab ring developed by Mr. Fraze was licensed to anyone and everyone for a very low, very modest royalty. The result was instant success. With royalties so low, it was not worth it for any single producer to buck the system and to try to design around the new flip-top system. Yet, the cumulative royalty stream from all the soft drink and beer manufacturers was enormous! The patent was eventually sold to Alcoa Aluminum some years later.

With nonexclusive licensing relationships throughout an industry or field, licensees can do as they please and use your invention as they please. You will have a relatively hands-off relationship with them. This arrangement may provide less incentive for a producer to push your invention. Your position would be to license anyone and everyone and hope they use the invention wisely. A good example of nonexclusive licensing is the "flip-top" can invention that allows the flip-top to remain attached to the can. It replaced all the throwaway flip-tops because of its positive environmental impact.

Anatomy of a License

You can become knowledgeable about licensing fairly quickly by reviewing the following license agreement and learning about the various clauses. Each element is explained so you'll know why they are in a license agreement and how to talk to a potential licensee and legal counsel.

The license agreement you use in your invention/patent-related matters really should be one tailored to the licensing of patents and the related intellectual party. It would be a big mistake to use a generic license agreement.

The example used herein is for an exclusive license to manufacture, sell, and use, the most common license of them all. An exclusive license is generally a bit more complicated than a nonexclusive license because it includes a performance clause. This license agreement has been successfully used in dozens of licensing transactions and has metamorphosed over the years to reflect the most modern, up-to-date clauses. This sample agreement has been also been reviewed and approved by patent attorney Michael Neustel (see appendix for more information) to ensure the legal content is accurate.

Should you be interested in using this license agreement or other similar agreements, such as a nonexclusive license agreement or the Sweetheart Agreement disclosed in the next chapter, you may purchase the *From Patent to Profit* CD ROM on our website, http://frompatenttoprofit.com. When purchasing the CD you will also be able to download future modifications and improvements to the agreements as they are introduced.

Licensing Parties

The first paragraph in a license agreement establishes the two parties entering into the license agreement. All legal agreements must be by the consent of two (or more) legal entities. Thus the names of the parties you put in the opening paragraph should be the legal names for the Licensor (you) and the Licensee (the company or entity you are granting the rights). If you're licensing a corporation, then the corporation's name should be inserted, not the officer who's signing the document.

You will also include your or your company's address and the licensee's address, usually the one where the license agreement will be kept and royalties paid. This is usually the company's headquarters, but it may be a regional office.

Note that in the document, there are no spaces between lines. This is common practice. The reason is to prevent a dishonest party to the agreement from making an insertion at a later date and trying to claim it was there all the time. Obviously one of the signed documents wouldn't have the extra line in it, raising a question, but then it becomes a legal issue as to which document is the valid one.

This paragraph is followed by the words, "Witnesseth," which is saying that the following terms and conditions are hereby going to be agreed upon by the two legal entities.

Ownership of Rights

The second one-sentence paragraph simply states that the licensor is the owner of certain intellectual property pertaining to a specified invention. If there is only one or two issued patents that are being licensed, insert them here. This section may also include any of the following intellectual property rights, which should be specified with a means of identification. Use a format like that shown here to identify your intellectual property rights:

1. **Patents issued:** U.S. Patent 6,095,687

2. **Patents pending:** U.S. Patent Application Serial No. 000–000

3. **Unregistered trademarks:** Mymark™ or you may add Application No. 000–000

4. **Registered trademarks:** Mymark®, Reg. No. 000–000

5. **Unregistered service marks:** Myservicemark^{SM} or you may add Application No. 000–000

6. **Registered service marks:** Myservicemark®, Reg. No. 000–000

7. **Registered trade dress** (as a mark): Reg. No. 000–000

8. **Copyrighted trade dress:** Copyright No. 000–000

9. **Copyrights:** Copyright No. 000–000

10. **Trade secrets:** Spell out the specific subject matter

11. **Know-how:** Know-how related to the invention

If you have a list of three or more intellectual properties you are licensing, it is usually best to simply add: "pertaining to [my invention] as described in the Protected Rights in Exhibit A." Exhibit A is an extra sheet that is added at the end of the license agreement that lists the various IP with their identifying numbers, descriptions, and so on.

This sheet should be titled simply "Exhibit A—Protected Rights." You may put a single sentence below that line reading: "Protected Rights" herein refers to the following:

Then list all the IP you are licensing.

Offer to License, Willingness to Accept

The third opening paragraph is one sentence basically saying that you, the licensor, are offering to license your property rights and that the licensee is accepting your offer. If the license is nonexclusive, you'd insert that word. In the event of a sales-only or a user-only license, those words may also be inserted here.

Definitions and Meanings

Paragraph 1 lists the various terms used in the license agreements and their meanings. The phrases in this license agreement are commonly used in patent-related licenses. These terms are capitalized throughout the document as a means of identification. More specifically, they are:

ANCIENT INVENTION

The folding stool was invented in Egypt 4000 years ago. It had a lightweight seat attached to two sides much like the director's chair used today. It was easily portable and therefore widely used. The only modern day improvement to the chair was the addition of a folding back support.

EXCLUSIVE LICENSE AGREEMENT

THIS AGREEMENT made and entered into in duplicate originals as of this _____ day of _____, 2___, by and between _____ (hereinafter referred to as Licensor) having its principal place of business at _____ _____; and _____ (hereinafter referred to as Licensee), having its principal office and place of business at _____ _____.

WITNESSETH THAT

Licensor is the owner of certain technology and know-how pertaining to _____ _____ (or "as described and claimed in the attached Exhibit A") and;

Licensor wishes to grant and Licensee desires to secure an exclusive license under the Protected Rights.

In consideration of the premises and of the mutual covenants set forth herein, it is agreed by and between the parties as follows:

1. For purposes of the present Agreement, the following terms shall have the meanings indicated:

A. "Protected Rights" shall mean those intellectual property rights listed herein, above (or in Exhibit A as the case may be).

B. "Licensed Products" shall mean any goods, materials, and products covered by or coming within the scope of the Protected Rights.

C. "Sale" shall include sale, gift, delivery, lease, or consignment, as the case may be, whether for domestic use or export, and goods subject to royalty shall be considered "sold" when shipped, delivered, invoiced, or paid for, whichever shall occur first.

D. "Sales Price" shall mean the price invoiced in the regular course of business in a genuine arm's length transaction and shall include a corresponding lease, rental, or consigned price.

2. A. Licensor hereby grants to Licensee on the terms and conditions set forth in this Agreement a personal, indivisible, non-transferable, exclusive license under the Protected Rights to make, use, and sell Licensed Products, with no right to grant sublicenses.

B. The exclusive license shall be for all sales made by Licensee in the field of, and specifically restricted to _____.

C. This license shall be effective as of _____, 2___, and shall continue for the duration of this Agreement.

D. Nothing contained in this Agreement shall be construed as conferring, by implication, estoppel, or otherwise, any license or other right under any patent or patent right except the license expressly granted herein.

3. A. Licensee shall pay royalties to Licensor on all Licensed Products sold by Licensee. The rate of royalty shall be _____ percent (____%) of the total Sales Price of Licensed Products sold by Licensee, less all returns and credits.

B. Within ten (10) days following each month's end, during the term of this Agreement, Licensee shall submit to Licensor a report certifying (i) the total number of goods subject to royalty sold during the previous month, (ii) the Sales Price or Prices of such products, (iii) the aggregate dollar volume of Sales for such period, and (iv) the payment to be made thereon under Paragraph 3A hereof. The first such report shall include the period from the effective date of this Agreement until _____. Each such report shall be accompanied by payment of the royalties due for a given period.

C. A similar statement shall be made within 10 days after termination of this license or of this Agreement for any reason, covering the period from the end of the last month reported until such date of termination.

D. In order for Licensee to maintain the exclusivity of the license, it must maintain minimum sales and royalties, which are calculated on a quarterly basis. If for any reason Licensee does not meet the minimum quarterly sales volume, it has the right to maintain the exclusive license by payment of the corresponding minimum royalty. Minimum sales are:

	1st Quarter	2nd Quarter	3rd Quarter	4th Quarter
Year 1:				
Year 2:				
Year 3:				
Year 4:				
Year 5 and each following year:				

E.   A non-refundable, one-time upfront fee of _____

($_____) is hereby acknowledged, attached to, but not credited against future royalty payments.

F.   In the event that all of the Protected Rights are ever deemed invalid in a court in the United States of America, then this license is null and void and Licensee is under no further obligation to continue with the terms and conditions of the license, unless, however; Licensor elects to contest the judgment of invalidity in a court of appeals, in which case the terms and conditions of the license will remain intact until Licensor has exhausted its last possible avenue of appeal.

4.   Licensee shall keep complete and accurate records of its manufacture and sale of goods subject to royalty sufficient to verify the accuracy of the reports and payments required to be made by it. Such records and other operations of Licensee shall be available during ordinary business hours for examination by an authorized employee or representative of Licensor for the purpose of verifying Licensee's compliance with all conditions and obligations of this Agreement.

5.   During the term of the license granted by the present Agreement, Licensee agrees to have marked permanently and legibly all Licensed Products manufactured and sold by it, with the number of the Licensed Patent under the Protected Rights, or the words, "Patent Pending" as the case may be, and in compliance with 35 U.S.C. #287, and in a form subject to the prior approval of Licensor. In such notices, licensee shall not use the name of Licensor in any literature, advertising, or other sales promotion.

6.   A.   Unless sooner terminated in accordance with the terms hereof, this Agreement shall remain in full force and effect until the expiration of the last of the Protected Rights.

B.   Should Licensee (i) fail or refuse to make any payment required under Paragraph 3 of this Agreement or be in breach or default of any other provision of this Agreement, or (ii) initiate any proceeding for dissolution or winding up its business, or (iii) apply for relief under any law relating to bankruptcy or insolvency, or (iv) become the subject of any proceeding relating to bankruptcy or insolvency, then Licensor, in addition to all other remedies available to it, shall have the right by written notice to Licensee to terminate forthwith this Agreement or any licenses granted hereby, as Licensor may elect.

C.   Upon termination of this Agreement for any reason, all rights and licenses granted hereunder to Licensee to sell and market Licensed Products shall terminate.

D.   Any termination of this Agreement shall not relieve Licensee from its obligation to make a final statement and report, or from its liability for the sales on goods subject to royalty, made or sold prior to the date of such termination, and shall not prejudice any remedy, cause of action, or claim of Licensor accrued or to accrue or which Licensor may have on account of any failure, refusal, breach or default by Licensee.

E.   Any termination of this Agreement shall not prejudice the right of Licensor to conduct a final audit of the records of Licensee.

7.   A.   This agreement shall be transferable by Licensee whether directly or by operation of law only together with Licensee's entire business and only with the prior written consent of Licensor, which shall not be unreasonably withheld. Upon transfer of this agreement, the transferee shall execute a written undertaking in a form provided or approved by Licensor. In the undertaking the transferee shall (i) assume all of the obligations undertaken by Licensee hereunder; (ii) acknowledge that it has satisfied itself that the Protected Rights and Licensor's right to license same are valid and incontestable; and (iii) covenant not to directly or indirectly contest, challenge, or deny the validity of any of the Protected Rights in any forum or for any purpose.

B.   This Agreement may be assigned by Licensor and shall inure to the benefit of Licensor and its successors and assigns, and all obligation of Licensee shall run in favor of the successors, assigns or other legal representatives of Licensor.

8.   A.   Licensee agrees to indemnify and hold Licensor harmless against any and all claims, expenses (including reasonable attorneys' fees), costs, charges, taxes and the like arising out of Licensee's operations under this License Agreement, and/or Licensee's manufacture or sale of Licensed Products.

B.   The Licensee shall procure and maintain comprehensive general liability insurance, naming Licensor as an additional insured, in an amount greater than or equal to $2,000,000 per incident, including:

a)   product liability;

b)   personal liability;

c)   personal injury;

d)   property damage; and

e)   broad form contractual liability coverage for all acts and omissions directly or indirectly associated with this Agreement.

Upon request of Licensor, Licensee shall provide written evidence, satisfactory to Licensor, of above-stated insurance.

9. A. Licensee (i) acknowledges that it is satisfied that the Protected Rights and Licensor's right to license same are valid and incontestable; and (ii) will not directly or indirectly contest, challenge or deny the validity of any of the Protected Rights in any forum or for any purpose.

B. No warranty or representation is given or made by Licensor that any goods sold are free from infringement of the patent rights of others.

C. A waiver of any of the terms or conditions of this Agreement shall not be considered a modification, cancellation, or waiver of such or any other term or condition thereafter.

10. A. All improvements in connection with the Licensed Products shall be the sole property of Licensor regardless if made by Licensor, Licensee, independent contractor, or in combination therewith.

B. If, during this Agreement, Licensor shall conceive any improvements in connection the Licensed Products, which come within one or more issued claims of any patents in Licensor's Patent Rights, then Licensor shall disclose the same to Licensee in writing within ninety (90) days of the conception of such improvement. Licensee shall elect in writing, within sixty (60) days after date of notice, whether they wish to have a license under such improvement. If Licensee so elects and notifies Licensor in writing, then any patent applications filed or to be filed for the improvements shall be added to and become a part of Licensor's Patent Rights under this Agreement with no additional compensation.

C. Any and all intellectual property rights resulting from any improvements made by Licensee during the term of this Agreement shall be the sole property of Licensor. Licensee shall, as soon as practicable after conception of an improvement, notify Licensor of such improvement within ninety (90) days. Licensee, its employees, and agents shall cooperate with Licensor and sign all assignments, applications, and other documents reasonably necessary to allow Licensor to prosecute any form of intellectual property protection or management chosen by Licensor in order to protect the improvements. Any improvements by Licensee shall be automatically included as part of Licensor's Patent Rights under this Agreement with no additional compensation.

11. All notices for all purposes of this Agreement shall be given in writing and shall be effective when either (a) served by personal delivery, or (b) deposited postage prepaid in the United States certified mails, or (c) delivered via Federal Express with verifiable signatures for receipt and, addressed to the respective parties at their given addresses or at such other addresses as either party may later specify by written notice to the other.

12. This Agreement shall be interpreted and construed under the laws of the State of _____.

13. This instrument contains the entire and only understanding between the parties and supersedes all prior agreements between the parties respecting the subject matter, and any representation, promise or condition in connection therewith not incorporated herein shall not be binding upon either party. No modification, renewal, extension or waiver of this Agreement or any of its provisions or any termination shall be binding unless in writing.

IN WITNESS, the parties hereto have caused this Agreement to be executed as of the day and years first above written.

Licensor:

By: _____
     Name   Title

Licensee:

By: _____
     Name   Title

- **Paragraph 1A:** "Protected Rights" is the term that will be used when referring to the licensor's intellectual property being licensed. If you wish, you may use the term "Protected Patent Rights" instead.

- **Paragraph 1B:** "Licensed Products" refers to the goods, materials, processes, and so on that fall under the Protected Rights.

- **Paragraph 1C:** "Sale" refers to the sale of Licensed Products and explains what qualifies as a sale. If your invention and patent are related to a process, you would typically use defining language such as " . . . any product produced by the process covered under the Protected Rights."

- **Paragraph 1D:** "Sales Price" refers to the price of goods sold. Frequently this may include additional clarifying language such as, " . . . less trade discounts and freight." In other words, the Sales Price is more specifically being defined as the "net" sales price. When calculating royalties, this may be the case.

At times there may be other specific terms you'll want to define in this section. For instance, you may include a definition of "Net Sales Price" as well.

Granting of License

- **Paragraph 2 (2A)** explains the granting of license based upon certain terms and conditions. It also says that the license is nontransferable, exclusive, and that the licensee may not sublicense the Protected Rights. The terms are as follows:

- **Paragraph 2B:** If there is a field restricting the exclusive rights, it should be defined here. Or, if there is a territorial restriction, that should be listed as well.

- **Paragraph 2C:** This paragraph establishes the effective date.

- **Paragraph 2D.** This paragraph states that no other rights other than those listed in the Protected Rights are being conveyed.

The language in this paragraph is fairly standard language.

Payment of Royalties

In Paragraph 3 the terms and conditions of the payment of royalties are defined. More specifically:

- **Paragraph 3A:** The royalty rate is defined. If royalties are based on a per-unit sale, then the language should reflect that. Royalties are almost always based on the Sales Price. This sentence includes the further explanation that royalties are not paid on returns and credits. This clarifying language may be in paragraph 1D instead.

- **Paragraph 3B:** Royalties in this agreement are paid on a monthly basis, which is preferred. It is equally common to pay royalties on a quarterly basis and less common to pay on a semi-annual basis. Also defined in this paragraph are the reporting method and the statement that the report should include a check for payment. Last, it defines the starting date for the first report to be submitted.

- **Paragraph 3C:** This paragraph states that a report and payment are also required upon termination of the agreement for any reason.

- **Paragraph 3D:** Performance clauses are essential in an exclusive license and should be clearly spelled out. This paragraph states the sales quotient or the minimum royalty the licensee must meet in order to maintain exclusivity. Since new products usually take years to ramp up, this provision gives the licensee five years to meet the predetermined expectations.

- **Paragraph 3E:** An up-front fee (also referred to as a "License Fee") is payable upon execution of the agreement. The amount is inserted here. While not usually credited against royalties, this may become a negotiating point, which is discussed in the next chapter.

- **Paragraph 3F:** This clause protects the licensee if your patent is ever invalidated. However, it also provides you with a means of reversing the decision and reinstating the license agreement should the appeal be in your favor.

Record Keeping

A licensee must keep accurate records as required in **Paragraph 4**. This is a two-fold obligation requiring both sales and manufacturing records. It also states that the records must be available for your review during normal business hours so an authorized individual, such as an appointed accountant, may review the records.

This paragraph may also include a provision for a penalty should the royalties be underpaid. Typically the penalty clause would require payment be made on the underpayment, plus 10% surcharge and reimbursement for your cost to audit.

Patent Marking

In **Paragraph 5**, the licensee is obligated to mark all products with the required patent number. It may also include markings for patent pending if patents have not yet issued. To be in compliance with 35 U.S.C. #287 essentially means that the products are clearly marked.

More specifically this means that the printing of patent numbers should be: 1) placed on the Licensed Product itself; 2) large enough to be easily read; and 3) in a contrasting color that makes it easy to see. For instance, $1/4$-inch letters printed in black ink on a white background for a medium-sized product, such as a basketball, a plastic grocery sack, or a ski rack, should suffice.

With very small items where placing a clear marking is not possible, such as a key chain ring or jewelry clasp, the patent marking may be placed on the sales tag or the packaging itself.

Patent marking gives constructive notice to all suppliers in the field that the product is patented. Failure to mark a new product, or to mark it clearly, may create problems later on. Should a competitor infringe, it may say it was unaware of the patentable matter and would most likely not be liable for infringement up until the time it was properly notified.

Termination of the Agreement

Paragraph 6 states that the license agreement will be in effect until the last of the patent rights expire, providing the licensee is not in default of the terms and conditions of the agreement. Shorter time periods may be used—for instance five years—but generally speaking it is unnecessary. Besides who wants to renegotiate the license agreement every five years?

Other provisions for termination of the agreement are as follows:

- **Paragraph 6B:** This paragraph says that the licensor may terminate the license agreement immediately should the licensee: 1) be in default of any provision in the license agreement; 2) be in default of royalty payments; 3) plan to go out of business, or 4) apply for bankruptcy protection or is forced into it. Lawfully you may not be able to terminate a license if a company goes into bankruptcy, but it's still best to have this clause as you may be able to at least try and put pressure on the other party. Should this happen to you, it would be wise to consult your attorney.

- **Paragraph 6C:** Essentially, all rights are terminated when the agreement is terminated.

- **Paragraph 6D:** When terminated, the licensee is still obligated to make a final report and payment.

- **Paragraph 6E:** A final audit may be conducted after termination.

Hopefully, termination will not be required. But if it is, you'll have a provision in place. The word used in this paragraph "forthwith" means "right away." A licensee may insist on a certain time period, such as ten days, to remedy any breach. As long as it is reasonable, you may have to agree.

Transferability

Paragraph 7 states that the license agreement may be transferred under certain conditions. Either you or the licensee may transfer the agreement. To expand:

- **Paragraph 7A:** The licensee may transfer the license but must have your written consent. You cannot be unreasonable about assigning the rights. However, if the new licensee is a competitive entity that may plan on shelving your technology and patents, then you may not want to consent. Or, at the very least, you may want to be paid off. In this paragraph, the transferee is also accepting all obligations in the agreement.

- **Paragraph 7B:** This says that you may transfer the license to any successor or assignees as you wish and all obligations of the licensee shall run in favor of the assigns.

Indemnification and Insurance

Paragraph 8 defines the licensee's obligations to maintain insurance and to hold you harmless should there be a claim. This covers several points. In summary:

- **Paragraph 8A:** This says that the licensee must indemnify and hold you harmless against any and all claims resulting from the manufacture or sale of the Licensed Products.

- **Paragraph 8B:** This requests that the licensee maintain product liability insurance of $2 million and that you, the licensor, are named as an additional insured. As you can see, this covers you under virtually all imaginable conditions. It also says that you may request a copy of the insurance policy should you request.

Patent Validity, Infringement Disclaimer, Waiver

In **Paragraph 9** of the license agreement, you want to make certain statements regarding your patent validity, infringement of other patents, and a waiver of any terms and conditions contained in the agreement. Specifically:

- **Paragraph 9A:** The licensee acknowledges that your patent rights are valid and will not contest, challenge, or deny the validity to anyone for any purpose.

- **Paragraph 9B:** It simply says that you are unaware of any other patents that infringe the licensed products.

- **Paragraph 9C:** Says that if any part of the license agreement is surrendered or unenforceable that the rest of it still remains in tact.

Improvements to the Invention

Paragraph 10 discusses improvements that the licensor may make to the invention and offers them for license under the same agreement. In detail:

- **Paragraph 10A:** This says that all improvements to the Licensed Product are the property of the licensor.

- **Paragraph 10B:** You must disclose your improvement to the licensee in a certain time frame and the licensee must decide whether or not it wishes to include them in the license agreement in a specified time period. This gives the licensee ample time to decide if the improvement improves marketability or not.

- **Paragraph 10C:** This says that any related intellectual propery rights resulting from an improvement are owned by the licensee. It also says that the licensee and any of its employees must cooperate in the filing and assignment of those rights. It also states that the rights then become part of the license agreement as well.

Notices to a License Agreement

Essentially this **paragraph 11** says that any notices sent from one party to the other shall be delivered either in person or by some form of verifiable means of delivery. Notices are usually matters such as change of address, phone number, and so on. It may also include a notice that a royalty payment has not been received or that the licensee is in default.

As a general rule, you don't need to send a certified notice that you have not received a royalty payment. Start by calling the accounts payable department, and if you experience problems, talk to one of the officers of the company, preferably the one who has signed your license agreement and see if he/she can fix the problem. It could be nothing more than a new individual in the department who did not understand the importance of maintaining royalty payments current.

Lawful Interpretation

Paragraph 12 says that you want the legal interpretation of the license agreement to be that of your state not the licensee's. There are two reasons for this. First, you'll be signing the license agreement after your attorney has reviewed it. Second, if there is a challenge, you want it to be based on your rights in your state should there be a legal issue later on. Since you're the licensor, the owner of the IP, any challenge is going to have to be in your state anyway. There are some exceptions to this rule; however, they are rare.

Clarification of Content

Paragraph 13 says that this is the only agreement regarding the Protected Rights and that any other agreements, oral or written, are not binding. It also says that if the two parties had any previous agreement, then this one supercedes them all.

Signing License Agreements

These last lines in the agreement are for the signatures of the licensor and licensee. If one or the other is a corporation or limited liability corporation (LLC), then an authorized officer of the corporation or LLC must sign. Otherwise, the signature is usually that of the president, CEO, or owner of a proprietorship.

One additional means of certifying a licensee's signature is to ask him/her to include the corporate seal. Only those of high standing should have access to the corporate seal and the ability to use.

Manufacturing Agreements

One other form of agreement that is used refers more specifically to the manufacture of a patented product. These agreements are commonly used with large corporate entities that want to protect their IP, but less commonly used with smaller entities. There are two forms of manufacturing agreements:

Authorization to Manufacture

This is essentially the same as the preceding license agreement except it is referred to as an authorization instead. The purpose of an Authorization to Manufacture is to make the producer aware of the patented nature of the innovation and that it may not produce the patented product for any other entities other than those you've licensed (or authorized). By using the previous license agreement and making the following modifications, you can turn it into an Authorization to Manufacture:

A. Change the words "License Agreement to "Authorization to Manufacture."

B. Change "Licensor" to your company name.

C. Change "Licensee" to "Manufacturer."

D. paragraphs 2B, change the field to read, "The Manufacturer is authorized to manufacture Licensed Product solely to those companies that have been licensed or authorized by [your company]. Such names shall be provided to Manufacturer by [your company] in writing.

E. Delete paragraph 3 in its entirety sine the manufacturer is not paying royalties. If it were, you should use the license agreement instead.

F. Add a confidentiality clause in your favor if you have not previously entered into such an agreement with them.

There are some additional minor modifications to make as you read over the agreement. But in essence, an Authorization to Manufacture is nothing more than making the producer aware of the patented nature of the product and holding it responsible for selling only to those entities you've previously authorized.

Manufacturing Agreement [for your company's use]

Similar to an authorization to manufacture, this agreement can usually be accomplished with a generally slimmed down version of the authorization to manufacture and may be as brief as one to two pages. It may be used with parts suppliers too, should they be privy to your product development.

If you're in business and would like to make sure your patented innovations are protected with your subcontractors, make a list of your concerns, based on the previous license agreement (or authorization to manufacture) and have your attorney draft a simple agreement.

LICENSING STRATEGY

A mirror image of your manufacturing,
marketing, and patenting strategy

Your license strategy should be based on several considerations. Before you run out and start contacting prospective licensees, make sure you put a plan in place so that you can communicate effectively to your prospects.

First, you have to know your market well. Who are the industry leaders? Who would be your most likely licensees? How do you plan to license them . . . exclusively or nonexclusively? If you believe you are short on knowledge in this area, you must conduct more research to better understand how your invention fits in the field and how it will be perceived by others.

With ample knowledge of your industry and your product's niche, you can then determine your licensing objectives. There are several approaches and several variables you can consider.

Last, you really should know your licensees as well. Your research will help you better understand what the company's objectives are for the short term and long term. Knowing this will help you explain to them the benefits and merits of your innovation and how they may profit from it.

Your Licensing Objectives

What are your present objectives? Have you been thinking only in terms of exclusive or nonexclusive licensing? Have you thought about producing the product for your own sales as well? Are there other patents or trademarks you can include in your license agreement that may improve the licensablity? You probably have some idea of your direction; however, you're probably best off keeping an open mind to alternatives and proposals you may receive from your potential licensees.

LICENSING LAUNCHES DISNEY

In 1928 Disney created Mickey Mouse. A woman approached Disney and asked if he would license the right to her to make Mickey Mouse stuffed dolls. Disney agreed and liked her work so much that he offered more characters for her to make. The licensing income from Mickey Mouse and eventually other characters and other products launched the Disney Kingdom we know of today. Even now, Disney's fortune is based in large part on royalties from licensing. Disney royalties are typically between 5% to 7.5%.

345

Your Licensing Focus

We know that licenses grant the manufacture, sale, and use of your intellectual property rights to other licensees. We also know that your underlying objective is to earn money from your intellectual property and gain security through patents. Licenses may be your means to accomplish both.

Many inventors seek to "license manufacturers," which in and of itself is usually the wrong perspective. Remember, nothing happens until something is sold. So by licensing a manufacturer that has no industry exposure from the marketing prospective, few orders, if any, will follow.

Remember that manufacturers are notoriously poor at sales marketing. They make products based on "what customers tell them they want." They don't usually create and promote new products well. Don't waste your time looking for or negotiating with potential licensees based on their manufacturing ability. There is an exception to this rule you'll read about later on.

The principal focus of your licensing activities should be to license "where the money is," or, in other words, with an entity that has market presence. You'll want to establish a mutually beneficial relationship between yourself (the owner of intellectual property) and these entities (licensees). In this way, both of you benefit. Until your invention is sold and begins producing income, you'll not be satisfied, so make sure your focus is on the marketing—where sales and profit is generated—not manufacturing. That is your chief objective. Your licensee is your marketing partner.

Licensing Out Only

Is your sole interest to license others? In other words, do you want to be an inventor and product developer and let other do all the manufacturing and marketing? There is nothing wrong with this approach. However, you must have a good understanding of the innovation's industry and know the players.

You should also have a good working knowledge of licensing, both from the legal side of what's contained in a license agreement and from the business side of how to introduce and conclude a license agreement. You really must understand royalty rates in your field and have a broad knowledge of competitive products and what royalties others may be charging for similar or competitive products.

Establishing good working relationships with your licensees should be one of your underlying objectives. You really want to be able to introduce new improvements and new product concepts to them at a later date. From this perspective, building personal relationships is invaluable long term.

When licensing out, your focus is almost always on the company that is marketing your innovation in the field, whether that is an existing supplier to the trade, a new supplier to the trade, or a private-label supplier to the trade.

Licensing as an Adjunct to Your Company's Sales

You may elect to license your patents and sell the invention through your company as well. This may not fit some strategies, but it can be a very good approach at times. There are different reasons for wanting to do this and a few different ways to accomplish it effectively.

First, you may license select companies to sell into markets that your company doesn't sell into. For instance, you've a presence on the Internet, but no presence in the commercial/industrial marketplace. Another reason you may want to consider this approach is that by licensing competitors in your field, you will be helping to standardize the industry. After all, there's nothing wrong with profiting a little from your competitors' sales if they can help expand the overall market for your innovation.

One other outcome from using this approach is that your ongoing market exposure through your company and your success in marketing will attract inquiries from additional potential licensees.

One notable success story using this approach is that of Shari's Berries®. It used this means to expand sales of the Shari's Berries line of chocolate-covered strawberries to the retail trade. It also profited by supplying the licensees with the ingredients on a wholesale basis and with the know-how to produce the Shari's Berries at the retail location.

Your Company as Supplier

If your company holds the patents and will be supplying various marketers or perhaps end users with the patented innovation, such as in an OEM situation, then no license agreement is required. All you have to do is start supplying the company based on its purchase orders.

The exception to this rule occurs if the marketers or companies you are supplying are selling or using your innovation in conjunction with one of your trademarks, for which you may want to specify certain quality standards. The primary purpose of the license agreement in such a case would be to protect the integrity of your brand name (trademark). Should this be your approach, consult with your attorney to have him/her prepare the appropriate agreement with the quality-control specifications you'll require.

JDAs and Licensing

A Joint Development Agreement may serve as an adjunct to licensing or as an alternative to licensing. JDAs are an excellent means of partnering, particularly for smaller companies that could use the support of larger entities in an industry. For instance, your company may be the creator and owner of the patents, while another company produces the product and yet another one markets it. There are many combinations and viable alternatives that may be pursued.

One approach may be to enter into a JDA and a license agreement. Since JDAs are usually for a specific time period, one of the parties to a JDA may also wish to retain license rights to preserve its position afterward. For example, the marketer may make a substantial investment during the JDA phase in order to test market and verify results, but it would want to continue its sales/marketing effort afterward. Or a manufacturing entity may wish to keep its position as a producer to the marketer or others as they may be licensed.

This may not be your primary approach to licensing, but you should leave the door open to JDA possibilities. They may work well with your overall licensing strategy and may serve as an excellent stop gap in your product development/licensing strategy.

Contractual Agreements

Many inventors and small businesses work with large entities via contractual agreements. These agreements may be used as a preliminary step toward licensing in a manner similar to a JDA or as a replacement for a license agreement.

PRICELINE V. MICROSOFT

One recent example of an eCommerce patent license resulted from the settlement between Priceline and Microsoft regarding Priceline's name-your-own-price business model. In October 1999, Priceline sued Expedia, owned by Microsoft, for infringement of U.S. Patent No. 5,794,207, "Method and Apparatus for a Cryptographically Assisted Network System Designed to Facilitate Buyer-Driven Conditional Purchase Offers." It resulted in Microsoft agreeing to pay royalties to Priceline to settle the dispute. There are three principle lessons learned from this suit. First, the present-day patent system works. Second, Priceline's licensing strategy was to license others, which it had succeeded in doing. Third, it provided a favorable rate to the licensee to boost the patent's strength and ultimately Priceline's licensing strategy.

As an interim step to licensing, a contractual agreement between the parties may be struck in order to develop and further qualify the marketability or manufacturability of an innovation. Since qualifying an innovation's potential may take months and at times years, this may serve both parties' needs.

Essentially, such an agreement would provide the patent holder with income during the qualifying period, while the two parties work together to develop the innovation into a viable product. This may be a particularly effective approach to take should your innovation require a substantial amount of engineering and further development. Remember, it takes months to assemble all parts suppliers, design and prepare packaging, build up inventory, and ramp up sales. A contractual agreement of this sort would typically provide the licensing company with the financial means to conduct some serious and, at times, expensive product development and testing during these final stages.

Licensing In and Sublicensing

At times, it may be valuable to license in other technologies or perhaps other existing patents, trademarks, and copyrights in order to improve the marketability of your invention. The term "license in" simply means you are licensing someone else's patent or intellectual property to use along with your innovation.

Licensing in other patents may expand the marketability of your innovation as well as the scope of the patent protection, which would be good for both you and the other patent holder. When doing so, you'll want to have the ability to sublicense the licensed-in patents so you may collect royalties on it. Generally speaking, royalties you'll be collecting on the licensed-in patent would be about twice that which you would pay to the patent holder. Here might be a typical scenario concerning how you would license your patent along with a sublicense on another party's licensed-in patent, assuming that the two patents are of equal value (50%/50% split) and have been licensed for a 6% royalty:

Royalty rate to your licensee:	6% payable to you
Sublicense royalty payment ($\frac{1}{2}$ of 3%)	1.5% you pay to the second patent holder
Net royalty you retain	**4.5%**

In the above example, both patents are considered of equal value and are being licensed at the same rate. Since you are doing the licensing, via a sublicense to your licensee, the second patent holder on the licensed-in patent receives $\frac{1}{2}$ of that part of the royalty. There is no real rule of thumb here other than you'll have to have a figure that makes sense. If the second patent holder is unreasonable, then it could kill the approach altogether.

It may also be valuable to license in certain trademarks or copyrights to help promote the sales of your innovation—for instance, should you receive a license from Disney for the use of Mickey Mouse® on your innovation's artwork. Companies such as Disney, Warner Bros., and many sports celebrities tend to have fixed rates. Thus, if your innovation includes artwork or designs from one of these entities, you'll just have to add it onto the cost of goods sold. Typical royalty rates are from 5% to 7%, but can go as high as 17% in the case of someone exceptionally famous like Michael Jordan.

Licensing in another party's trademarks gets a little trickier because the other party will want to make sure the integrity of the marks are upheld with adequate quality control. If this is your approach, you'll need to discuss these issues with the licensing executives of the trademark holder.

Who Do You License?

Your licensing focus is to license "where the money is," or, in other words, those companies that are directly involved with actual sales of the innovation. That's where the profit is and that's who will want protection. Usually this is your marketing partner or marketing partners if the license is nonexclusive.

So, your primary focus for licensing is on those marketing partners that are actively selling product in the field of your invention, not the manufacturers or subcontractors who may be making the product for the marketers. Now, let's be more specific.

In retail applications, the supplier is the company with the brand name you see on the retail shelves. It's not a distributor, with the exception of one that may have it own private label products, such as with Sysco Products Company in the restaurant trade, or a supermarket chain that distributes its own private label. With industrial/commercial sales, the licensee is typically the same—that is, the company whose brand can be seen on the supply shelves in the purchaser's plant or on the shelves of a distributor. Again, the marketing partner you license is not the distributor, unless you are licensing it under its own private-label product line.

There are applications where you'll be licensing the manufacturer or the end user itself, such as may be the case with an industrial/commercial product or process. A good example of this would be the flip-top aluminum can invention licensed to the various canning and bottling companies, or a manufacturing process licensed to a manufacturer in a given trade.

However, keep in mind that you can also consider licensing "products produced by the patented manufacturing process," instead of the manufacturer itself. For instance, if your patented process used in tortilla manufacturing equipment in a food facility improves output by 10%, you could ask for a small royalty on the future sales of tortillas coming off the patented process. Since an increase of 10% represents a significant profit potential, a small royalty per ton of tortillas manufactured may make sense.

Licensing Both the Manufacturer and Marketer

There is one strategy you may consider in which you license and collect royalties from a manufacturer (or multiple ones), instead of the marketers. If you have granted one or more sales-only licenses to your marketers, you can also license the supplier of the patented product with a manufacturing-only license. In the latter type of agreement, the manufacturer is going to be shipping product to end users, but only through the licensed marketers.

The licensed manufacturer may ship to and bill the licensed marketer or the marketer's customer. It really doesn't matter which method is used to get the product to market. With this strategy your royalties are paid by the manufacturer. If the marketer is billing its customers, then they have their set profit margin. If the manufacturer is billing the customer, then it would also be paying a commission to the marketing partner.

BEING INFRINGED?

Patent infringers make great licensees! If you are infringed, REJOICE! Work with a smart patent enforcement company and have them structure a profitable license to your patent.

–Andy Gibbs, CEO, PatentCafe.com

Applying this double licensing strategy requires that both parties submit regular quarterly or monthly reports on sales. You can then compare the reports and see if they corroborate.

International Licenses

Many innovations have international marketing potential; however, searching for international licensees can be extremely costly and frustrating if you or your company is not familiar with overseas marketers and suppliers. Also the cost of patenting overseas can be very costly. However, there are two primary approaches you may take to license your invention for the international marketplace.

1. **License an international entity.** You may license one single entity that has international exposure and capabilities. When doing so, you have the option of asking it to pay for all international patent filings and prosecution in exchange for a reduced royalty. This type of approach may also include the ability to sublicense your technology to some of the licensee's established contacts. In such a case, you would share in the royalty stream.

2. **License individual companies overseas.** If you take this approach, it will usually be more costly, since the responsibility and expense of finding licensees is up to you. With a foreign licensee, you may ask it to pay for the patent costs in its country or marketplace at a reduced royalty rate.

With foreign licensees, you may either allow them to market the patented innovation in their designated country or region or the entire world. Thus, you may protect the interests of other companies in other countries should you desire.

The Right Strategy for You

There are several points to consider before applying a licensing strategy and determining if exclusive or nonexclusive licensing best fits your innovation. Above all, your licensing mindset should be as follows:

Make all your decisions based on what is best for your innovation's future

Of course, what is best for your innovation's future is to have a lot of customers, sales, and profit. When licensing, forget about personal preferences and predetermined ideas you might have. You want to apply a strategy and then see where it goes once you're active in the process. Here are some points to consider to help you formulate your strategy:

1. **Niche-oriented products.** All products start out as niches, but the narrower the niche, the more likely you will license a single, exclusive company. For instance, a toy such as a Star Wars® Laser Gun would be marketed through a single toy company, not several. So, the narrower your niche, the more likely the license will be exclusive.

2. **Commodity-oriented products.** Inventions that will become broadly used by all companies in a field will be nonexclusively licensed. In fact, licensing such an innovation such as this on an exclusive basis may kill the sales and may promote design-arounds and patent infringement.

3. **Based upon fields.** If your innovation may be used in various fields, you'll want to

ONE WORLD

Taking their inventions overseas to offshore markets is the next big frontier for American inventors. But, with costly national patent system redundancies, translation requirements, globe-wrapping red tape, and startling differences in national laws and court procedures, inventors will find the swim to be both long and treacherous. Independent inventors and large corporations need to find common ground and ways to work together to influence international patent rules and policies that will enable seamless patent protection across national borders.
–Don Kelly, Intellectual Asset Management Associates, LLC

license it separately in those fields. For example you may license a patented plastic water-soluble trash bag in the retail field to one exclusive marketer; to several nonexclusive licensees in the industrial/commercial field, and to two exclusive licensees to the hospital trade.

Do your research ahead of time so you may develop a strategy that fits the invention. It makes no sense looking for a single licensee or focusing only on one field if your innovation must be sold through different sales channels to get market penetration. Besides, if you have more than one field in which you can license an innovation, you'll have more potential partners to pursue.

Exclusivity, Nonexclusivity, and Variables

The concept of an exclusive or nonexclusive license may take on a multitude of variables and forms depending on several factors. Nonexclusive licensing follows a more straightforward format and is at more of an arm's length transaction. At times, you may use both an exclusive and nonexclusive license, all wrapped into one. Last, there is the strategy of staging in licensees over a period of time. You'll want to use the strategy that fits your innovation best and works for the industry.

The following are some guidelines you can use to develop your strategy.

Exclusive Considerations

You should know that a license may be exclusive in one field and nonexclusive in another. Exclusives may also be for a certain time period, for instance, to give a first producer a jump on the competition. This is sometimes referred to as "favored nation status." Last, exclusive licenses may be limited to a certain number of exclusive licensees—for instance, five licensees in an industry that may have twenty suppliers. Here are some considerations for exclusive licensing:

1. Niche-oriented products are usually exclusively licensed to a single licensee.

2. Although commodity-oriented products would usually be nonexclusive, if you are trying to carve out a specific commodity niche, it is probably advisable to have a dedicated, exclusive relationship for at least a certain time frame—from one to five years or so. The narrower the niche, the longer the exclusive time frame may be required, up to the life of the patents.

3. A limited exclusive license at start up serves as an excellent incentive to a first licensee. It gives the licensee a jump on the marketplace and the all-valuable first-to-market position.

4. Exclusive licenses must be predicated on performance and guaranteed earnings year after year. Make sure that your guarantees and your earnings begin early on.

    Don't spend time talking to a potential candidate who wants an exclusive license but will not provide any long-term performance guarantees.

Nonexclusive Considerations

Generally speaking, these are easier license agreements to strike because the exclusive language and performance clauses are not usually required.

REALISTIC EXPECTATIONS

You and your licensee should determine performance requirements based on conservative, realistic expectations. Hopefully, this will dovetail with your earnings expectations. Unrealistic expectations may result in failure for you and your licensee. Hopefully, your marketing study will be accurate enough to give you both a good idea of earnings potential.

351

1. If your invention is a commodity-type product, it may be best to grant a simple, nonexclusive license. This is, of course, assuming that it is the type of invention that would apply to all commodity products in a given field. An example is the aluminum flip-top can.

2. There may be less incentive for a company to support your invention if "anyone can make the product."

3. If you have a major breakthrough that is of great benefit to mankind, you may want to consider only nonexclusive licenses.

4. Supply items for industry (raw materials or otherwise) is best nonexclusively licensed as purchasers will typically want multiple suppliers.

If you are not sure of your marketing and manufacturing approach, you will not be sure of your licensing approach either. If this applies to you, learn as you go and be cautious.

Exclusive and Nonexclusive

There are two forms that this kind of license strategy may take on. First, a license may be exclusive for a certain time period and then become nonexclusive as the case may be when providing favored nation status.

The second form would be a single license agreement to a company that is exclusive in one field and nonexclusive in another field. In other words, you may grant exclusivity in the field of strength for the licensee, but nonexclusivity in a field in which it is weak.

One other variable on this theme is that a licensee may want to consider converting an exclusive license into a nonexclusive license should it not be able to reach its sales minimums in a performance clause.

Staging-In Licensees

One approach that is effective in overcoming objections the nonexclusive licensees may have is to stage them in on an orderly basis. For instance, you may have a first licensee that begins in January of year 2005, a second one that begins in January 2006, a third one in January 2007, and then in January 2008, you may license any and all others. This type of licensing approach may be actually considered a form of exclusive and nonexclusive license. This approach was used with the licensing of Direct TV's rights. It started with RCA; three months later Sony was licensed; and every three months thereafter, a new TV supplier was added on.

When staging in licensees, there may be competitive issues that might be of concern to a licensee. In almost every industry, there are companies that tend to get along well as competitors and others that may not. You do not want to do anything to cause antitrust problems, such as blacklisting certain companies from being licensed. However, you can include a list of "other companies" that "may be staged in" at a future date. Your patent attorney can help you prepare this language in a legal format.

Royalty Rates

How do you determine a royalty rate when licensing a patented invention? This is a fairly difficult question to answer since it depends on several factors. Generally speaking, royalties are based on what the market will bear. But that perception is sometimes wildly different between an inventor and a licensee. Inexperienced inventors all too often get greedy and expect high royalty rates. Many think that just because they have a patent, it is worth millions. We know that's not true. But how can we figure out a royalty rate?

Think about this—if you are asking for a 10% royalty, you are essentially asking for an amount of income equal to that of the licensee's profit margin. A licensee will usually have a 10% profit after all costs are deducted, if even that. And the licensee is the one that must invest substantial money, time, people resources, and so on, as it creates the market and builds sales and inventories. It has all the risk, not you. From that perspective doesn't it seem absurd that the inventor would want as much profit as the company that is really taking all the risk? Asking for high royalties usually—and irretrievably—kills the deal. So, what's reasonable?

Keep in mind that royalty rates tend to vary depending on the industry and type and number of intellectual properties being licensed. In the toy industry, royalty rates may vary from 5% to 15%, but it is trendy so the earnings may be short-lived. The royalty rate for housewares, appliances, electronics, and hardware may vary from 1% to 7%. The packaging industry tends to fall between 1% and 5%. Tools and sporting goods tend to be between 3% and 10%; automotive aftermarket about 3% to 8%. Gimicky infomercial products may be associated with much higher royalty rates, as great as 20%, but they're frequently dead in a matter of months. The pet supply industry and baby goods tend to be between 5% and 10%, with sales in lower volumes. Computer software tends to have royalty rates as high as 25%, but the life cycle is usually very short-lived. In most every field, there are too many variables to have "one standard rate for all."

Some well-known royalty rate examples include Disney's and Warner Brother's copyrights, which have a 5% to 7% royalty for most of their products. Michael Jordon gets 17% for his product endorsements. But he is a well-known name. Are your creations as sought after as theirs? On the other hand, licensing something for as little as $1/100$ of 1% may be extremely profitable due to the high volume. An example is the flip-top opener on aluminum cans that stays attached to the can. The annual royalty earnings were well into the tens of millions. Or, how would you like to have been the inventor of the blinking cursur used on computer screens at 1 cent per unit? Generally speaking, products that sell in high volume will have much lower royalty rates than those that sell in lower volumes. But the earnings can get quite interesting. In contrast to the high volume of the flip-top can is the Ant Farm, which can support a 20% royalty with its year-in, year-out, annual sales of about $1 million. The anticipated sales volume of a new invention/product then becomes a key reason why royalty rates tend to vary so widely.

There are several other factors to consider as well. For instance, how easy is it to design around your patent? How broad is its scope? Will it support a validity challenge? Do you have one patent or several? The more patents you have, usually the more secure your position and perhaps the more you may be able to charge. If patent protection is weak or easy to design around, no one will want to license your inventions/patents . . . at any royalty rate. The same goes for any patent that is only a minor improvement over existing products.

Another factor is based on the basic economic theory of price elasticity. A 2% higher price on a commodity product is usually enough to put that commodity out of the 99%

PROFIT SHARING

Asking for profit sharing instead of royalties is usually a bad idea. First of all, the term profit is a nebulous term, and in many cases a good certified public accountant (CPA) is going to do his/her best to minimize it so that taxes will be minimized as well. Taking profit instead of royalties is the perfect way for a dishonest licensee to take advantage of an inventor. Second, any profit sharing will be based on the innovation's performance, not the company's overall profit and loss (P&L) statement. Thus, it may take years to share in the return, if ever. Should you ask for royalties and profit sharing then? This kind of request reeks of greed. You're asking the other company to invest its time, money, and resources and take all the risk and you're telling it to pay you both royalties and profit? This is a deal killer. The only realistic approach to profit sharing would be to have it buffered by an underlying minimum royalty. In other words, let's say you agree to take 25% of the profit on the sales of the product line. You should also have a royalty payable to you in a certain minimum percentage; this royalty is then credited back against profit sharing. That way you'll not become vulnerable to the ineffectiveness of a licensee should it not perform or should it take years to realize an honest profit.

sales category. In other words, the product that costs 2% more is likely to be virtually shut out of the mainstream market in which it competes. This holds true for many products, such as plastic resins, well-established wood and paper products, metals, trash liners, fruits and vegetables, bread and milk, automobiles (sold by different dealers), and even computers (to a lesser degree). So it is important that a new invention and its accompanying patent(s) do something wonderful to the product that will make customers want to pay more. And keep the royalty rate low . . . very low with commodity products, as low as $\frac{1}{2}$% or less with the flip-top can patent. Higher royalty rates for a commodity-oriented product may only be warranted if you have developed a new concept (for instance, a new high-output process) that can replace an existing commodity product and reduce costs by say 10% or more.

DO NOT ASSIGN

If you are selling your patent as a condition of a purchase agreement or a license, never assign it until you have been paid in full. It is like the mortgage to the ranch. You wouldn't want to deed it over to someone after they made only the down payment!

The result of interviewing dozens of inventors says the median royalty rate is probably about 3%. However, one of the best ways to find out what a viable royalty rate would be is to ask a potential licensee what it thinks would be fair. If the company has licensed other technologies, it will be able to tell you what some of the ranges are. One thing you may also consider is a sliding royalty rate. This is a good way to perhaps appeal to your desire for a higher royalty and the licensee's request for a lower one. Start with a higher rate for lower volumes and a declining rate as the sales volume increases. For instance, you might have a rate of 5% for sales up to $1 million, then 4% for sales from $2 to $4 million, then 3% for sales from $4 to $8 million, and 2% for sales over $8 million. This also serves as an incentive for the licensee to really go out there and make things happen.

Another factor that plays a part in determining a royalty rate is based on the up-front money you receive. It is just like a mortgage on a house—you can buy down the interest rate with up-front payments. So can a licensee buy down royalty rates with a one-time up-front fee.

Also keep in mind that an exclusive license will probably command a higher rate than a nonexclusive license. And if there is more than one licensee, you may have a more favorable rate for the first licensee with the favored nation status.

In addition to up-front money influencing a royalty rate, consideration for a lower royalty can also be given to a licensee if it is willing to pay all your costs for R&D, patenting, traveling expenses, and so on. With exclusive licenses, further consideration may include the licensee's defense of the patents if infringed or there is an invalidity challenge. You'll read more about this approach in the coming section on Licensing Incentives.

What is usually more important than royalty rates is to make sure your licenses have performance clauses that the licensee has to meet, or the license—if exclusive—is voided or perhaps made nonexclusive. And of course, the license should include a clause for immediate retribution if royalties and their accompanying reports are not received on time.

Do some research in your field and you'll be able to get a good idea what royalty rate you can charge for your innovation. But whatever you do . . . don't get greedy!

Up-Front Fees

Just like royalties, up-front fees may vary tremendously. Some licenses may have no up-front fee, some $25,000, and others may be into the millions of dollars. It depends on several factors, which are essentially the same factors that affect royalty rates. And just like with royalties, don't get greedy and ask for the moon on an unproven innovation. Most entities that ask for large up-front fees have a proven track record. If you're new to the field, be flexible.

A higher royalty rate typically reflects a lower up-front fee. Likewise a large up-front fee may buy down or reduce the royalty rate. In many industries, up-front fees are common, yet in other industries, they may be uncommon. Some research will bear this out.

As a rule of thumb, with multiple patents and broad scope, you can ask for a higher up-front fee, providing the invention is considered highly marketable. If the product has been successfully launched and established, and you are nonexclusively licensing companies in the field, you may also ask for higher up-front fees.

In most cases you should ask for an up-front fee even if you are inexperienced in your field. If the patent protection is solid, ask. Regardless of whether the license is exclusive or nonexclusive, up-front fees are usually justified. The only question is "how much?"

One word of caution with up-front fees: don't let them kill a license deal. If your patent protection is sufficient and the licensee is willing to spend a lot of money to ramp up, be flexible.

Licensing Incentives

Since licensing should be reviewed as a form of partnership, providing incentives for licensees makes a lot of sense. Here are some incentives you may consider:

1. **Royalties.** A new product with added benefits to end users can usually bring a higher price. But this does not mean a dramatically higher price, as price elasticity becomes an issue. Keep in mind, the more price competitive your innovation is, the higher the sales. Try to keep royalties low providing your licensee agrees to keep his sell price competitive with other products. The difference in 1% on a royalty may mean the difference in millions in sales. What's better? 2% of $10 million in sales or 3% of $500,000?

2. **Sliding scales.** Royalty rates may decline as sales volumes increase. A sliding scale is an incentive for a producer to promote and sell more product, resulting in lower costs. A typical sliding scale may be:

Less than $2,000,000 in sales	3%
$2,000,000–$5,000,000	2%
Over $5,000,000	1.5%

The sales volume may be based on annual sales volume, which resets every January of the calendar year. It may also be based on total sales volume. In other words, once sales volume reaches a certain point, the royalty rate goes down and stays at the lower rate. There are other approaches as well, which you may discuss with your licensee and attorney.

3. **Start-up exclusives.** Even in a commodity market in which you may ultimately license many entities, it is usually a smart move to grant a temporary exclusive to a single start-up company that is willing to get the innovation launched. In exchange for the start-up exclusive, the licensee must invest its capital to make sure the product launch succeeds. Being first to market is just as important as having proprietary

DON'T GET GREEDY!

When considering royalties, keep in mind that most companies earn about 10% profit after all costs are paid. By requesting a 10% royalty, you are asking to make as much as they make. Can you justify this amount? Did you put in as much time and effort and financial risk as they have? Most likely you have not. Licensees are the ones that must spend their money—earned from the sales of other products—in order to launch your innovation. Don't get greedy! Very few high-volume products have large profit margins. An overly aggressive royalty will kill a new idea fast. Besides, a small percentage of a multimillion dollar market is a lot better than 10% of nothing.

LOST ROYALTIES OR LOST SALES?

Assume that your licensee is exclusively licensed and is being infringed on by another company, which has been notified of patent infringement to the degree of $10 million a year. If for a period of three years (the statute of limitations on patent infringement is 6 years), no action was taken, the potential loss is $30 million in sales. This means a lawsuit can be filed for $90 million since willful infringement gets treble damages! Compare this to the amount of royalties you could sue for. At a 3% rate, you could only sue for $900,000 x 3 or a total of $2.7 million. If your licensee is suing for $90 million and you share in 20% of the award, you stand to gain $18 million. Your licensee is also footing the bill. Which would you prefer?

product protection. And frankly, what should you care if your licensee has an exclusive for twenty years or more? If it is based on guaranteed sales and royalty income to you, you should be satisfied.

4. **Sublicensing others.** You can include a clause in a license agreement with a start-up licensee that allows it to sublicense others in the future. This would typically be done only if the licensee is a company with sufficient clout and has the ability to establish the product line in the marketplace. The structure of such an agreement would be one in which both parties share the income from the sublicensing. This may be an attractive incentive based on your licensee's success in launching the innovation. The long-term benefit is that it helps standardize the industry as well.

5. **Participation in patent infringement awards.** If a licensee has an exclusive license and your patents are infringed, it may be in a position to sue the infringer for the loss of sales and income. This is a great deal more than you can sue for . . . which is lost royalties. In such an event, your licensee should also be defending the patents against invalidation. Usually this clause is agreeable to a licensee as it has much more to gain than you do. If you include this clause in your agreement, ask for a percentage of any infringement award the licensee receives; 15%–25% is a reasonable request.

6. **Right of first refusal on new developments.** You can also give a licensee the right of first refusal on any improvements to the licensed invention. The alternative to this clause would be to automatically include all future improvements to the innovation. In either case, it should be predicated on the license being in good standing and the licensee achieving its objectives. This clause is not one you'd normally offer a licensee; however, if you have established a very good working relationship, it makes a lot of sense.

7. **Trade dress protection.** If your innovation has the potential for trade dress protection, make it part of your license agreement. Doing so helps establish your trade dress rights. Licensing trade dress is usually in the best interests of your licensee, too. Long term the licensee stands to gain from continued intellectual property protection after the patents have expired.

There are other incentives that may unfold as a result of the licensing process. Be open-minded and realistic. You don't want to give away your income or control of the patents without receiving something in return. Incentives must be earned.

The Simplified License Strategy

Here is a licensing approach that provides great incentives for a licensee and puts you in a more hands-off position. It's sometimes called a "sweetheart agreement." A sweetheart agreement would be used for a special company that is willing to give solid guarantees on sales. It exploits most of the previous incentives discussed. Here's how it works:

1. The license is exclusive in the licensee's trade, country, industry, etc.

2. You give the licensee a substantially reduced royalty rate—for instance, 1.5% instead of 3%.

3. The licensee guarantees an agreed-upon monthly income from royalties. Set up a reasonable timeline to ramp up.

4. The licensee pays for all patenting, R&D expenses, training, travel, etc.

5. The licensee pays for prosecution of infringers in the event of infringement.

This type of agreement with ·guaranteed payments invariably ensures that the licensee will aggressively take your product to market. Likewise, it gives you security and allows you to focus your efforts on what you do best—invent.

A simplified license agreement written with single-line spacing and about two pages long is shown on the following two pages. The license agreement contains the essential elements of a license agreement. Should you elect to use this approach, make sure you review your licensing strategy and the agreement's content with your patent attorney.

Overview of the Simplified License Agreement

This license agreement is essentially like that of any other agreement, but it is greatly slimmed down and more concise. The real difference is that with single line spacing, it is reduced to two pages and is perceived as a fairly easy document to read. Here are the details:

The first paragraph begins with the basic information: names, addresses, date, etc.

Paragraph A lists the intellectual property rights you're licensing.

Paragraph B states that the licensee wishes to receive those exclusive rights and the licensor wishes to grant them in the United States. You can apply this agreement to any other country or part of the world by changing the territory.

Paragraph 1 defines the additional intellectual property rights covered under the agreement. This agreement includes the licensing of trademarks and trade dress as well.

Paragraph 2 grants the exclusive license to the licensee, which may or may not be limited to a certain field.

Paragraph 3 sets the royalty rate. It may be payable on a monthly or quarterly basis (or otherwise).

Paragraph 4 states that the exclusivity is based on performance. This agreement allows for a five-year ramp-up period. Exclusivity is maintained by either maintaining sufficient sales or paying the minimum royalty. One guideline you can use to establish a five-year ramp-up period is to establish an acceptable sales volume target for five years down the road and then back it off from there. For instance, have a target of about 20% of this amount in year 1, 40% in year 2, 60% in year 3, and 80% in year 4. However, it really comes down to whatever figures and means the parties agree upon to reach the targeted goal.

This paragraph also includes a clause saying that the license agreement may become nonexclusive should the licensee not wish to pay the minimum royalties. In this case, the royalty rate is then renegotiable. However, you may decide to completely void the agreement should the licensee decide not to continue.

Paragraph 5 requests normal record keeping of the manufacture and sale of licensed products sold.

Paragraph 6 is a statement establishing a marketing right.

Paragraph 7 is the clause that says that the licensee will pay all of licensor's costs

ROYALTY INCENTIVES

Think about this. If you receive a 3% royalty and you have to pay your R&D costs, attorney's fees, etc., you would be much better off accepting a lower rate and having your licensee pay those costs instead. Your licensee will most likely have a budget to pay for some of those expenses. With lower royalties, they can be much more competitive in the marketplace. The outcome of this approach is obvious . . . it is a win-win situation for both.

LICENSE AGREEMENT

THIS AGREEMENT is made this _____ day of _____, 2___, between _____ located at _____ (Licensor) and _____, a _____ corporation located at _____ (Licensee).

Recitals

A. Licensor is the owner of certain technology, know-how and intellectual property pertaining to _____ in which Licensor has filed a patent application (U.S. Patent Application No._____ entitled _____ filed on _____) or has been granted a patent (U.S. Patent No. _____) revealing the invention (the "Invention").

B. Licensor desires to grant Licensee the exclusive right to manufacture, sell, and market the Invention in the United States and Licensee desires to license the rights to manufacture (including subcontract via designated subcontractors) and market the Invention in the United States (the Territory).

NOW, THEREFORE, in consideration of the premises and of the mutual covenants set forth herein, it is agreed by and between the parties as follows:

1. *The Intellectual Property Rights.* Licensor defines its intellectual property rights (Rights) as those falling under the above referenced patent application [or patent] and also includes the trademark, _____™ and any trade dress that results from the manufacture and marketing of the Invention. It may also include the following alternate marks owned by Licensor: _____™, _____™ and _____ ™.

2. *Grant of License.* Licensor represents that he has full and complete authority to license the Rights to manufacture, sell, and market the Invention in the Territory and Licensor grants to Licensee the exclusive Rights to manufacture, sell, and market the Invention in the Territory. This exclusive license granted by Licensor to Licensee is restricted to the manufacturing, selling, and marketing of the Invention in the field of _____. Licensee understands that Licensor also intends on granting exclusive licenses of the Invention to other entities in at least two other unrelated fields, one of which is in the field of _____ and the other of which is in the field of _____, both of which are specifically not in Licensee's field. Exclusive licenses in the other fields will specifically include clauses excluding the ability to manufacture and market in Licensee's field.

3. *Royalty.* In consideration of this Agreement, Licensee agrees to pay Licensor a monthly royalty based upon _____ % of the gross sales revenue derived from the sales of the Invention. For purposes of this Agreement, gross sales revenue will be calculated based on the amount stated on customer invoices issued by Licensee and is payable on the 20th of the month following the close of any previous month. The royalty check will be accompanied by a sales report summary.

4. *Maintaining Exclusivity.* The exclusive Rights herein will remain in force based upon Licensee maintaining minimum sales and minimum royalty payments which are calculated on a quarterly basis. If for any reason Licensee does not meet the minimum quarterly sales volume, it has the right to maintain the exclusive license by payment at the end of each quarter of the corresponding minimum royalty. Minimum sales required to maintain the exclusive license are:

	1st Quarter	2nd Quarter	3rd Quarter	4th Quarter
Year 1:				
Year 2:				
Year 3:				
Year 4:				
Year 5 and thereafter:				

If for any reason Licensee decides it no longer wishes to maintain its exclusive license, then this agreement shall become nonexclusive, this paragraph 4 is then stricken, and Licensee's royalty rate will be subject to renegotiation.

5.   *Records.* Licensee agrees to maintain complete and accurate records of its manufacture and sale of goods subject to royalty and will request the same of any sublicensee. Such records and other operations of Licensee shall be available during ordinary business hours for examination by an authorized employee or representative of Licensor.

6.   *Pricing.* Licensee shall have the right to establish the price at which it will sell the Invention.

7.   *Reimbursement of Costs.* Licensee agrees that in lieu of the favorable royalty rates it will reimburse Licensor for any related patenting costs, trademark or trade dress registration costs, now or in the future, and any other R&D costs incurred by Licensor on behalf of Licensee.

8.   *Defending the Rights.* Licensee agrees to defend all the Rights owned by Licensor in the event of infringement in the field of Licensee's exclusive license. In the event of an award (which may also include damages) being granted Licensee for its lost sales and profits, Licensee agrees to pay Licensor _____% of the net award.

9.   *Parties and Assignment.* This agreement shall be binding upon and inure to the benefit of and be enforceable by the parties or their successors or assigns. If Licensee wishes to transfer the license with the sale of its business, it must have the prior written consent of Licensor, which shall not be unreasonably withheld. Upon transfer of this agreement, the transferee shall execute a written document declaring it will: (i) assume all of the obligations undertaken by Licensee hereunder; (ii) acknowledge that it has satisfied itself that the Licensor's right to license same are valid and incontestable; and (iii) covenant not to directly or indirectly contest, challenge, or deny the validity of any of Licensor's Rights in any forum or for any purpose.

10.   *Insurance.* The Licensee shall procure and maintain comprehensive general liability insurance, naming Licensor as an additional insured, in an amount greater than or equal to $2,000,000 per incident, including product liability, personal liability, personal injury, property damage, and broad form contractual liability coverage for all acts and omissions directly or indirectly associated with this Agreement. Upon request of Licensor, Licensee shall provide written evidence, satisfactory to Licensor, of above-stated insurance.

11.   *Indemnification.* The Licensee agrees to indemnify and hold harmless the Licensor and its heirs successors, assigns and legal representatives for liability incurred to persons who are injured as a consequence of the use of any Invention manufactured by the Licensee or as a consequence of any defects in the Invention design.

12.   *Reasonable Efforts.* The Licensee agrees to utilize all reasonable efforts to manufacture, promote, market, distribute, and sell the Invention.

13.   *Amendment and Termination.* This agreement may be altered, amended, or terminated only in writing and signed by both parties.

14.   *Governing Law.* This Agreement shall be interpreted and construed under the laws of the State of _____.

IN WITNESS WHEREOF, the parties hereto have caused this Agreement to be executed as of the day and year first above written.

Licensor

By: _____
      Licensor's authorized signature

Licensee

By: _____
      Licensee's authorized signature

There are no rules of thumb to know how much a royalty should be. Try asking your licensee what he/she thinks is fair. Frequently they will have a larger amount in mind than you! Never assign a patent for money up front unless it is considered "payment in full." Once assigned, a patent is like a deed conveyed. In other words, don't expect any more payments. Don't get greedy. If your licensee is taking all the risk, then it should get the largest share of the profit. Greed is the #1 deal-killer for inventors. Experienced negotiators, licensors, and licensees never let attorneys negotiate for them. That is why attorneys carry the nickname, "deal killer." You and your licensees reach a mutual win/win agreement, and then afterward, run it by your attorneys.

pertaining to patenting and other IP registration and R&D expenses. You may also include costs such as travel for training and trade shows. Regardless of how many other costs are identified, you will always have the option of requesting reimbursement.

Paragraph 8 establishes the licensee's responsibility to defend the patents. This includes defending the validity of the patents and includes prosecuting infringing parties. With this clause, you should ask for a percentage of any judgment that is awarded the licensee, up to about 25%.

Paragraph 9 provides for the standard terms of assignment.

Paragraph 10 states that product liability insurance is required on the product line.

Paragraph 11 indemnifies you from any and all claims made by other persons on the licensee.

Paragraph 12 states that the licensee will use all reasonable efforts to achieve its objectives.

Paragraph 13 is a basic termination and amendment clause requiring the signature of both parties.

Paragraph 14 is a statement about the state in which the laws governing the agreement will be construed. Make sure it is your state.

Last, the signatures of you and an officer of the corporation make it official.

Using this type of agreement doesn't necessarily mean it'll be easier to close a deal, but it usually is. With the understanding of who is going to pay for what and a reduced royalty rate established up front, there's just less to talk about and hopefully less conflict and fewer objections. When using this strategy, you'll still want to have your attorney cross the t's and dot the i's after you've reached an agreement with the licensee.

The Best Time to License

Usually the best time to start searching for licensees is as soon as possible. However, make sure you've done enough research to qualify the marketability of your innovation. There is no need to wait for patents to issue. A patent is not required to license an invention. It is usually best to begin your licensing activities during the patent pending phase and use confidentiality agreements with prospective licensees. The exception to this rule may be if you plan to go into business and use licensing as a means to generate additional income. Thus, you may elect to wait until your company has successfully launched the concept, so you can bring on additional licensees later on.

As an independent inventor, you don't want to spend a lot of money pursuing patents on something that doesn't sell, so the earlier you can qualify the innovation and the sooner you can secure a licensee, the further ahead you'll be. Besides, a smart, aggressive marketing partner doesn't want to wait around either.

If a company wants to wait to license your technology until after the patent issues, it will just have to miss out on the opportunity. Those really are not the words of a progressive company anyway. Smart marketing experts will see the opportunity at hand and aggressively go for it. They are willing to "bet on the come" instead of waiting to see what happens. Smart marketers and innovators know that you can't bet on the horse after it races.

Your Negotiating Strategy

Generally speaking, there are two negotiating strategies you can use. You can negotiate the license agreement through an attorney, or you can negotiate it yourself.

Using a patent attorney for negotiating can be extremely costly. It can also result in "building a fence" between the parties, which is not conducive to building a partnership. Many attorneys are known as deal killers when it comes to negotiating license agreements; this results from their aggressively representing your position instead of looking at it from a business perspective. It's best to use a patent attorney to negotiate a license agreement only in the event of infringement and when the matter has been turned over to your attorneys to pursue. Whether the attorney is pursuing the case in court or looking for an out-of-court settlement is part of your strategy.

Negotiating your own license agreements is an alternative and is doable once you have experience. However, how can you get experience? Mistakes can be costly if you're inexperienced and can result in a one-sided license that is worthless.

The best licensing approach is one that builds trust between the parties and strives to conclude a satisfactory business relationship for both parties. A trust relationship should not require the involvement of an attorney, at least not during the initial stages. The most successful approach is usually a two-fold approach with the two principals building a relationship first and then having the attorneys address the legal details later.

This approach also works well with first-time inventors. Your licensees will be your partners for years to come. With the objective of having mutually attainable goals, you'll be on the right track. With a partnership approach to licensing, it is a lot easier for the licensee and the licensor to consummate the majority of a license agreement. Once you've drafted a business-wise agreement, the two parties may then have their attorneys tie up loose ends. This process will be expanded on in detail in the next chapter.

THE BEST STRATEGY ... NO NEGOTIATING

The drive for partnerships is a powerful influence in business. Let a potential licensee know that you prefer not to negotiate a license but want to have a partnership instead. This reflects a team-building approach. Your willingness to share your IP and a strong licensee's ability to manufacture and market can be powerful.

Chapter

16

THE LICENSING PROCESS

Here's a straightforward process you can follow
to license your intellectual property.

Y ou're ready to license your innovation! Before you just go out there and start making phone calls or sending letters in the mail, prepare a licensing plan you can follow. Use the following plan to license your intellectual property. You'll find it easy to follow and cost effective to employ.

This process is based on years of experience and on techniques used by successful licensors and attorneys. The approach is flexible enough to be used for most innovations and technologies from the simple ones up to and including those that are relatively complex with multiple patents.

It works well with licensing small businesses or large corporate entities. After all, the strategy is essentially the same for all. The licensees are looking for opportunity, profit, and protection, and that, we hope, is what you are offering. With a licensing plan in place, you are displaying professionalism and will be taken seriously. You simply must be ready to answer and anticipate questions and objections a licensee may have.

Start with a plan and don't go out half prepared. You'll only be setting yourself up for failure. Once you've prepared your plan and completed the initial stages, you'll be on your way.

The Licensing Plan

When preparing to license your innovation, the initial stage is nothing more than preparing a strategy and conducting an analysis of your intellectual property and perhaps potential partners. With that understanding in place, you'll be ready to start looking for and qualifying licensees.

Here are the essentials of a licensing plan:

Initial stage:

1. **Develop and define your strategy.** Essentially you want to determine your approach to licensing. Will it be an exclusive or nonexclusive license? Will it be limited to various fields? Or, will you be licensing after your company launches the product? Will you be licensing a limited number of exclusive licensees? Staging them in? What incentives have you considered? The answers to these questions are discussed in detail in Chapter 15.

2. **Do a SWOT analysis.** This term, used by licensing experts, means you need to analyze and evaluate your intellectual property's Strengths, Weaknesses, Opportunities, and Threats. This is important, because your licensees will do this and if you're not prepared, you'll not be able to handle their questions.

Second stage:

1. **Find licensees.** Use New Product Summaries and the methods discussed in detail in Chapters 10 through 12.

2. **Research interested candidates.** Once a candidate is identified, you'll want to research it to better understand its position, its needs, and its direction.

3. **Use a term sheet.** Send a term sheet outlining the basic details of a license agreement.

4. **Prepare a license agreement.** Once the term sheet is accepted, the formal agreement should be prepared.

5. **Send the agreement to the licensee for discussion.** Using the methods covered later in this chapter, send the license agreement to the prospect.

6. **Handle objections.** This is where the rubber meets the road. During the course of discussing the term sheet and/or the license agreement, objections will arise. You should be prepared to handle them. They can make or break an agreement.

7. **Modify the agreement accordingly.** Once you and your licensee have reached an acceptable business relationship, make the changes to the document.

8. **Send it to attorneys.** Once you and your licensee have agreed upon a working agreement, it is time to send the agreement to your respective attorneys to cross the t's and dot the i's.

9. **Sign it.** With the details tidied up, sign it!

There are other considerations you'll want to be familiar with such as when and how to use a letter of intent or alternatives to licensing. What's important is to plan ahead, be ready, be professional, and be receptive to offers.

SWOT Analysis

We know that a SWOT analysis is to identify strengths, weaknesses, opportunities, and threats relevant to your invention and intellectual property. This is also referred to as an IP audit and is what most licensees will conduct before licensing in, so you should be prepared and do it yourself ahead of time. Having completed this analysis you'll be in an excellent position to handle objections from prospective licensees.

Before meeting with a prospective licensee and before determining your royalties, you'll want to have your SWOT analysis completed. It will have a direct impact on the royalty rate you may ask for. Make sure this is done before offering your innovation, because licensees will expect qualified answers. You must demonstrate your knowledge of the trade and that you are an expert in your field. If you haven't done this, then you're not ready to license.

The primary purpose of a SWOT analysis is to determine the viability of your licensable matter. You'll have to be ready to anticipate objections a licensee may have. Your challenge is to take the weaknesses and threats and turn them into strengths within the presentation so that you'll have the most viable concept possible. At times, this may even require improving on your innovation to increase its value and potential for licensing.

Look at it from the licensee's perspective. If patent and IP protection are weak, and a major competitor can easily design around your innovation, where's the opportunity, where's the protection? Where's the strength in your position? Strong patents represent better protection and an elevated royalty potential. Weak patents represent poor protection and low royalty potential, if any at all.

According to Andre Lake Mayer, a world renowned licensing expert, you may need anecdotal research, qualitative research, and expert testimonials to prove your point. You have to qualify your proprietary advantage. If you can't do that, who will want to pay royalties on a new concept that will put the licensee at a market disadvantage?

Andre challenges every licensor to qualify the following:

- What is the value and potential of your idea?
- Is it one product? Is it a line of products/collection and is it sustainable?
- Is it a trademark that could evolve into a far-reaching brand across multiple categories and platforms?
- How will it gain consumer awareness?
- Why does the marketplace need this?
- Is your timing right?
- What differentiates your innovation from others?
- How do you get an edge on the competition?
- How will the retailer or distributor have a competitive edge?

The answers to these questions need to be clear and concise, because that's what a licensee will be expecting.

If you've done your homework throughout, your customer-driven innovations will be well-received and highly licensable. Now, you have to put them down into concrete words and language that shows the licensee the merits of the opportunity.

Licensing IP is "licensing money" from the perspective of the licensee. Can you "show him/her the money?"

Finding Licensees

The process of finding licensees uses essentially the same methods as finding marketing partners (and at times manufacturing partners) as outlined in Chapters 11 and 12. Only your focus is very specific: you're seeking marketing experts in the field of your invention to license. Two of your most important aids are your New Product Summary so you may quickly qualify a company's interest and your Invention Prospectus (Chap-

---

ONE CHANCE

You only have ONE chance to make a first impression with a licensee—so EVALUATE your invention one more time before presenting it to a prospective licensee. Their job is to find why NOT to license, so find the problems—and fix them—before your licensee does!
–*Andy Gibbs, CEO, PatentCafe.com*

ter 10) so you may explain the opportunity in a qualified, professional manner. If you've not prepared these two documents, you're probably not ready to license your innovation.

If you're exclusively licensing your innovation, which is the case more often than not, you'll want to base your interview questions after those outlined in Chapters 11 and 12 and will want to include some additional considerations such as those listed in the following checklist. For the most point these seven points are based on the qualities you have pursued in selecting potential manufacturing and marketing partners, as explained in Chapters 11 and 12.

Consistent with your strategy to base your licensing focus on the ability of a licensee to market your innovation, not manufacture it, the following is a checklist to use when evaluating a potential exclusive licensee.

Checklist for Exclusive Licenses

1. **Size.** Is the potential licensee's present sales and marketing network the right size and the right fit to handle your innovation's needs, whether it is being considered as a long-term exclusive or a start-up exclusive? If not, does the prospect have the ability to expand sales?

2. **Cutting edge capabilities.** Is its marketing approach really dedicated to innovation? Does it have the graphics and marketing abilities your new product needs to succeed? Is its manufacturing ability state of the art? Are they a low-cost supplier/producer of the required quality?

3. **Financial clout.** Does it have the financial ability to invest in machinery, dies, molds, and so on in order to ramp up the innovation in the targeted field, to the target market?

4. **History and reputation.** Has this company launched other innovations successfully? Does the company have similar licensing or product development relationships with other companies? Is it now selling or has it ever sold a product that has infringed a patent? Does it have an ethical reputation?

5. **Testing and trials.** With the added responsibility of being exclusive or being the start-up exclusive, is this candidate prepared to handle hands-on product tests and trials? It is more important than ever that an exclusive licensee does this well as it spearheads the effort to arrange customer trials and monitor results.

6. **Trust.** The trust issue is crucial. You are trusting this entity with an important part of your financial future. Do you believe they'll do what they say?

7. **Leadership.** Conceptually, does the candidate have leadership with vision and great determination to succeed, or does it just want to get in on the action and try to corner your invention to "see if it sells?"

Exclusive licensing relationships can make or break your innovation. If a licensee has been successful in the recent past, it will probably be able to make things happen with your invention. As long as your exclusive license agreement has performance requirements and a means of canceling the agreement in the event of nonperformance, you'll be protected. However, it only makes sense to get off on the right foot with the right licensee from the onset.

WHAT'S THE INVENTION?

If you can't talk about your invention for 30 minutes WITHOUT disclosing your invention, you're not ready to present to a licensee. You should be able to talk about the licensee's market share, profitability, manufacturing process, sales channels, and other ways they can make money. Convince them that they will make money without discussing your invention, and you will most likely have a licensee regardless of what you invented.

–*Andy Gibbs, CEO, PatentCafe.com*

Researching the Licensee

When you've talked with a prospective licensee that shows interest, you should do some research on the company to better qualify it as a prospective partner before sending out term sheets and license agreements. This will be particularly important if you are talking about some sort of exclusive arrangement. Here's what you can do and what to look for.

Start by searching on the Internet at the company's website, google.com or hoovers.com and try to find the following information:

1. **Learn about the company's market.** Is it national or worldwide? Is it sufficient for your innovation's needs? Will you need another licensee in another field or in other countries?

2. **Look for strategic announcements.** What kind of news releases has it made recently? Has it announced expansion plans? Has it purchased any companies recently? How does your innovation fit? Has it fallen on hard times?

3. **Get a copy of its annual report.** This can also be very informative. You can see how it's doing now and what plans it has for the future. Read the notes at the end of the report to look for any surprises they may be obligated to report. They'll usually be buried there.

4. **Get a copy of its 10Ks or 10Qs.** These are the annual and quarterly reports required by the Securities and Exchange Commission (SEC). Usually you can find these on the companies website or through the SEC itself at www.sec.gov.

5. **Contact the individual tracking the stock.** Almost all public companies are being tracked by some expert at one of the major stock brokerages. You can ask your stock broker to contact the individual to find out its present condition and what its plans are for growth.

Should you search a company on the Internet and not find it, remember that it may be a privately held company, which makes learning about it a bit more difficult. Nevertheless, be diligent and you'll be able to find the information you need.

With this background information on a licensee, you'll be able to have meaningful conversations about your innovation, how it fits with the company, and how the prospect may view its potential. Later on, when you're discussing the details of a licensing deal, this information may be valuable in helping you understand how the prospect sees the potential of your innovation and why it may be interested. You'll be able to talk the same language and about the same potential.

The Licensing Mindset

In many respects licensing is just like any other business negotiation, and in many respects, it is quite different. If you have the right frame of mind with the right attitude, you can do well and become very successful in your licensing endeavors.

On the other hand, if you approach licensing without sufficient knowledge in your field and with an attitude that your invention is the best thing since sliced bread, you're probably in for a big surprise.

Having the right mindset will help you build some excellent working relationships, which may provide a home for your future innovations. Keep in mind that there are sev-

---

THE YO-YO

The yo-yo is most likely the second oldest toy in the world, after dolls. Ancient Greek yo-yos were made of terra cotta, and yo-yos are pictured on the walls of Egyptian temples. Even warriors such as Napoleon and the Duke of Wellington were avid yo-yo enthusiasts. The modern yo-yo got its start in the 1920s when a young man from the Philippines, Pedro Flores, drew large crowds when playing with his traditional pastime, the "yo-yo," which means "come-come." He started a company to make the toys, calling it the Flores Yo-Yo Company. One entrepreneur, Donald F. Duncan, saw Flores playing with his yo-yo during a business trip to California. A year later in 1929, he returned and bought the company from Flores, acquiring the unique toy and the name "yo-yo." Shortly afterward Duncan invented the looped slip-string, which allows the yo-yo to sleep.

eral keys to being a successful negotiator, but with licensing it is usually wise to change your mindset to one of partnering instead.

Partnering Instead of Negotiating

What's the difference? It's primarily in your attitude. Fortunately, the attitude of partnering is well received in the modern business world regardless of whether you are a small business or a very large corporation. Partnering makes sense to both entities. Large companies may have excellent market exposure, plenty of money to finance an innovation and ramp up, but they may be lacking in the innovative spirit and know-how of a smaller entrepreneurial business.

Regardless of your company and its size, here are a few tenets you may apply to licensing in order to establish a good working relationship and the making of a partnership:

1. People like to do business with friends. So, be friendly, be polite, show your expertise and enthusiasm, but be realistic in what you say.

2. Don't be greedy; be realistic and understanding in your expectations.

3. Listen carefully to the other party's position and viewpoints. His/her experience in the marketplace may differ from yours and, in most cases, is probably right. If nothing else, it's probably the right viewpoint for the individual's company.

4. Make all decisions based on what's best for the innovation. In fact, those are words that you may—should—repeat throughout your discussions.

Let the partnership theme pervade your discussions and you'll find that 90% of the time it will be well received. If it's not, you should question if the prospective licensee is really a good match.

Power, Information, and Time

Master negotiator Herb Cohen, author of *You Can Negotiate Anything*, talks about the three primary elements of negotiating, power, information and time. These also apply to partnering and are helpful when preparing for your discussions.

Having strong patents and knowing that your invention is highly marketable gives you substantial power (potential power we might say). However, you don't want to flaunt it. Instead, it's best that you position your power as an advantage for a licensing partner as well. Patents will protect the investment and the marketability potential represents profit potential.

All during your product development process you have been building your knowledge base on your innovation. You have also qualified marketability and potential in your prospectus. Now you've completed a SWOT analysis and have researched a potential licensee. Those three components represent a substantial amount of information. The prospectus may be shared with the right licensee. The information you've gathered with a SWOT analysis and company background will assist you in your licensing discussions.

Time and timing is also important. If there is any one area in which an innovator has a disadvantage, it is this element. Successful innovators tend to have a strong sense of urgency in their activities. This may not be an attribute that a careful, deliberate licensee may have. Nevertheless, you want to show your sense of urgency in getting the innovation to market. You want to partner with the right company that is willing to take on the project now, help get out the final bugs, and take the innovation aggressively to market.

You do not want to have your sense of urgency reflect an individual who is pushing to get a deal signed! That is a sign of weakness and may appear desperate.

Again, if you make your decisions based on what's best for the innovation—which is a prompt (as opposed to delayed) market launch—then you're making the right decisions. That's the message you want to convey with all your power and knowledge supporting it.

Using Term Sheets

When a prospective licensee shows interest based on a New Product Summary or more likely, a prospectus, the next step is usually to send the party a term sheet with the basic elements of what would be in the license agreement. This is usually a one- or two-page summary that shows the licensee what you are looking for. In a way, you can think of it as a simple offering, but by no means should it be considered a "take it or leave it document."

Invariably there will be some negotiation, modifications, and additions. Just like when you list a house for sale, some bargaining takes place afterward. When listing a house, you do some research to make sure your asking price is reasonable and accurate; otherwise, you'll scare away buyers. Just like offering a house for sale, a term sheet should be carefully prepared so it doesn't scare away prospective licensees.

The term sheet on the following page is typical of one you may use in your licensing activities. Here's what a term sheet may include:

Title: State the title "Term Sheet" with a description of the specific innovation.

Summary paragraph: Include some basic expectations of the licensee. In this case, you're looking for a start-up licensee that will aggressively go out and ramp up the innovation.

1. **Protected rights:** List the patents issued and serial numbers of the published patent applications. If your patents are pending and unpublished, then you may choose to list your patent applications by title and include an attorney docket number instead of the actual serial number. Usually you will be required to reveal unpublished patent applications so the licensee may review the patentable subject matter. Thus, until that time, the serial numbers and filing dates are confidential. They may then be revealed under a confidentiality agreement.

2. **Field of the license:** In the example given, it includes all applications, but it may be specific to a given field.

3. **Territory:** It is based on licensing a U.S. patent; the territory is the United States.

4. **Term:** The life of the patents is the term (almost always recommended).

5. **License fee:** The term "standard" is being used here as a means to suggest that it is negotiable.

6. **Royalties:** Again the term "standard" is used in the example shown to suggest it's negotiable. Discussions on this paragraph and paragraph 5 set the stage for favored nation status of the first licensee.

7. **Royalty accounting and reports:** You may elect quarterly or monthly royalty payments.

8. **Marking of bags:** This is a standard requirement.

THE FRISBEE

During the depression of the 1930s, people had to find inexpensive ways to have fun. Kids discovered that metal pie plates flew well, which became a popular pastime. However, pie plate tossing had its risks. The plates made a strange unpleasant noise when they flew and could injure your hand if you didn't catch them just right. In 1948, Walter Morrison and Warren Franscioni got the idea to make the flying pie plates out of plastic. They called them "Flyin' Saucers" in hopes to cash in on the UFO craze in America. The plastic version also flew further and straighter than its metal prototype. In 1955, Wham-O, the makers of the Hula-Hoop, licensed the rights to "Flyin' Saucers." The name Friesbie came from the Friesbie Pie Company in Bridgeport CT, where the word was shouted out to warn bystanders a plate was sailing by. Thus the toy was renamed Frisbee®. More than 200 million Frisbees® have been sold in last 50 years, more than baseballs, footballs, and basketballs combined! Mattel Toys owns the rights to the Frisbee today.

TERM SHEET – SQUARE BOTTOM PLASTIC BAG

An exclusive start-up license is being considered for the plastic square bottom bag, providing certain requirements are met by an acceptable licensee. This includes having the financial means, technological ability, a strong customer base in the field, and sufficient production capacity to ramp up the innovation.

The terms and conditions of the license agreement include:

1. **Protected rights:** U.S. Patent 6,095,687 and U.S. Patent Ser. #20020009575.

2. **Field of the license:** Includes all applications and uses.

3. **Territory:** All products manufactured or sold in the United States.

4. **Term:** The life of the patents.

5. **License Fee:** The standard one-time license fee, $50,000.

6. **Royalties:** The below rate is paid on the net sales price, which is the actual sales price, less freight, returns, rebates, and trade discounts, but not sales commissions. The standard royalty rates for annualized sales are:

Up to $2 million in annual sales:	six percent (6%)
$2,000,000.01 to $10,000,000:	four and one-half percent (4.5%)
$10 million:	to $25 million three percent (3%)

7. **Royalty accounting and reports:** Reports and payments due on a quarterly basis.

8. **Marking of bags:** Includes relevant patent numbers and patents pending.

9. **Default:** Licensor may terminate with 30 days notice if licensee is in default of royalty payment, 10 days for all other breaches.

10. **Insurance:** $2 million is required.

11. **Licensor enforces patents:** This is the obligation of Licensor and will be diligently pursued as desirable or required.

9. Default: It goes without saying that a license may be cancelled, but you can put some general language in the term sheet.

10. Insurance: You want the licensee to know what acceptable insurance standards will be required.

11. Patent enforcement: This is your responsibility; however, if you're pursuing a sweetheart licensing strategy, then this paragraph should reflect that language.

Term sheets may be a valuable interim step, but they are not always needed. Many times you may jump directly to submitting a license agreement for review with the licensee.

Term sheets (and licenses too, for that matter) may or may not be considered confidential. That's up to you and your approach. However, if you have unpublished patents pending and the review of those patent applications is going to be discussed further to evaluate them for possible licensing, you should have a confidentiality agreement in place. This is also true if you are going to be discussing any other sensitive material with a prospective licensee.

Usually this stage of discussions take place after the licensee has already signed a confidentiality agreement prior to review of the invention, as outlined in Chapter 10.

Whatever your approach, if you've done your research and your homework well, the next step is to tailor a standard license.

Tailoring a Standard License

With your marketing and manufacturing objectives determined, and with or without input you've received from a prospective licensee, you can tailor your own standard license agreement to fit your needs. You may want to consult your attorney to help with your license draft or you can use an existing license agreement and adapt it to your needs.

The most widely used pre-published standard license agreements (exclusive, nonexclusive or sweetheart) are in the *From Patent to Profit Resource Guide*. They were drafted by a patent attorney and are specifically for licensing innovations and technologies with patents pending or patents already granted. The Resource Guide CD, available at www.frompatenttoprofit.com, allows you to modify the documents on your own computer.

Using a standard license agreement is a valuable tool when introducing a license agreement to a prospective licensee and opening discussion on the finer points. It doesn't matter how much experience you have, a standard license agreement represents an excellent starting point. As you'll see, it will be discussed, modified, and improved upon by both parties until a business agreement is reached. Then, it'll be sent to attorneys for final approval.

Introducing a License to a Potential Candidate

Usually a license agreement is introduced to a prospective licensee when it requests one. Sometimes you may take the lead and initiate this step by asking the question, "Shall I send you a copy of my standard license agreement and we can go over it?" Regardless of approach, this is an important step because it shows serious intent on the licensee's behalf and a willingness to enter into some form of agreement.

Whether or not you are inexperienced with licensing, here is an approach that is very

---

SLICED BREAD

You've heard the expression, "it's the greatest thing since sliced bread," but have you ever wondered who invented it? Bread was first made around 12,000 years ago. The dough was made by mixing crushed grain and adding water. Most likely it would have been baked by covering the dough with hot ashes. It was probably tough, flat and chewy, probably similar to pita bread. The Egyptians discovered that if you let the dough ferment, gases will build up inside the loaf and cause it to rise when baked. Otto Frederick Rohwedder is considered the father of sliced bread who started working on a bread slicer in 1912. When Rohwedder introduced the concept to bakers, they had no interest and told him that pre-cut loaves would quickly go stale. So, in order to overcome the problem Rohwedder designed a machine in 1928 that would slice and wrap bread keeping it moist and keeping it from going stale.

SUPPORTING ROLE

Never negotiate a patent license without the assistance of an attorney. Your attorney can help provide a legal background and assistance when needed. Having any attorney to contact during licensing talks can make or break a deal. It shows your professionalism as well.

–Michael S. Neustel,
U.S. Patent Attorney

straightforward, friendly, and disarming. Almost always it is well-received by prospective partners. By this time you should reasonably understand the anatomy of a license as outlined in Chapter 14.

1. **Offer to send a standard license agreement.** With the understanding that there is interest on behalf of a potential licensee, tell the prospect that you will send him/her a draft of your standard license agreement so that you and he/she may "fill in the blanks" together. If the party agrees, this would be an appropriate step to take.

2. **Set the attitude up front.** Tell your prospect that before you send the standard agreement to him/her that you'd like to have an understanding up front. Explain that *your philosophy views licensing as developing a partnership* and not as a negotiation. Tell the prospect that an acceptable agreement must be good for both parties.

3. **Next, set the conditions for discussion.** Tell the prospect that you would like him/her to agree on one condition in your discussions. That one condition is that *the two of you agree to work out as much of the agreement as possible in order to make a good business decision* for both parties. Then, only after you've done that, will the agreement be turned over to your attorneys so they may cross the t's and dot the i's. The two of you should make the agreement ahead of time that "the license will not be negotiated by our attorneys."

4. **Send the standard license agreement.** The standard agreement should include several blanks you'll need to fill out together, unless some of those terms and conditions were part of the term sheet and have been already agreed upon by the licensee. Usually you will send this document via overnight mails or via email. The two most obvious blanks that you'll be discussing are the royalty rates, up-front fees and sales guarantees if it is an exclusive license.

5. **Call the prospective licensee.** Once the potential candidate receives the document, you each can "fill in the blanks" over the phone. Better yet, meet in person, talk it over, and complete the agreement. Take your time and listen. You'll want to know those items that are important to your partner. It is usually during these discussions that objections will arise. This step may very possibly take several phone calls to complete.

6. **Prepare another draft.** Based on your discussions you'll want to prepare another draft that includes all the terms and conditions you've discussed with your counterpart. After one last review, it should be ready to turn over to your attorneys for a final editing.

7. **Send to your respective attorneys.** Now your attorneys may cross the t's and dot the i's. They may also work out acceptable language for any details you've left up to them to figure out. When having attorneys review your draft, make sure that any changes they suggest are from a legal point of view and not from the business side. Remember, you previously made an agreement that the business decisions are up to you and your prospective licensee and not the attorneys.

You Go First

Here's a simple approach you may take when licensing your patents to companies that will have a permanent or start-up exclusive. Once nonexclusive royalty rates are established, this approach would most likely not be used.

The most important consideration with a license agreement is usually the royalty rate. Obviously with stronger patents, you can ask a higher royalty rate. But aside from that, what is the best way to establish a fair royalty rate?

First of all, you don't want to negotiate against yourself. If you've used a term sheet, it included some semblance of a royalty rate and a starting point. But almost invariably a question will be asked if the rate is negotiable or a statement will be made saying it's too high. What do you do?

When you encounter this question, the best response is to ask the prospective licensee to give you a proposal. Don't say you'll accept a lower rate or "we can work it out." It's best just to ask the other party to give you a proposal.

In other words, "you go first." Look, the reality is that you will not know what makes sense or what doesn't until the other party makes some form of offer. Just like when you're selling a house, you have a list price (similar to a term sheet), but the buyer has to make an offer. Nothing happens until that's made.

Similarly, with an exclusive or start-up exclusive license, nothing happens until you receive a proposal from the licensee. If you don't do this, you'll end out negotiating with yourself.

Objections From the Licensee

After setting the stage for a partnership and submitting a term sheet and your standard license agreement, there will invariably be some objections and some requests from a prospective licensee. Here are several issues that may arise, along with ways you can handle them.

Short-Term Exclusive

When offering a short-term exclusive, you're giving the licensee an opportunity to be first to market. This is an exceptionally valuable position to have. However, the amount of time the licensee may want may be much longer than the amount of time you believe is necessary.

There are a few way you can resolve this objection. But money talks . . . and hot air walks. Here's what you can do:

1. **Offer guarantees during ramp up.** In a way, it doesn't matter if the start-up exclusive is longer than what you thought was necessary as long as it is backed up with real time guarantees and royalties. A typical ramp up may take from one to four years depending on the complexity of the innovation and the technology. So if the licensee is willing to make a powerful commitment, then why not?

2. **Limit the number of start-up licensees.** Instead of making the license nonexclusive after a short time frame, offer to limit the number of licensees during the start-up period. This has a two-fold positive effect. First, licenses can be limited to a select list that would be considered favorable competitors and not ones that may destroy market prices. This is usually one of the biggest concerns to start-up licensees in nonexclusive relationships. So a preset list will help alleviate anxieties during the start-up months and years.

The time period for the limited number of start-up licensees may be anywhere from

HOT DOG!

Hot dogs were invented by German immigrants in the 1860s when they began selling sausages from carts in New York City. They peddled their sausages in a different way, by putting them inside a bun topped with sauerkraut. They were easy to carry, easy to eat and provided a way to keep the mustard and relish from sliding off. A butcher, Charles Feltman, popularized them at Coney Island, giving birth to the famous Coney Island hot dog. During a major league baseball season, fans eat over 26 million hot dogs. The world's largest hot dog was 608 meters in length . . . before it was eaten.

one to five years or so. This type of arrangement may also be (in many cases "should be") tied to some sort of minimum guarantee on behalf of the licensees.

3. **Stage in licensees.** You can also offer to stage in licensees over a certain time period. This is similar to limiting the number of licensees, but you're establishing time periods in which a new staged-in licensee may begin manufacturing and selling product.

With any of these plans, consult with your attorney to prepare the appropriate language and make sure you are protected.

Determining Performance Clauses

There are several ways you may structure performance clauses. The most common way to establish a performance clause for an exclusive license is to have quarterly sales objectives for a five-year ramp-up period, after which sales may be flat. In the case of a start-up exclusive, the ramp-up period would typically be less, as short as a few months to one to two years. The five-year ramp-up is what's illustrated in the two license agreements in Chapters 14 and 15.

There are other ways to establish performance clauses. Here are some suggestions:

1. **Units produced.** Instead of guaranteed sales, you can base performance on numbers of units produced. This is similar to having sales guarantees except that an inventory buildup precedes sales and almost guarantees sales will follow.

2. **Machinery purchased.** At times, you can tie in an exclusive to production machinery required to make the innovation.

3. **Market demand.** If there is some form of measurable market demand you can track, you can tie your exclusive into the company's ability to meet the demand. For instance, this may be predicated on back orders and lead times. If they reach a certain mark, the nonexclusive clause kicks in. The only problem with this is that it would be difficult to monitor.

4. **Customer demands.** If an end user requests an alternate supplier, then that too should be way to end the exclusive. Usually with this type of clause the request would come through a competitor.

5. **Increase up-front fee.** Another approach is to ask for a larger up-front fee in lieu of performance guarantees during a start-up exclusive.

There is another area of performance clauses that can become challenging as well—that is, the quarterly or annual estimates. While you are looking for large numbers, the licensee is typically going to be conservative and will want to give much lower numbers. Here are a few suggestions:

1. **You go first.** Have the licensee submit his/her sales figures to you instead of your presenting it with your expectations.

2. **Be prepared.** If the prospect's figures are much lower than you anticipated, you must be able to provide market data supporting otherwise.

3. **Meet with the National Sales Manager.** If you can arrange a meeting with the company's national sales manager and discuss why the figures are so low and perhaps provide your supporting market data, it may allow you to resolve the issue.

4. **Ask your attorneys to figure it out.** As a last choice, ask your attorneys to give you input on what may be acceptable. You and your licensee will still have the last say regardless of what the attorneys come up with.

Or course in most performance clauses you will want to give the licensee the ability to keep it in force by maintaining the equivalent of a minimum royalty.

Objection to Royalty Rate

This is usually the most difficult of all areas to agree on, but not always. If you and the licensee have differences in this area, here are some approaches you may consider to break the stalemate. But first, make sure you have carefully listened to your licensee's position and its argument for a lower royalty. Lower rates mean more competitive products and usually much larger sales. If you're still at an impasse, here are some suggestions:

1. **Accept a sliding scale.** Offer to accept the licensee's lower rate when it meets certain sales objectives. The sliding scale may kick in on annualized sales or may be based on total sales. It may also be based on a formula that provides for the next year's sales to be at the same rate established by the previous year's sales volume. Your attorney can write this clause. An example of a sliding scale clause is illustrated in Chapter 15.

2. **Increase the up-front fee.** If a licensee wants a lower royalty rate, you can agree providing the licensee agrees to pay a higher up-front fee. Remember, this is just like buying down a mortgage rate. With a large up-front fee, you can accept very low royalties.

3. **Offer a paid-up license.** This may be a bit of a bold step, but ask the prospective licensee if it would prefer to pay a one-time lump sum for a paid-up license. Or, as an alternative, perhaps it would prefer to pay lump sum payments on an annual basis. The lump sum payments could be based on several factors, including sales volume, number of units produced, number of production lines employed, and so on. This would most likely require a regular audit and would have to be written by your attorney.

4. **Apply marketing credits.** If the concern of the licensee is that it will be making a substantial investment to ramp up the innovation, offer to reduce the royalty in exchange for its marketing commitment. You may lower the royalty rate based on a certain dollar expenditure commitment.

Above all, there must be some common ground between the two of you. You can't expect high royalties without substantial justification from a marketing and patent protection viewpoint. On the other hand, you have to be rewarded for your inventive effort. Finding that common ground is very challenging at times.

Resistance to Up-front Fee

This is another common area of disagreement. There is a rule of thumb you can apply regarding up-front fees. Generally speaking, the more experienced you are in the field and the more patents you have, the easier it is to ask for up-front fees. Nevertheless, it's a good idea to always ask for them in every licensing application. Just make them reasonable.

Before you counter a licensee's objections, you should let it know that you've invested a substantial amount of time and money to get this project where it is (typically

#1 LICENSE KILLER

The demand for high royalty rates is the number one killer of a licensing deal. You simply don't want to fall into this category. Be creative and listen. Strive to find a solution that makes sense for all parties. If you go into licensing talks with demands of high royalties or become stubborn and unwilling to budge, more often than not, it will end the partnering opportunity. Worse yet, once the opportunity is lost, it is unlikely that the potential licensee will be interested later on . . . even though you offer lower rates.

far more than what you are requesting in the up-front fee). Without your time investment, there would not be an opportunity. Based on this, an up-front fee is justified.

If after discussing this, there are still objections, here are a few ways you can deal with them:

1. **Split out the up-front fee payments.** Instead of having the entire amount due at signing, you can agree to have it payable over a certain time period, for instance, two payments over 90 or 180 days.

2. **Apply part of the up-front fee against future royalties.** You may consider applying 50% or more of the up-front fee against future royalty payments.

3. **Offer to reduce the up-front fee.** Don't let an up-front fee kill an otherwise good deal for the parties. If you offer to reduce it, make sure you get something in return like leaving your royalty schedule untouched. However, if the company wants no up-front fees, low royalties, and no performance guarantees, you really must pass.

THE TOUGH PROBLEMS

In the later stages, if there are any details you two just can't agree on, it may be best to have your attorneys work them out instead. This is more disarming than getting into a disagreement over some minor point.

Usually the amount of the up-front fee is weighed against the royalty rate, but if both are unreasonable, there won't be any resolution to consider as the prospective licensee will walk and nothing will happen.

What if a Patent Doesn't Issue?

A licensee may not be comfortable entering into an agreement that it believes may later penalize it. This might happen if it were locked into your technology and no patent issues, thus allowing competitors to enter the marketplace. If this happens, you may offer to include a clause that says the license agreement will be terminated if no patent issues in five years. This should be ample time to prosecute the application even if an appeal is required.

The risk of this type of clause is minimal. First, five years is better than none. Second, if you are having a difficult time getting your patent to issue with sufficient scope, you can personally visit the U.S. Patent Office and, with your attorney, plead your case. Having an existing licensee in place, who is relying upon the issuance of patent, is a compelling argument. In essence, having a licensee helps validate the value and usefulness of the invention. That these licensees are experts in the field and consider your invention novel enough to license it helps, too. And protecting U.S. commerce has always been one of the primary purposes of the U.S. Patent Office.

This is not a common occurrence, but have your attorney add this simple clause to the agreement.

Assignability

With corporate America, assignability of a license agreement is usually not an issue since corporations are perpetual organizations. However, at times smaller companies may insist they would like to assign the license agreement without any need to have the licensor's approval.

When this happens, there are a few things you can do. First, make sure that your license agreement includes the clause that the assignee signs off on the validity of the IP being licensed. Second, add language to the effect that it is assignable only if in good standing and only if the assignee is not considered a hostile entity to the IP. Last, you should consult your attorney to write these clauses and to make sure you are protected.

In such a case, your primary objective is really to make sure that your license agreement is not transferred to a hostile company that may cease production or that promotes competitive technologies and products.

International Law

At times you may licensee a company that may have facilities overseas or may be a foreign entity altogether. Most overseas companies are fairly familiar with licensing based on the laws of the U.S. However, a question may arise regarding certain legal issues and potential conflicts.

The best way to avoid conflict and misunderstanding is by using World Intellectual Property Organization (WIPO) guidelines. Your attorney will be familiar with them. Having WIPO be the means of conflict resolution is usually acceptable to most international entities.

You may learn more about WIPO and even find the various clauses you may want to include at http://www.wipo.org.

Modifying Agreements

It is uncommon, but acceptable, to use a license agreement such as ones on the preceding pages and insert hand-written names, addresses, products, and so on. It is also legally acceptable to modify an agreement by crossing out words or inserting additional language. Frequently this may happen at the eleventh hour when a simple modification is made. If you modify an agreement in written form, here are three guidelines to follow:

1. **Crossing out.** Cross out words with a single, bold horizontal line. Use a ruler, line it up over the words, and strike a line directly through the center of the unwanted words.

2. **Adding words.** A license agreement is usually double-spaced. This makes it easy to add words and/or additional language between the lines. Neatly print the new language between the lines with a lead line indicating the point of insertion.

3. **Initial changes.** If you cross out words or add language, both signors of the agreement should initial next to the change. The changes may not be valid unless both initials acknowledge them.

Letters of Intent

When a potential licensee is interested, but believes it needs more time to perform due diligence before licensing, you may find yourself in a temporary stalemate. The problem at hand is that you want things to happen now and you don't want to "take the invention off the market," waiting for a company to express due diligence. Likewise, the licensee must be responsible and thoroughly qualify the opportunity.

In such a case, there is one particularly good alternative you can pursue. That is a letter of intent, which is similar to an option. Letters of intent are not uncommon. Many businesses use them for various purposes, including as a temporarily means of securing the rights to an innovation or product offering.

With licensing, you can say that a letter of intent allows a marketing company to test drive your innovation during a short time frame. With a letter of intent, some specifics should be laid out and there should be some form of monetary compensation to you dur-

---

CANADIAN INVENTORS

A surprising number of successful inventions and technologies have come from Canada. Here are just a few prolific inventions from Canadian inventors:
1. Basketball
2. Hydrofoil boat
3. Electron microscope
4. IMAX
5. Paint roller
6. Snowblower

ing the time frame. A sample letter of intent from a potential licensee is given on the next page.

A letter of intent comes from the potential licensee. However, you're most likely going to be the one to recommend it. When you do, you may use a format like the one given. It sets up the basic conditions of a license agreement (similar to a term sheet) and has a time frame to complete certain tasks. Here's what's included:

- **The heading** is addressed to you, the licensor.

- **The first paragraph** expresses the company's desire to enter into a license agreement on your protected rights based on completing certain tasks.

- **Exclusivity** is defined in the next paragraph and is based on U.S. patent rights.

- **Royalties** are outlined in this example with a clause stating that more negotiation will take place for a sliding scale. It may also include language regarding up-front fees payable upon execution of a license.

- **Costs** after the license agreement is in place are established in the next paragraph.

- **Timing** refers to the time frame in which the parties must fulfill their tasks, which is 90 days in this example. This is a fairly common and reasonable time period.

- **Undertakings of prospective licensee** lists the various tasks the prospective licensee will perform. In this example it is three-fold: to test market the innovation; to qualify market demand; and, to pay the licensor a certain sum of money.

The payment or payments is/are essentially in exchange for the time frame the innovation is being kept off the market. There are no guidelines for the amount or number of payments with a letter of intent. A not unreasonable figure for an innovation with solid patent protection is $5,000 per month. Certainly high technology requests would be far greater, into the tens of thousands to hundreds of thousands of dollars. In the example, the payments are credited against future royalties.

- **Undertakings of licensor** is a list of your obligations under the agreement. The example requests the licensor to secure a valid U.S. patent and help the licensee market the innovation.

- **The last paragraph** closes the document with the clause that the undertakings of the prospective licensee must meet its approval.

There is no specific science to letters of intent, other than they should include the following: specific tasks to perform, a form of consideration such as money, and a time frame to complete.

Signing and Archiving Your Agreements

Make sure you prepare and execute your important agreements in duplicate originals. One is for your archives and the other for your licensee's. These documents should be kept in metal, fireproof containers to help protect them against not only fire but also insects, rodents, and the like.

When signing license agreements, it is not necessary to have them notarized or have a corporate seal placed on the documents. However, if a licensee or attorney insists, then by all means do it.

Make a copy to keep in your files for ongoing reference. And by all means keep a

LETTER of INTENT

Dear _____ (Licensor),

We, _____ (prospective licensee), are prepared to enter into a licensing agreement for the _____ protected rights owned by _____ (licensor) as follows:

Protected Rights: List them here.

Exclusivity: In the territory of the United States.

Royalties: _____ percent (____%) of net sales. Net sales are defined as invoice price less freight, discounts, and rebates. The royalty will be reduced with higher volume sales, as we will agree upon in the license agreement.

Payment of Costs: All costs for patenting in the United States as well as all R&D costs incurred by Licensor will be paid by _____ (Potential Licensee).

Timing: The License Agreement will be executed within 90 days of the signing of this Letter of Intent.

Undertakings of Prospective Licensee:

- To test market the _____ during the 90 day period. This will entail any equipment conversions required and, assessment of the economics to manufacture as well as the evaluation of value-added pricing in the market.

- Following the above assessments and determination of market demand, _____ (Potential Licensee) will broaden supply capability and expand sales.

- _____ (Potential Licensee) agrees to pay Licensor an advance against royalties in the amount of $_____ as follows:

  $_____ upon signing.
  $_____ in 30 days.
  $_____ in 60 days.

Undertakings of Licensor:

- To provide a valid U.S. patent within _____ years of signing this agreement.
- Work with _____ (Potential Licensee) to develop a marketing plan.
- Work with _____ (Potential Licensee) to develop future improvements.

It is our desire for the parties to enter into a mutually beneficial agreement providing the undertakings of the prospective licensee meet its approval.

Sincerely,

Signed (name)

copy in your computer so you may pull it up and use it as a format for future licensing activities.

Human Nature and Licensing

Licensing is best initiated as early as possible after you've conducted sufficient research to qualify manufacturability, marketability, and patentability. The first license is the toughest one. Once you've built up established relationships, you'll find the process much easier.

Always remember that licensing is a big step for many companies and a big commitment, too. It must be of mutual benefit to the parties. Human nature plays a big part in how well received you'll be and how willing the licensee may be. Here are a few points to remember:

- **Be friendly.** People like doing business with friends, with people they like. Smart businesspeople know that if they can become friends with the other party at the beginning, they are more likely to have a successful business relationship. You can do this too. Try to discover your mutual interests and, see if you can spend some time apart from business to get to know one another better.

- **Be positive.** Keep in mind that licensing may be intimidating to a potential licensee. Maintain a positive approach with your licensing activities and solve problems as they arise. Your attitude will encourage others to be positive as well.

- **Be understanding.** Not everyone is willing to enter into a license agreement nor has the understanding of what it takes to launch innovative new products. A licensee is naturally conservative with guarantees, since launching your innovation may put them in unfamiliar territory.

- **Don't be greedy.** This may be the number one deal killer in licensing patents. You really must be flexible and think in terms that will be acceptable to your licensee.

- **Be happy.** When you have succeeded in signing a license agreement, go out and *celebrate!* Be happy and remember the feeling, because that feeling will help you strive for many more successes in the years to come.

Licensing is the culmination of your invention activities. With one success, you will be motivated to create many more. Continue to invent and have the time of your life. Now, go out there and make it a better world! –Bob DeMatteis

SMILE AND SAY THANKS

There's one thing to remember. When someone tells you "no," make sure you say, "thank you!" Why? Well, this only means that you are that much closer to hearing a "yes!"

CONCLUSION

I have been inventing for some time now and have seen some incredible things happen. By reading this book, you've taken the most important first step in the right direction. As you proceed, take your time, don't rush, and be determined to see it through. Your creativity and commitment will get you started and well into the development process. When you find yourself on unfamiliar ground, I hope you'll find this book a road map to answer those questions. I hope you'll find it will give you the direction you need to overcome obstacles and to take you to the next level.

I've been fortunate in my invention career. I've had many opportunities to profit from my inventions, but it didn't happen without learning from all the difficulties and obstacles I encountered along the way. You'll invariably encounter some as well. When you feel something is not right, you spot red flags, or your invention development is stymied, remember:

1. Don't be discouraged.

2. Brainstorm, test and evaluate all the alternatives.

3. Always keep records and remember what you've learned for future reference.

You see, I know if you are determined to see your project through, and continually evaluate and improve your invention based on what the customer wants, that you'll be successful. The only necessary ingredient is to take action.

Should you have any questions regarding this book, you may find it helpful to visit our website at www.frompatenttoprofit.com. You'll find a list of our seminars and workshops and other inventor resources of interest.

Now go out there and make it a whole new world!

APPENDICES

From Patent to Profit ^{SM}

The following four processes should be pursued as one continuous, yet somewhat flexible, timeline. Begin at phase 1 at the far left column until the four processes are simultaneously completed at phase 5 at the far right column. Next, move to the launch phase.

	1	2	3	4	5
1 INVENTING	**Get an Idea** You have an idea or made a discovery. Now you take action. ☐	**Brainstorm** Ask yourself, "How can this be turned into a dazzling, high-impact product? How should it be engineered to accomplish this?" ☐	**Quality** *Qualify marketability.* Start by asking friends, then ask engineers, department managers or other experts in the field. **(3)** ☐	**Prototype** Determine best prototype—a model, CAD drawings or computer-generated design? First model does not have to function. Follow-up with functional prototype to show. **(5)** ☐	**Engineer** Refine prototypes by carefully engineering the best attributes, until you have one that can be thoroughly tested in-house and fix it as required. ☐
2 PATENTING	**Learn/Protect** Learn about IP, patents, and patent protection, including products, processes, and systems. *Write an invention disclosure* **(1)** illustrating and describing your invention. Have it validated. ☐	**Search** Do a *thorough patent search* after you understand manufacturing and marketing basics. Begin on the Internet or use appropriate software. **(2)** ☐	**Consult/Hire** Consult with your legal counsel to confirm patentability. If you do not have a patent attorney, start interviewing now. ☐	**Learn/Write** Learn about provisional and permanent patent applications. Start writing one right away. ☐	**Write/Review** Complete writing of provisional patent application. Patent attorney reviews. You re-review. *File the provisional patent application.* **(7)** ☐
3 MANUFACTURING	**Learn** There are several kinds of manufacturing processes with different limitations and variables. ☐	**Brainstorm** Determine the best, most cost-effective process for your invention. If you are unfamiliar with manufacturing processes, visit a local **SBDC** for help. ☐	**Interview** Ask managers and engineers in your manufacturing or procurement department about the best, low-cost means to produce. ☐	**Qualify** *Qualify that manufacturing cost and expected quality.* Verify engineering, start-up costs, and anticipated production output. **(4)** ☐	**Commitment** Get a commitment from manufacturing or enter into a supply agreement with an outside supplier. ☐
4 MARKETING	**Learn** Learn the marketing essentials for your invention. ☐	**Brainstorm** Consider the various ways your invention can be sold and used in the market. Determine the best few sales approaches. Or identify outside sales experts in the field of your invention. ☐	**Interview** Meet with your company's national sales manager to verify marketing potential. Or, interview outside prospects starting with a New Product Summary. ☐	**Qualify** Invention meets company's sales manager's expectations. If outside company, its ability to take to market is qualified. ☐	**Commitment** The *marketing department makes a commitment* or you enter into an outside sales agreement or a commitment to **license** with a marketer in the field of your invention. **(6)** ☐

Shaded boxes are benchmarks

Continuing the same concurrent timeline, follow through as you take your invention to market. The launch phase begins in phase 6 at the left hand column and moves through to completion at phase 10 at the far right hand column.

6	7	8	9	10
Test	**Perfect**	**Anticipate**	**Prove**	**Improve**
Test prototypes until one functions close to what is expected. Manufacturing uses it as guide to make working models. ☐	You work with the manufacturing and marketing departments (or partners) to perfect the working models. ☐	You tighten up all the details, anticipate problems, and formulate solutions. ☐	Assisting your marketing team, you stand close by while customers prove the product. ☐	You expand upon the invention's potential. With new improvements, you start process all over again. ☐
Other IP	**File provisional**	**File IP**	**Scope**	**Permanent Ap.**
Develop other IP such as copyrights, trademarks, trade dress. Have attorney qualify so there is no conflict or confusion. ☐	Application is filed before public disclosure to preserve worldwide filing rights. ☐	Any additional filing of trademarks, copyrights, or other IP should be in place. ☐	Is market potential international? Evaluate international cost vs. market potential. ☐	Before end of year, the permanent application or an international PCT is filed. ☐
Working model	**Test**	**Gear up**	**Ship**	**Expand**
Manufacturer employs its engineers and gears up to make working models to prove quality and qualify performance. ☐	Working models are thoroughly tested and proven. ☐	Manufacturer gears up for production. ☐	Ship first orders on time. ☐	Manufacturing expands production and inventory levels. ☐
Packaging	**Test/Verify**	**Sell**	**Oversee**	**Expand**
Marketing has prepared packaging, using either internal department or outside experts. Prepares spec sheets, product brochures, and so on. ☐	Marketing works with manufacturing to ensure quality and acceptance. Beta tests are conducted now. Results verified. ☐	First customers are sold for initial test marketing. Market ramp-up is underway. ☐	Sales team follows up to address all customer, distributor, and retailer concerns. ☐	Marketing expands sales. ☐

INVENTOR RESOURCES

■ INVENTOR BOOKS, MATERIALS, AND SOFTWARE

IPT Co. (Inventions, Patents and Trademarks Co.) is an ever-expanding resource for inventor-related books and materials. Check them out on the web where you may purchase inventor start-up kits, home study courses, and many more specialized tools for inventors and product developers. This list includes:

- *From Patent to Profit:* Purchase additional copies of the book.

- The *From Patent to Profit Resource Guide*: This is a CD that contains all of the patent filing forms, confidentiality agreements, invention disclosure forms, and license agreements you'll need for your personal use. Modify and copy them as many times as you want. Free updating of forms is provided.

- The *Scientific Journal*: This Inventor's log book includes guidance to protect all your rights, develop your invention in the proper order with the Strategic Guide, and track your day-to-day activities.

- *From Patent to Profit* Home Study Courses: Learn all you need to know to commercialize your invention fast. This is the leading home study course in the world.

- *PatentWriter*: Your personal patent writing book walks you through all the steps you'll need to take to write a successful patent application. Get patent pending fast!

- *PatentWizard:* This software guides inventors in writing a provisional patent application.

- *PatentHunter:* This software helps you professionally search patents internationally in just minutes and download unlimited numbers of patent document copies.

- *Essentials of Patents*: This book exposes the various aspects of invention development and patenting in the corporation and the various departments. It is an excellent book to read to understand how a licensee may view your inventions and patents.

For more information about these products, visit IPT Co. at www.frompatenttoprofit.com or call IPT at 1-888-53-PATENT (1-800-537-2836). Bookmark the site to stay abreast of this ever expanding educational and information resource for inventors.

Patents in Commerce DVDs and Workbooks. Twenty individual DVDs covering every aspect of invention—protecting your ideas, patent strategy, prototyping and manufacturing inventions, marketing, and licensing—may be purchased along with the corresponding workbooks, which will guide you in becoming proficient in the subject matter. Or, purchase the entire set for your library and for future reference. Visit www.patentsincommerce.com to learn more.

■ INNOVATION AND INTELLECTUAL PROPERTY WORKSHOPS AND TRAINING

***From Patent to Profit* Workshops.** These workshops are given in many parts of the country and are based on the book's teachings. The workshop presenters are Bob DeMatteis or other highly qualified experts in the field. *From Patent to Profit* workshops are sponsored by prominent colleges, universities, SBDCs, the SBA, and inventor organizations. They show small businesses and first-time and experienced inventors invaluable insight into the field of invention and the *From Patent to Profit* system. To find a workshop near you, visit our website at www.frompatenttoprofit.com/workshops.htm. If you are unable to attend one of the seminars, take a look at out *From Patent to Profit* Home Study Course on our website.

Patents in Commerce Series. This is ideal for home study or remote learning at the office. *Patents in Commerce*™ (PIC) is a complete invention development, patent, and intellectual property training series consisting of DVDs, Research Workbooks, invention and patent books, and other materials and software. The authors and presenters are a team of highly experienced inventors, marketers, attorneys, prototypers, and government specialists who show inventors and small business owners how to successfully commercialize their innovations. More than twenty world-class invention and small business experts present a process-driven approach that shows inventors and business innovators the various aspects of product development and design, patenting, prototyping, manufacturing, invention financing, licensing, and product marketing that are needed to successfully commercialize an invention. Visit the Patents in Commerce website at www.patentsincommerce.com and see how you may become an expert in your field and learn at your own pace. *From Patent to Profit* is the official textbook of the Patents in Commerce series.

NCIIA (National Collegiate Inventor's and Innovators Alliance). Sponsored by the Lemelson Foundation, this tremendous resource supports colleges and universities that teach invention, innovation and entrepreneurship. The NCIIA provides grants for the establishment of E-teams (Excellence teams) in order to develop some of the student projects. It is also a resource for curricula, patent advice and services, technology transfer, and assistance in locating business incubators and enterprise development organizations, industry experts, and sources of capital. Visit their website at www.nciia.org.

■ GOVERNMENT ASSISTANCE AND INVENTOR ORGANIZATIONS

The *From Patent to Profit* way of inventing and patenting is one that requires inventors to take action. While most SBDCs can be very helpful in your initial start-up efforts, not all inventor organizations can. Some may even be invention promotion companies fraudulently parading as inventor organizations. Likewise, we cannot honestly recommend an inventor organization just because of its national association. The reality is that many inventor organizations are inexperienced hand-holding groups that don't have experts on staff and don't promote taking action. You want an inventor organization that has experienced inventors and marketers on staff and has the ability to help members prepare their innovations and to take action . . . to get them commercialized. Thus, the only reasonable assurance we can give is to keep in touch with the Patents in Commerce Affiliate list on its website, as it qualifies only those inventor organizations that have proven their ability to fulfill the principal obligation to its members—that of commercializing their inventions.

Patents in Commerce Affiliate Inventor Organizations. Patents in Commerce has identified several progressive inventor organizations throughout the United States that help inventors with a strong focus on assisting their members in the essential elements to successfully launch and commercialize an innovation. This includes counseling, seminars, an adequate library, and much more. For more information on Patents in Commerce Affiliates, visit the web page www.patentsincommerce.com/affiliates.htm.

SBDC (Small Business Development Centers). Big Brother can help! The SBDCs are almost always willing to provide strong counsel to inventors/innovators. The SBDC helps you out in the early stages and its big sister, the SBA, will take over from there. They have more than 1,000 offices throughout the United States. If you are starting out, this might be one of the most valuable first steps you take! For more information, go to their Internet website at http://www.sba.gov/gopher/Local-Information /Small-Business-Development-Centers/.

U.S. Patent and Trademark Depository Libraries. These libraries are helpful for conducting patent and trademark searches. Some allow teleconferences directly with the U.S. Patent Office. They are always helpful and are interesting places to visit. Some also provide seminars and workshops for inventors. For more information on the U.S. Patent and Trademark Depository Library near you, call 1-800-PTO-9199.

U.S. Copyright Office, Library of Congress. For copyright information, contact the U.S. Copyright Office, Library of Congress, Washington DC 20559-6000. Their Internet site is cweb.loc.gov. You can download copyright forms online.

Federal Trade Commission. If you want to verify that a company you have contacted is not under scrutiny by the FTC and project Mousetrap, call 202-326-2222. To find out the latest news as it is announced, call the FTC at 202-326-2710 or visit their website at http://www.ftc.gov.

HELPFUL INVENTOR LINKS AND RESOURCES

http://www.frompatenttoprofit.com/Helpful_Invention_Links.htm is the place to stay up-to-date on the most important invention resources, patent office news, and so on.

LICENSING AND MARKETING CONTACTS

You are encouraged to follow the teachings in Part Three and find, interview, and evaluate prospective partners in the field of your invention. As you do this, you'll find some companies that actively seek inventions for license. The following is a list of companies that are receptive to outside inventions. We've listed their websites and the key pages for submitting your inventions. However, companies may alter their course and methods of doing business, job responsibilities may change, phone numbers and addresses may change, and at times, companies get sold or go out of business. Thus, the following list should be viewed as a starting point to identify a potential licensee's names and current information for the types of products they seek. If the URL has changed, try going to the home page and searching from there.

As Seen on TV. Enormous volume of infomercial products, go to www.asseenontv.com.

As On TV. Go to http://www.asontv.com/. You'll recognize their many products.

Aztech. Seeks inventions on aging and persons with disabilities. Their website is: www.wnyilp.org/aztech.

Baby Net. You can locate 400+ manufacturers of baby, toddler, and juvenile products. Go to www.thebabynet.com/manufacturers/manulist_all.htm. Many will be receptive to hearing about your invention.

Bench Dog, Inc. Manufactures tool-related inventions. Visit www.benchdog.com.

Benjamin Obdyke. Home improvement products. Go to www.benjaminobdyke.com/.

Big Idea Group. Toys, stuffed animals and more. Visit www.bigideagroup.net.

Bobcat, Inc. Visit www.bobcat.com/faqs/attaidea.html for inventions related to the Bobcat tractors, lawnmowers, and snow blowers.

Buske Industries, Inc. Seeks automotive auto body-related inventions. Visit www.buskeindustries.com.

City of Hope. Medical inventions. Visit www.cityofhope.org/ott/?DROP=%23\ &I1.X=29\&I1.Y=14.

CNS Corp. Personal health care and equine products. Visit www.cns.com/inventors /submit.html.

Continental Window Fashions. Bedding, window, and drapery innovations can be submitted at www.ambrosiawindowfashions.com/newproductsubmission.htm.

Corning Consumer Products. Consumer products company's website is http: //www.coring.com.

Crawford. Shelving, brackets, ladders, and so on. Visit www.lehighgroup.com/product_list.asp.

Department of Energy. Visit www.osti.gov/dublincore/gencncl/ to learn what the U.S. Department of Energy is pursuing for license.

DeWalt Industrial Tool Co. Tool-related products sought by one of America's premier marketers. Visit www.dewalt.com/us/service/company/invention.asp.

Dirt Devil. Owned by Royal Appliance, Dirt Devil seeks vacuums, household appliances. Visit www.dirtdevil.com or www.productlaunchpartners.com.

Erico Products Inc. (Caddy Fasteners): For electrical fasteners for the commercial/industrial trade, visit www.erico.com.

Ethical Pet: For pet-related innovations, go to www.ethicalpet.com/Password/inventor_designer.htm.

Excel Development Group, Inc. Places new toys worldwide. Website is www.exceld.com.

Fujitsu-Siemens, Inc. Visit www.fujitsu-siemens.com/rl/aboutus/intelprop/rules.html for submitting concepts related to computers, notebooks, handhelds, and so on.

Fundex. Game inventions by the dozens! Visit www.fundexgames.com/products/?c=17&g=2633.

Gadget Universe. All kinds of gadgets! Visit www.gadgetuniverse.com.

Genius Babies. Baby and children's goods. Go to http://www.geniusbabies.com/new-product-submissions.html.

Great Northern Corporation. Point of purchase displays and packaging. Visit www.greatnortherncorp.com.

Guthy Renker. For health and women-related innovations used in infomercials, visit www.guthyrenker.com/index.php?intSectionID=11.

Hammacher Schlemmer. Annual competitions in search of new inventions. Visit www.hammacher.com/sfi/rules.asp.

Harvard University. Visit www.techtransfer.harvard.edu/ReportingIntro.html.

Hearlihy & Company. Seeks technology and curriculum-related products for middle level schools. Visit www.hearlihy.com.

Homax. Home improvement products. Visit www.homaxproducts.com.

Inno-pak. If you have a packaging-related invention that can be used in bakery, deli, produce applications, and supermarkets, visit http://www.innopak.com.

Intel Corp. A wide variety of products related to its core business of computer chips. Visit www.intel.com/intel/company/corp7.htm.

Interordnance of America. For gun- and rifle- related innovations, visit www.interordnance.com.

Izzo Golf. For golf industry products, visit www.izzo.com/suggest.asp.

Johnson & Johnson. Patented ideas will be considered in the field of house wares and health and beauty aids. Visit www.jnj.com/contact_us/new_product_ideas.

Johnson Outdoors, Inc. For boats, motors, diving and outdoor sporting equipment, visit www.jwa.com/corp/idea.html.

KDI Precision Products, Inc. In the field of communication, visit www.kdi-ppi.com.

Kraco Enterprises. Looks for aftermarket auto accessories. Visit www.kraco.com.

Lehigh. For cordage, rope products, Bungee cords, etc., visit www.lehighgroup.com/product_list.asp.

L.H. Dottie. For electrical and plumbing fasteners and fittings for the residential and commercial trade, visit www.lhdottie.com.

Lisle Corporation. Automotive products. Visit www.lislecorp.com.

Long Short Golf, Inc. Submit your golf-related innovations at http://www.longshotgolf.com/msg8.htm.

Manufacturers' Agents National Association. Commonly referred to as MANA, it is a professional organization of manufacturers' representatives in virtually every industry. Since manufacturers' reps frequently make great marketing partners, you may advertise in their monthly publication to find them. Their website is www.manaonline.org/.

Marra Design Associates. For toys on a commission only basis, visit http://www.marradesign.com.

McClean Hospital Research. Visit www.research.mclean.org/administration/inventions/ for medical-related innovations.

MedCatalog.com. For a wide variety of medical inventions, visit www.medcatalog.com/.

Mediacorp. Direct response TV leaders. Visit www.mediacorpworldwide.com.

Michigan State University. Visit www.research.unc.edu/otd/faq/procedure.html.

NASA. Visit www.invention.nasa.gov/nasa_employee.html to submit your inventions to space travel and NASA.

National Presto Industries, Inc. For mass merchandise house wares, visit www.gopresto.com.

Natural Bambinos. Eco-friendly, organic, and cruelty-free baby innovations are considered at: www.store.naturalbambinos.com/info.html

New Funtiers. Paul Lapidus heads this company that has licensed more than 100 toy inventions. The website is: www.newfuntiers.com.

Nicholas Webb Consulting. For inventions in the medical field, visit www.healthcareinventions.com/ or www.nickwebb.com.

Nordic Track. Seeks new inventions in the in-home fitness field. The website is www.nordictrack.com.

Northwestern University. Visit www.northwestern.edu/ttp/investigators/ttpservices.html.

Optima Chemical. Develops an interesting array of chemical substances. Visit www.optimachem.com.

Oregon State University. Visit www.oregonstate.edu/research/TechTran/invdiscl.html.

Party America. For party-related innovations, visit www.partyamerica.com/contact.html.

PatentCafe. www.2xfr.com/ offers inventions for license and exposes to licensees as well.

Pelham West Associates. New product scouts for a variety of companies, representing a variety of items. Visit their website at www.pelhamwest.com.

Penn State University. Visit www.research.psu.edu/ipo/faculty_staff/disclosures2.html.

Perrin Mfg. Co. This company seeks kitchen and bath, hardware, stationery and automotive aftermarket accessories. Visit www.perrin.com.

Petcology. For ecologically sound pet inventions, visit www.petcology.com /docs/about.shtml.

QVC. Start with the QVC web page at www.QVCproductsearch.com.

Rectorseal Corporation. Visit www.rectoseal.com if your innovation is related to plumbing.

Rutgers University. Visit www.ruinfo.rutgers.edu/external/views/view1.asp?serial=2518.

Safety Components International. For safety and security related technologies, visit the website: www.safetycomponents.com.

Samson. For marine related rope products, visit www.lehighgroup.com/product_list.asp.

SnS International. Specializes in the home and bath industry. Visit www.cleanimals.com

Stanford University. Visit www.otl.stanford.edu/inventors/.

Starret, Inc. For precision tools, saws and blades, calculators and calibrators, meteorology concepts, visit www.starrett.com/pages/57_if_i_have_an_invention_or_an_idea_how_do_i_go_about _presenting_it_to_starrett_.cfm.

Storehouse. For portable plastic products, visit www.lehighgroup.com/product_list.asp.

Strength Tech, Inc. Visit www.jwa.com/corp/idea.html for weight lifting-related innovations.

Stride Tool, Inc. For tool and hardware related inventions, visit www.stridetool.com/.

Thane International. Leading infomercial partners. Visit www.thane.com/.

The Sharper Image. Visit www.sharperimage.com.

Think USA. Located at www.thinkusa.com, Murray Ansel handles technology transfer in several high-tech and low-tech fields and has excellent connections in New Zealand and Australia.

Tufts University. Visit www.techtransfer.tufts.edu/tufts/disclosures.html.

Tulane University. Visit www.tmc.tulane.edu/techdev/inventors/disclose.html.

Twister Sisters. Visit http://www.twistersisters.net/ for innovations related to entertainment, animation, music, toys, design, food, and fashion. Andre Lake Mayer is a highly skilled licensing agent.

Ultra-Hold. For storage-related devices, visit www.lehighgroup.com/product_list.asp.

United States Department of Agriculture (USDA). For agriculture-related inventions suitable for government grants, visit www.ars.usda.gov/business/docs.htm?docid=1129.

Universities and Colleges. Many colleges sponsor technology transfer and will want to participate in the commercialization and licensing of inventions in their field. Contact your local universities and discover which technologies they may be pursuing. Since we cannot list every college and university, the ones we've included in our list are some of the more prominent ones.

University of California Davis. For agriculture-related inventions, visit www.ovcr.ucdavis.edu /TTC/Default.html.

University of Connecticut. Visit www.cstc.uchc.edu/inventors/disclosure.html.

University of Illinois. Visit www.uic.edu/depts/ovcr/otm/com-discl.htm to submit your technology.

University of Mississippi. Visit www.olemiss.edu/depts/HR/handbook/HB_sec10.html.

University of North Carolina. Visit www.research.unc.edu/otd/faq/procedure.html.

University of Pittsburgh. Visit www.tech-link.tt.pitt.edu/inventors_inventiondisclosureinstructionsv3.htm.

University of Tennessee. Visit www.tennessee.edu/.

Utah State University. Visit www.usu.edu/techcomm/invent.htm.

Ventura Distribution, Inc. For video and DVD products distribution, visit www.venturadistribution.com/corporate/submissions.phtml.

WestBend Co. For household appliances, visit www.westbend.com.

Wine Appeal Products. Products related to the wine industry, corporate gifts and golf. Visit www.wineappeal.com.

WirthCo Engineering Inc. Seeks automotive related inventions. Visit www.wirthco.com.

Wow Shopper. For lifestyle inventions, visit www.wowshopper.com.

TRADE MAGAZINES

There are thousands of trade magazines and trade journals at your fingertips. The best way to find those in the field of your invention is to go to www.google.com and search the words, "trade journal +<your field>" and you should find all the related trade literature you'll need.

PATNEWS. A worldwide news service with topics including Patent Office news, stories about who is suing who, interesting new patents, styles of patent drafting, statistics on issued patents (focus on software), reviews of patent books and computer programs, and biting commentary if Greg happens to be in a lousy mood that morning. To subscribe, visit http.www.patenting-art.com/clients/partnews.htm.

www.ferro.com/industry+information/trade+journals/default.htm. Lists dozens of trade journals.

Inventor's Digest. Contains interesting articles on inventing. Visit www.inventorsdigest.com/.

www.penton.com. Lists trade magazines in several industries.

■ PROTOTYING AND MANUFACTURING

Titoma. Contact Matt Berg at www.titoma.com.tw/for most prototyping needs, including rapid prototyping technologies.

Local colleges and universities. Most local colleges and universities have metal, plastic, and woodworking shops that can help you convert your inventions into prototypes. Some will charge you and some won't. The quality is generally excellent.

www.thomasregister.com. The famous Thomas Register is available on line, although not as complete as the volumes you can find in the library.

■ PRODUCT DESIGN AND MANUFACTURING ASSISTANCE

Studio Red. Superior product design in many fields. Visit www.studiored.com and contact Phil Bourgeois.

The Gibbs Group. Located at www.productbuilders.com/, they specialize in domestic and import products with an emphasis on converting an invention into a real time, high-quality product.

Teim Inc. Visit www.teiminc.com to learn about start-up manufacturing and streamlining manufacturing processes and methods. Teim means "Total Excellence In Manufacturing."

SBDC. Don't forget to contact your local SBDC should you have any questions about how your innovation is made.

■ PATENT ATTORNEYS AND AGENTS

Go to www.frompatenttoprofit.com/patentattorney.htm for an up-to-date list of patent attorneys that understand the workings of independent inventors and small companies.

U.S. Patent Office. Provides a publication called, "USPTO's list of Attorneys and Agents Registered to Practice before the USPTO" that is available at many Patent and Trademark Depository Libraries or can be purchased from the Government Printing Office. Call the U.S. Patent Office at 1-800-PTO-9199.

■ PATENT SEARCHING

IPT. The Best Patent Search on Earth. Visit www.frompatenttoprofit.com/patentsearch.htm for the most recent information on patent searching, the Patent Search Handbook, PatentHunter software and special offers including complementary patentability opinions. Search domestic and worldwide in minutes.

U.S. Patent and Trademark Depository Libraries. Located throughout the United States, these experts will help show you how to patent search (but they won't do it for you!). For more information on the U.S. Patent and Trademark Depository Library near you, call 1-800-PTO-9199.

■ INTERNET NEWSGROUPS

alt.inventors. If you have the time, you can post your questions on this website.

misc.int.property. This group is similar to the alt.inventors group but more focused on the legal side of patents, trademarks, and copyrights and visited more frequently by patent attorneys.

ABOUT THE AUTHOR

Bob DeMatteis

Bob DeMatteis is an inventor extraordinaire and is living proof that the American dream can be realized. Bob had an idea he wanted to pursue. He pursued that dream, patented it, and today is a successful inventor with more than twenty U.S. patents and several international patents. Sales of his inventions exceed $1 billion and have been sold to companies throughout the world. You have most likely used some of Bob's innovations at one time or another . . . plastic grocery sacks, plastic merchandise bags, and more recently, plastic fast food bags and concrete bags.

When Bob saw plastic bags could economically replace paper bags in stores, he begin developing bagging systems so the bags would self-open. The result is that plastic bags are now the preferred bag by retailers and consumers. Today, he is still one of the leading innovators in his field.

Bob knows how to develop an invention and get it marketed and patented. He also knows the importance of creating a team effort with manufacturing and marketing partners. His team-building approach to inventing gives him a 100% success rate. All of his patents have been licensed either before or during the patent pending process.

In 1995, Bob had another dream…that is, to teach others how to realize their dreams through invention. He enjoys teaching as much as inventing and patenting. Bob developed the seminar series entitled *From Patent to Profit*®. These exciting seminars help inventors, entrepreneurs, and small businesses develop their ideas and patent, commercialize, and license them. Colleges and universities, SBDCs (Small Business Development Centers), U.S. Patent and Trademark Depository Libraries, and inventor organizations sponsor the seminars.

Out of demand, Bob became one of the most published authors in the field of inventing, patenting, and intellectual property. He is the author of *From Patent to Profit*, *Patent Writer*, *Essentials of Patents*, the *Scientific Journal*, with others on the horizon. Bob is also a founder and host of *Patents in Commerce*, a training series for small business and independent inventors. Read more about his books, seminars, and training at www.frompatenttoprofit.com.

Today Bob is a sought after lecturer, a member of IPO and LES, and an advisor to many small and medium-sized businesses. He and his family live in northern California.

INDEX

Liquid Paper, 105

Literal interpretation, 123, 145–146

LOC, see "Letter of Credit"

Log,
    book, see "Scientific Journal"
    records, 25, 28

Lo-Jack, 210

Long-term trend, 9, 12, 33, 34, 86

Looksmart, 214

Low cost producer, 220, 290

Low sales volume, 135

Lyondell, 292

M

Machine patent, 129, 130

Machinery suppliers, 233

Macintosh, 73, 130

Macintosh mouse, 128

Macy's, 241

Made in USA, 280

Madison, James, 203

Magic words (of a marketer), 307

Mail order companies, 212, 213

Maintain logs (and records), 27–29

Maintenance fees, 114, 116, 131,
    144, 187, 196

Make money,
    from invention, 16
    from patent, 16
    objectives, 9, 10

Make, use, sell, 193, 326, see also
    "License to manufacture,
    use and sell"

Malsed, Helen, 328

Mankind, inventions good for, 60, 61

Manufacturability, 78, 84, 85, 150,
    348, 380

Manufacturer, as a team member,
    40, 41

Manufacturer's cost, 25, 55, 72, 273,
    300, 384

Manufacturer's representatives, 297,
    302–305, 308, 309

Manufacturer's sell price, 216, 218

Manufacturers, finding, see "Find
    manufacturing partners"

Manufacturers, interviewing, see
    "Interviewing manufacturers"

Manufacturing,
    agreement, 342, 343
    complex inventions, 278
    expert, 91
    objective, 274
    partner, 20, 271–296
    partnership, 290, 291
    process, 22, 37, 87, 93, 94,
    275, 276
    records, 339
    simple inventions, 278
    (the wrong approach), 275

Manufacturing/marketing strategies,
    227–230

Manufacturing-only license, 327, 328

Margolin, George, 93, 138, 144,
    148, 315

Market,
    advantage, 236, 242
    creation, 202–206, 299–301
    demand, 33, 80, 374, 378, 379
    description, 227, 269
    durability, 11, 86
    readiness, 17, 302, 268
    share, 35, 52, 203–205, 298,
    300, 301, 305
    size, 11, 85, 244, 270
    trends, 8, 267, 269

Marketability, 30, 165, 228, 242
    assessment, 25, 85
    methods to determine, 30–32

Marketing,
    credits, 375
    expert, 208, 227, 239, See also
    "Marketing partner"
    expert, definition, 297–289
    objectives, 298, 299
    opportunity, 210–214, 269
    partner size, 202–207
    partner, 20, 31, 233, 297–311
    partners, finding, see "Find
    marketing partners"
    philosophy (for inventions),
    201–214

plan, 30, 218, 269, 274, 279

potential, 14, 140, see also
    "marketability"

strategy, 299, 300

warfare, 98

Marking of product, 133, 211, 229,
    339

Mark-up, 215, 275, 276

Masked terms, 242–244

Master cartons, 221, 223, 293

Masterson, Kelly, 281, 283, 290

Mattel, 369

Maytag, 71, 106

McDonald's, 56, 108

Measuring,
    assessments, 89 90
    performance, 93

Mediocrity perpetuated by
    standardization, 92

Medium corporations, 206

Mega-giants, 205, 206

Mephisto, 63

Mercedes Benz, 62, 70, 214, 290,
    305, 307

Metal,
    forging and casting, 277
    stamping, 277

Metamorphosis (of a product), 19, 24,
    29, 37, 38, 83, 294, 312

Method patent, 128, 129

Microsoft, 34, 49, 51, 53, 54, 64, 71,
    163, 205, 208, 213, 319, 347
    Front Page, 213

Miller Brewing, 327

Mindset,
    of an inventor, 54, 55
    licensing, 367–369

Minimum,
    quality standards (for trademarks)
    326
    royalties, 326, 333, 358
    sales volume, 326
    training, 72

Misconceptions about patenting,
    121–123

Mistake Out, 105

Need More Help?

From Patent to Profit
Resource Center

Our Resource Center provides inventors with legal forms and materials, and includes an elite team of experts in a variety of fields to answer your questions. These experienced professionals know the From Patent to Profit system and share the same winning philosophy.

Instant online help only a click away

We understand that decision-making along the way can be challenging and difficult, if not downright confusing. You may wonder where to get forms and legal inventor journals, who to hire, when to file a patent or trademark application, whether to go into business or license, etc. Answers to these and many more questions are at your fingertips when you visit our website.

Our materials and services include:

- **JOURNALS, FORMS, AND INVENTION-RELATED BOOKS.**

- **HELP WITH WRITING AND FILING STRONG PATENT APPLICATIONS—PROVISIONAL AND REGULAR.**

- **SOFTWARE FOR FORMS, PATENT WRITING, AND SEARCHING.**

- **FILING TRADEMARKS, COPYRIGHTS, AND ESTABLISHING TRADE SECRET PROTECTION.**

- **REFERRING ATTORNEYS FOR PATENTING AND INFRINGEMENT.**

- **DESIGNING AND PROTOTYPING INVENTIONS THAT ATTRACT PARTNERS.**

- **CREATING HIGH-EXPOSURE WEBSITES THAT ARE IDEAL FOR INVENTORS AND NEW PRODUCTS.**

- **TV EXPOSURE FOR INFOMERCIALS AND DIRECT RESPONSE (DR) ADS.**

- **PRESENTING YOUR INVENTION TO PRODUCT SCOUTS (WE CURRENTLY SCOUT FOR OVER 100 COMPANIES).**

- **INTERNATIONAL EXPOSURE AT INNOVATIVESHOWCASE.COM.**

- **START-UP GUIDANCE FOR GOING INTO BUSINESS (INCLUDING HOME-BASED), LICENSING, PARTNERING, FINANCING.**

- **LICENSING ASSISTANCE.**

Take action now. Make it happen and let us help you achieve your dream!
Visit us at www.FromPatentToProfit.com